Springer Tracts in Mechanical Engineering

Springer Tracts in Mechanical Engineering (STME) publishes the latest developments in Mechanical Engineering - quickly, informally and with high quality. The intent is to cover all the main branches of mechanical engineering, both theoretical and applied, including:

- Engineering Design
- Machinery and Machine Elements
- Mechanical structures and stress analysis
- Automotive Engineering
- Engine Technology
- Aerospace Technology and Astronautics
- Nanotechnology and Microengineering
- Control, Robotics, Mechatronics
- MEMS
- Theoretical and Applied Mechanics
- Dynamical Systems, Control
- Fluids mechanics
- Engineering Thermodynamics, Heat and Mass Transfer
- Manufacturing
- Precision engineering, Instrumentation, Measurement
- Materials Engineering
- Tribology and surface technology

Within the scopes of the series are monographs, professional books or graduate textbooks, edited volumes as well as outstanding PhD theses and books purposely devoted to support education in mechanical engineering at graduate and post-graduate levels.

More information about this series at http://www.springer.com/series/11693

Junjie Zhang · Bing Guo
Jianguo Zhang
Editors

Simulation and Experiments of Material-Oriented Ultra-Precision Machining

Springer

Editors
Junjie Zhang
Center for Precision Engineering
Harbin Institute of Technology
Harbin, Heilongjiang, China

Bing Guo
School of Mechatronics Engineering
Harbin Institute of Technology
Harbin, Heilongjiang, China

Jianguo Zhang
School of Mechanical Science
and Engineering
Huazhong University of Science
and Technology
Wuhan, China

ISSN 2195-9862 ISSN 2195-9870 (electronic)
Springer Tracts in Mechanical Engineering
ISBN 978-981-13-3334-7 ISBN 978-981-13-3335-4 (eBook)
https://doi.org/10.1007/978-981-13-3335-4

Library of Congress Control Number: 2018961689

This Springer imprint is published by the registered company Springer Nature Singapore Pte Ltd.
The registered company address is: 152 Beach Road, #21-01/04 Gateway East, Singapore 189721, Singapore

Contents

Fundamentals of Nanometric Cutting of Nanotwinned Copper

Junjie Zhang, Yongda Yan and Tao Sun

Abstract Nanotwinned (NT) Cu containing a high density of growth twin boundaries (TBs) is one emerging precious metal for its extraordinary properties of high strength, intermediate ductility, and high electric conductivity. In the present work, we elucidate the deformation mechanisms of nanotwinned Cu subjected to the diamond cutting-based nanometric cutting by means of molecular dynamics simulations, with an emphasis on examining the influence of intrinsic microstructural parameters and extrinsic machining parameter on the nanometric cutting processes. The underlying deformation mechanisms of the materials are further correlated with the evolution of machining forces and the formation of machined surface and chips. Our simulation results indicate that dislocation slip, dislocation–TBs interactions, and TBs-associated mechanisms work in parallel in the plastic deformation of the NT Cu. In particular, dislocation–TB interactions and TBs-associated mechanisms are strongly dependent on rake angle of cutting tool, TB inclination angle, TB spacing, and grain size, which leads to strong anisotropic cutting response of NT Cu that originates from the heterogeneous localized deformation.

1 Introduction

An interface is a common boundary separating two different phases and/or crystals of different chemical compositions and/or crystallographic misorientations. Interfaces have a strong impact on the chemical, physical, and mechanical properties of materials particularly with high surface-to-volume ratio, due to the chemical and/or crystallographic discontinuity. Furthermore, the impact of GBs is heavily influenced by the structural type of GBs. For instance, while nanocrystalline metals with average grain sizes down into the nanometer regime are strong but brittle due to severely confined space for dislocation motion, the introduction of aligned coherent twin boundaries (TBs) leads to a superior combination of ultra-high strength and interme-

J. Zhang (✉) · Y. Yan · T. Sun
Center for Precision Engineering, Harbin Institute of Technology, Harbin 150001, China
e-mail: zhjj505@gmail.com

J. Zhang et al. (eds.), *Simulation and Experiments of Material-Oriented Ultra-Precision Machining*, Springer Tracts in Mechanical Engineering,
https://doi.org/10.1007/978-981-13-3335-4_1

diate ductility in NT metals. The high strength of NT metals is a result of TBs acting as effective obstacles to dislocation motion, while the eminent ductility is owing to the enhanced dislocation nucleation sites provided by TBs [1, 2]. The NT metals thus have become emerging precious metals for the fabrication of advanced functional nanostructures for its extraordinary mechanical properties.

The diamond cutting-based nanometric cutting is one prominent technique for the fabrication of ultra-smooth surface as well as precise surface micro-/nanostructures for its nanometer-scale machining accuracy and shape accuracy [3, 4]. To facilitate the development of nanometric cutting technique, a fundamental understanding of the nanometric cutting mechanisms at the nanometer scale is essentially required. Since the tool edge radius is comparable with depth of cut in the diamond cutting process, the properties of workpiece material play an important even dominant role in the cutting process, which introduces great challenges in elucidating mechanisms of NT metals for its complex deformation behavior. TB-associated mechanisms also play an important role in the plastic deformation of NT metals, in addition to dislocation slip that dominates intragrain plasticity. Specifically, dislocation–TB interactions in terms of blocking, accommodation and transition, as well as TB migration and detwinning, are common deformation modes of NT metals under external loads [5–10]. More recently, Wang et al. presented direct evidence of interaction of dislocation with a single nanotwin by performing in situ nanoindentation [11]. Although deformation behaviors of NT metals under tension, torsion, compression, nanoindentation, and friction have been reported in enormous literature, the dislocation–TB interactions, particularly its role in determining the material removal and evolution of machining force in nanometric cutting of NT metals, are largely unknown.

A typical cutting route on NT metals involves multiple grains, which are of different crystallographic orientations as well as the inclination angles of aligned TBs with respect to the surface. Specifically, the inclination angle is also of great importance in understanding of the nanometric cutting mechanisms of NT Cu, as inclined internal twin planes with varying Schmid's factors possess distinguishable dislocation–TB interactions. Stukowski et al. [12] reported that the inclination of dislocation to TBs is strongly dependent on Schmid's factors of individual grains. Brown et al. [13] showed that the placement of TB orientation with respect to the loading axis greatly affects the plasticity reversibility of NT Cu. Furthermore, Wei et al. demonstrated that the plastic deformation of NT Cu nanowires can be governed purely by TB migration through designing the inclination angle of TBs [14]. Jang et al. reported detwinning-governed plasticity in NT nanopillars containing inclined TBs [15]. Zhang et al. demonstrated that the anisotropic frictional response of NT Cu originates from the heterogeneous localized deformation, which is strongly influenced by crystallographic orientation, twin boundary orientation, and loading condition [16]. Therefore, it is intriguing to explore the effect of inclination angle on the nanometric cutting of NT Cu. It has been demonstrated that the TB-related impact on the mechanical properties of NT metals is strongly dependent on the intrinsic sizes of internal microstructures. Recent experimental tests and atomistic simulations have shown that the strength of NT Cu has a strong dependence on TB spacing. Specifically, the strength of NT Cu first increases with decreasing TB spacing in a Hall–Petch manner. However, after reaching the maximum strength at the critical TB spacing, strength softening occurs when TBs further decrease [17]. In addition, it is found that the critical TB spacing depends on grain size; i.e., the smaller the grain size, the smaller the critical twin thickness [2]. Therefore, it is also intriguing to study the effect of TB spacing and grain size on the nanometric cutting of NT metals.

In light of these findings reported in the literature, in the present work we perform MD simulations to investigate the nanometric cutting of NT Cu using the diamond cutting tool. The influence of rake angle, inclination angle of TBs, TB spacing, and grain size on the nanometric cutting in terms of elastic and plastic deformation of the material, variations of machining force, machined surface quality, and chip formation is further studied. Without loss of physics, we first consider single NT Cu samples in the absence of grain boundaries (GBs) but containing twin planes that are parallel, inclined, and perpendicular to the free surface. To characterize the effect of cutting tool geometry, we perform nanometric cutting simulations on the same NT Cu sample containing 26° inclined TBs with respect to the free surface using diamond cutting tool with three different rake angles, which shows that the rake angle of 45° results in smaller energy dissipation and better machined surface quality than the other two rake angles of 0° and −45°. Then with the optimized rake angle of 45°, we address the effect of inclination angle of TBs by performing cutting along the same scratching direction but on different surfaces with varied inclination angles of the TBs cutting the surfaces. We found an inclination angle of 0° for the best machined surface quality. More practical polycrystalline NT Cu is then simulated, in which the effects of twin spacing and grain size on the cutting processes are studied. The transition in deformation mechanisms, the evolution of cutting force, and the cutting-induced microstructure change are closely investigated. Based on such atomistic simulations, we demonstrate that cutting of NT Cu exhibits strong heterogeneous characteristic, originating from the strong competition between individual deformation mechanisms. The chapter is organized as follows: In Sect. 2, we describe details about the models of NT Cu samples, the nanoscratching procedure, and the utilized defect analysis and visualization techniques. The results of nanometric cutting of NT Cu with emphasis on the aspects of rake angle, inclination angle, TB spacing, and grain size are then presented and discussed in Sect. 3. Finally, in Sect. 4 we summarize the results.

2 Simulation Method

The utilized MD model of nanometric cutting of NT Cu is composed of a NT Cu substrate and a diamond cutting tool, as shown in Fig. 1. The common neighbor analysis (CNA) [18] is utilized to identify types of lattice defects, and the coloring scheme is as follows: Blue stands for face-centered cubic (FCC) atoms, red for hexagonal close-packed (HCP) atoms, and gray for other atoms including surface atoms and dislocation cores. A single HCP-coordinated layer identifies a coherent

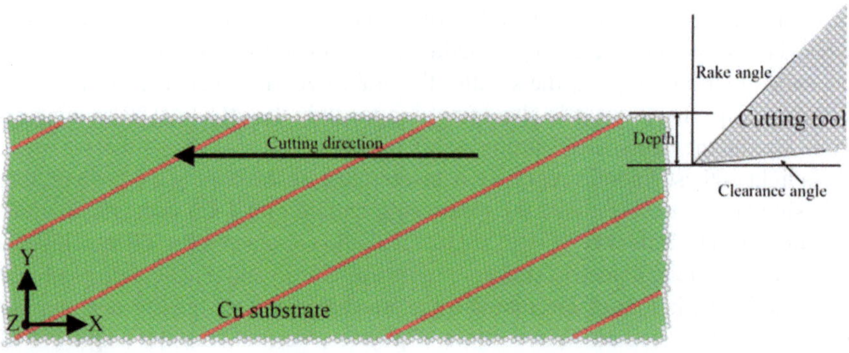

Fig. 1 MD model of nanometric cutting of NT Cu. Atoms are colored according to their CNA values

TB, and two adjacent HCP-coordinated layers indicate an intrinsic stacking fault (ISF) [19].

The NT Cu substrate has a dimension of 30, 18, and 10 nm in horizontal, longitudinal, and normal directions, respectively. Periodic boundary condition is only applied in Z direction, and the bottom of the substrate is fixed during the nanometric cutting process. Figure 2 presents atomic configurations of single NT Cu substrate that contains different inclined TBs with respect to the top free surface. In particular, five inclination angles, such as 0°, 26°, 45°, 64°, and 90°, respectively, are considered. For each NT Cu, the TB spacing is the same as 3.75 nm. The substrate containing inclined TBs is obtained by extracting the desired geometry from rotated bulk material that contains parallel TBs. Twin-free single crystal substrates that have the same crystal orientations with the five NT Cu substrates are also considered.

The fully three-dimensional polycrystalline NT Cu substrate is prepared by using the Voronoi construction. To address the influence of TB spacing, five substrates composed of nanoscale TBs with different TB spacing embedded in grains, from 0.63, 1.25, 1.67, and 2.92 to 3.75 nm, are considered, as shown in Fig. 3. The grain structure for each NT Cu sample is the same, and each sample contains 12 grains with an average grain size of 10 nm. In addition, for the TB spacing of 2.92 nm, four grain sizes, such as 6, 8, 10, and 14 nm, respectively, are considered to investigate the influence of grain size, as shown in Fig. 4.

The diamond cutting tool in diamond structure with a sharp edge has a relief angle of 10°. To investigate the influence of rake angle on the cutting process, three rake angles as −45° (referred to as negative), 0° (referred to as zero), and 45° (referred to as positive) are considered, as shown in Fig. 5. Given the ultra-high hardness of diamond as compared to copper and ultra-short cutting distance, the wear of diamond tool during the cutting process is not considered. Therefore, the diamond cutting tool is set as a rigid body, i.e., the coordinates and velocities of all the atoms in the cutting tool are updated every time step, in such a way the cutting tool moves as a single entity without any deformation.

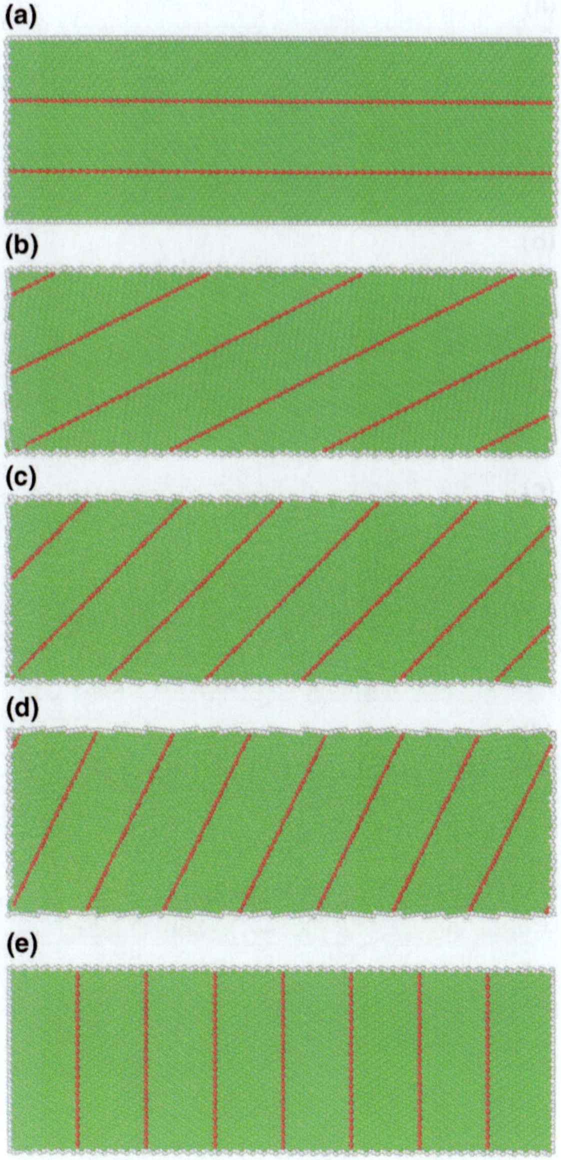

Fig. 2 Atomic configurations of single NT Cu containing inclined TBs. Inclination angles—**a** 0°, **b** 26°, **c** 45°, **d** 64°, and **e** 90°. Atoms are colored according to their CNA values

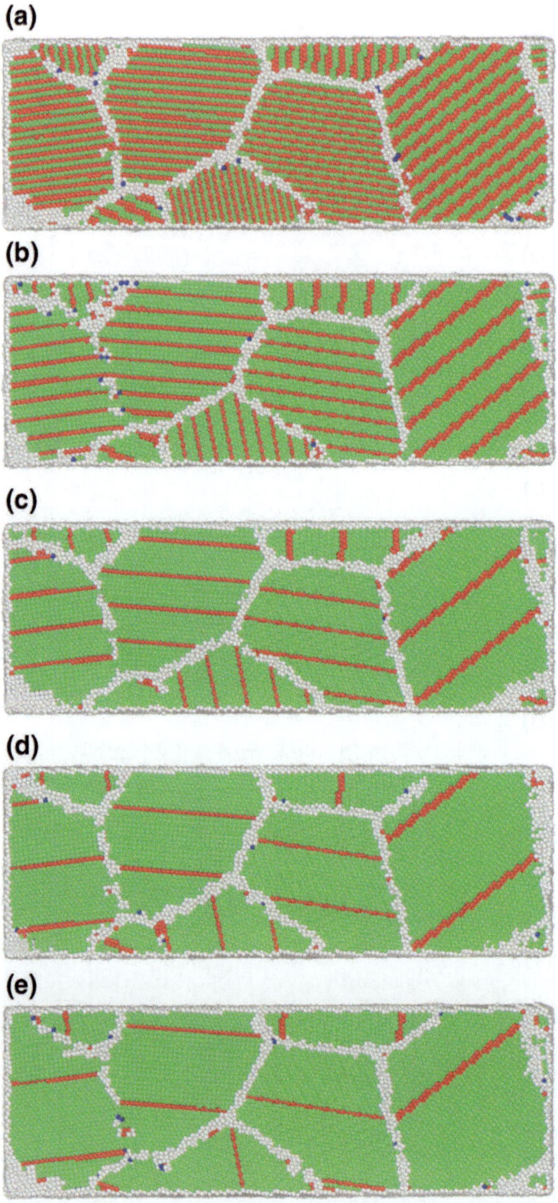

Fig. 3 Atomic configurations of polycrystalline NT Cu with different twin densities. TB spacing—**a** 0.63 nm, **b** 1.25 nm, **c** 1.67 nm, **d** 2.92 nm, and **e** 3.75 nm. Atoms are colored according to their CNA values

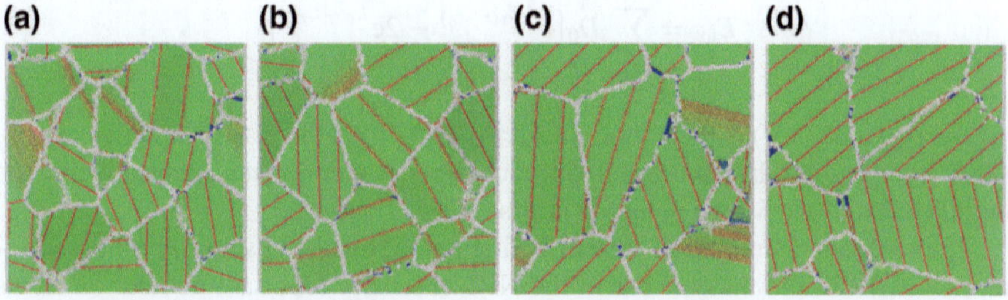

Fig. 4 Atomic configurations of polycrystalline NT Cu with different grain sizes. Grain size—**a** 6 nm, **b** 8 nm, **c** 10 nm, and **d** 14 nm. Atoms are colored according to their CNA values

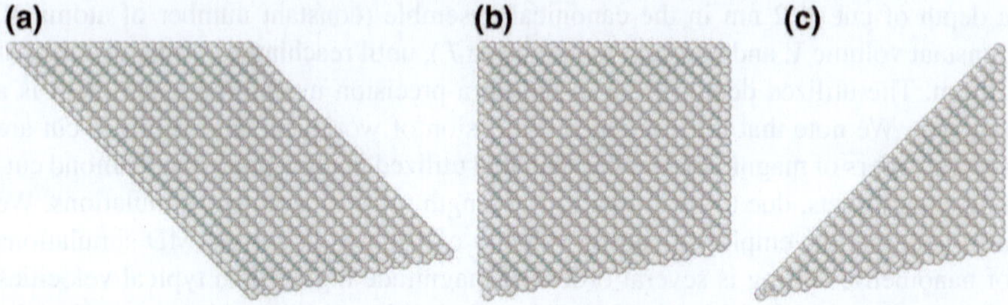

Fig. 5 Atomic configurations of diamond cutting tool with different rake angles. Rake angle—**a** −45°, **b** 0°, and **c** 45°

There are three types of atomic interactions in the simulated system, as Cu–Cu in the Cu substrates, Cu–C between the Cu substrate and the diamond cutting tool, and C–C in the diamond cutting tool, respectively. The C–C interactions are omitted as the diamond cutting tool is treated as a rigid body without any deformation in the cutting process. The embedded atom method (EAM) composed of interacting pair potential and electron embedding energy has been widely used to describe metallic systems, which can be expressed as:

$$E_{\text{tot}} = \frac{1}{2}\sum_{i,j}\phi_{ij}(r_{ij}) + \sum_{i} F_i(\rho_i) \tag{1}$$

$$\rho_i = \sum_{j \neq i}\rho_i(r_{ij}) \tag{2}$$

where r_{ij} is the distance between atoms i and j, ϕ_{ij} is the pair potential between atoms i and j, F_i represents the embedding energy that is generated when an atom i is embedded, ρ_i is the electron density at i atom generated by all atoms except atom i, and ρ_j is a function of the electron density of atom j at atom i. The EAM parameters for copper developed by Mishin et al. are utilized to describe the Cu–Cu interactions [20]. The Morse potential is utilized to describe the Cu–C interaction, which can be expressed as:

$$E_{\text{tot}} = \sum_{ij} D_0\left[e^{-2\alpha(r-r_0)} - 2e^{-\alpha(r-r_0)}\right] \tag{3}$$

where D_0 (0.087 eV) is the cohesive energy, α (5.14) is the elastic modulus, and r_0 (2.05 Å) represents the equilibrium distance between atoms *i* and *j*, respectively. The cutoff radius of the Morse potential is chosen as 1.0 nm [21].

The as-created simulation system is first equilibrated to its equilibrium configuration at 30 K and under 0 bar in the NPT ensemble (constant number of atoms *N*, constant pressure *P*, and constant temperature *T*). The low temperature of 30 K is chosen to minimize the thermal fluctuation influence. Then the equilibrated substrate is subjected to the diamond cutting with a constant velocity of 100 m/s and a depth of cut of 2 nm in the canonical ensemble (constant number of atoms *N*, constant volume *V*, and constant temperature *T*), until reaching a cutting distance of 20 nm. The utilized depth of cut in the ultra-precision machining experiment is a few μm. We note that the simulated dimension of workpiece and depth of cut are several orders of magnitude smaller than that utilized in ultra-precision diamond cutting experiments, due to the limitation of length scale in atomistic simulations. We also note that the employed cutting velocity of 100 m/s in current MD simulations of nanometric cutting is several orders of magnitude higher than typical velocities of tens of μm/s utilized in ultra-precision diamond cutting experiments, giving the intrinsic requirement of the integration time step to be of the order of femtosecond (fs). All the MD simulations are performed by using the LAMMPS code with an integration time step of 1 fs [22]. And the OVITO is utilized to visualize MD data and generate MD snapshots [23].

3 Results and Discussion

3.1 Nanometric Cutting Mechanisms of NT Cu

There are three force components acting on the diamond cutting tool, as the cutting force along *X* direction, the normal force along *Y* direction, and the lateral force along *Z* direction, respectively. Figure 6a, b plots variations of the cutting force and normal force with cutting length during the cutting process of the NT and single crystal Cu substrates with a rake angle of 0°, respectively. The NT Cu substrate contains 0° inclined TBs with respect to the free surface, and the TB spacing is 3.75 nm. When the cutting tool starts to contact with the left side of the NT Cu substrate, the material first undergoes elastic deformation, accompanied with a rapid increase of the cutting force. However, the normal force decreases precipitously after the initial increase, which is caused by the accommodation of elastic strain by the TBs. After increasing to a local maximum value, the scratching force decreases dramatically due to the yielding of the material. However, the normal force continues to increase at the onset of plasticity due to the positive value of the clearance angle. Upon further cutting,

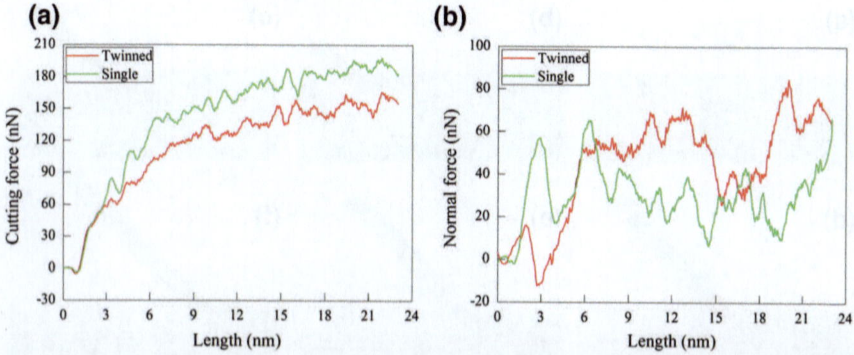

Fig. 6 Variations of machining force during nanometric cutting of twinned and single crystal Cu. **a** Cutting force and **b** normal force

both the cutting force and normal force increase with strong fluctuations caused by continuous yielding events. Figure 6 shows that the normal force is smaller than the cutting force during the cutting process.

It is also found from Fig. 6 that in the initial elastic deformation regime, both the cutting force and normal force are higher for the single crystal Cu than the twinned Cu. Furthermore, both the local maximum values of the cutting force and normal force associated with the yielding point are higher for the single crystal Cu than the twinned Cu, indicating that the introduction of TBs lowers the yield strength of the copper substrate in the cutting process. However, in the later period of plastic deformation regime, the normal force for the NT Cu is higher than the single crystal Cu due to the strengthening of the material by TBs.

Figure 7a–f presents instantaneous defect structures in the NT Cu at different cutting lengths. Atoms are colored according to their CNA values. Since the cutting tool first penetrates into the left side of the substrate, the plasticity is initiated by the nucleation of partial dislocations from the left surface and their subsequent glide on adjacent {111} slip planes in the NT Cu. Figure 7b shows that nucleated dislocation mainly glides inclined to both aligned TBs and rake face of the cutting tool. Furthermore, dislocation activity is mainly confined in the formed chip volume, and there is no dislocation observed beneath the machined surface. Upon further cutting, the severely confined dislocations in the chip annihilate in the upper surface of formed chip, as shown in Fig. 7c. Figure 7c also indicates that there are dislocations gliding along the rake face of the cutting tool, which leads to easy flow of removed material. With advance of the cutting tool, plastic deformation mainly localizes in the cutting zone, and dislocation motions are dominantly confined in the chip volume. Furthermore, no dislocation–TB interaction or TB-associated mechanism is observed. After the completion of cutting, a straight chip is formed, as shown in Fig. 7f.

Figure 8a, b presents representative cross-sectional views of instantaneous defect structures after nanometric cutting of the two Cu substrates, demonstrating that dislocation patterns in single crystal Cu are significantly different from the NT Cu.

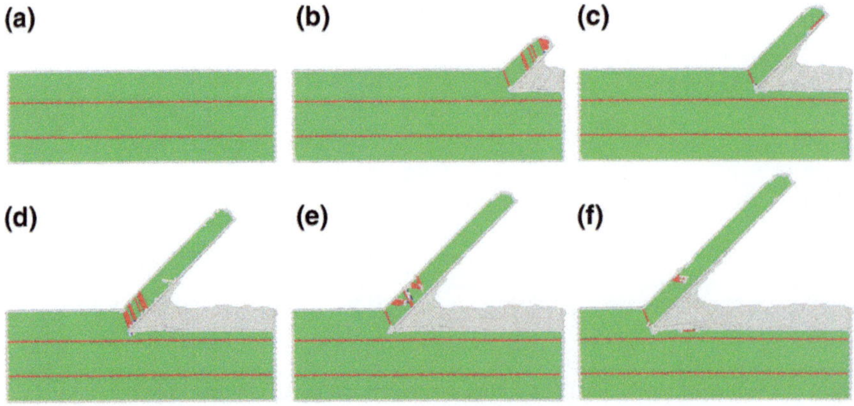

Fig. 7 MD snapshots of defect structures at different cutting lengths. **a** 0.0 nm, **b** 6.9 nm, **c** 11.5 nm, **d** 17.3 nm, **e** 20.0 nm, and **f** 23.0 nm. Atoms are colored according to their CNA values

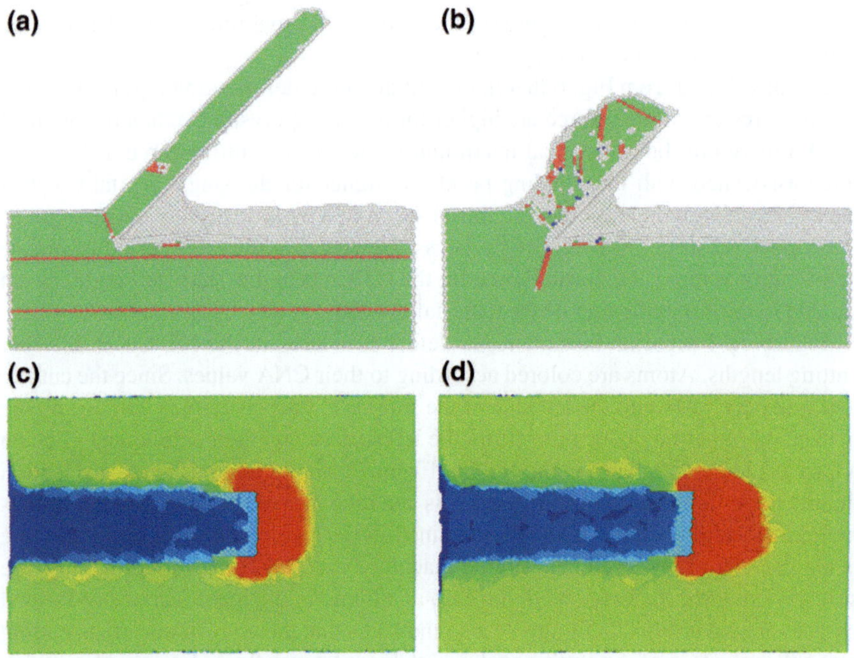

Fig. 8 Comparison between twinned Cu and single crystal Cu. Top row: MD snapshots of defect structures after cutting processes. **a** Twinned Cu and **b** single crystal Cu. Atoms are colored according to their CNA values. Bottom row: Machined surface morphology after cutting processes. **c** Twinned Cu and **d** single crystal Cu. Atoms are colored according to their heights in the range from 2.0 to 6.3 nm

Figure 8a shows that the plastic deformation is mainly confined in the chip volume, and there are rather limited dislocations below the machined surface. In contrast, the plastic deformation in the single crystal Cu substrate is more severe than the NT Cu substrate for its higher dislocation density. Furthermore, there is an emission of lattice partial dislocation from the edge of cutting zone and glide toward bottom of the substrate. Resultantly, both the chip profile and machined surface morphology of the NT Cu substrate are more uniform than the single crystal Cu substrate, as shown in Fig. 8c, d. In the NT Cu, dislocations with low density glide freely in the chip volume with long distance, as indicated by Fig. 8a. In contrast, the motion of nucleated partial dislocations with high density in the single crystal Cu is blocked by each other in the formed chip volume, which leads to strengthening of the material. Consequently, smaller machining force is required for the NT Cu than the single crystal Cu.

3.2 *Effect of Rake Angle*

Figure 9 plots the cutting force–length curves and normal force–length curves in the cutting processes of NT Cu substrate with different rake angles. The NT Cu substrate contains aligned 26° inclined TBs, which gives the lowest yield strength and the highest friction coefficient, and due to that the plasticity is dominated by TB migration and detwinning [16]. It is seen from Fig. 9a that the variation of the cutting force for each rake angle has similar characteristics: It first increases rapidly in the elastic deformation regime, then drops dramatically when the plasticity in the substrate initiates, and finally fluctuates around a constant value when the cutting process is stable. However, it is found from Fig. 9a that the variation of cutting force strongly depends on the rake angle. Specifically, the larger the rake angle, the smaller the period of elastic deformation regime and the smaller the cutting force. The normal force also has similar rake angle dependence, as shown in Fig. 9b. For the rake angles of −45° or 0°, the normal force first increases in elastic deformation regime and finally fluctuates around a positive value in lateral plastic deformation. However, the normal force for the rake angle of 45° mainly fluctuates around 0, indicating that the normal force acting on the cutting tool is negligible [24].

Figure 10 presents MD snapshots of instantaneous defect structures in the NT Cu substrate after nanometric cutting with different rake angles. It is found from Fig. 10 that for each rake angle, in front of the cutting tool the formed chip mainly flows along tool rake surface. There are considerable dislocations accumulated in the vicinity of the TB, indicating that the TB acts as a barrier to dislocation motion. Being similar with conventional strengthening mechanisms by high-angle grain boundaries (GBs), the TB blocking dislocation motion also leads to an increase in strength of the material. In contrast to conventional GB-associated strengthening strategy that is accompanied with a compromise in ductility, the NT Cu does not lose its ductility because of the dislocation nucleation sites provided by the migrated TBs. The nucleation of dislocations from the migrated TBs leads to softening of the material.

It is found that during the nanometric cutting of the NT Cu, TBs can act as sinks and sources to dislocation motion, and there is TB migration observed, which are consistent with previous tension studies. However, the rake angle has a significant impact on either the chip profile or the internal microstructure in the chip. Specifically, the larger the rake angle, the more compliant the chip profile. For the rake angles of −45° and 0°, Fig. 10a, b shows that dislocations are mainly inclined to tool surface. In contrast, dislocations mainly glide parallel to the tool surface for the rake angle of 45°, as shown in Fig. 10c [24].

Figure 11 presents machined surface morphologies of the NT Cu substrate after nanometric cutting with different rake angles. Atoms are colored according to their atomic heights. Figures 10 and 11 jointly demonstrate that the rake angle has a strong influence on the machined surface quality. Figure 11 shows that for each rake angle, there is an accumulation of dislocations parallel to TB formed in front of the probe. However, the dislocation density decreases with increasing rake angle. Consequently, Fig. 11c shows that the deformation of the TBs for the rake angle of 45° is negligible, while TB migration is pronounced for the rake angles of −45° and 0°. While the rake angle of 45° or 0° has a larger chip volume than the rake angle of −45°, the surface pileup accumulated on both sides of as-fabricated groove is less pronounced. Figure 11 shows that a smooth machined surface can be obtained for each rake angle. However, the machined surface is smoother for the rake angle of 45° than that for the other two rake angles, although there is material accumulated in the end of the machined surface for the rake angles of 45° or 0°. Furthermore, the depth of subsurface damage layer is the lowest for the largest rake angle of 45°. The rake angle dependence of nanometric cutting of NT Cu can be mainly attributed to the stress state between the tool surface and the substrate, which undergoes a transition from compressive stress to shear stress with the increase of rake angle [24].

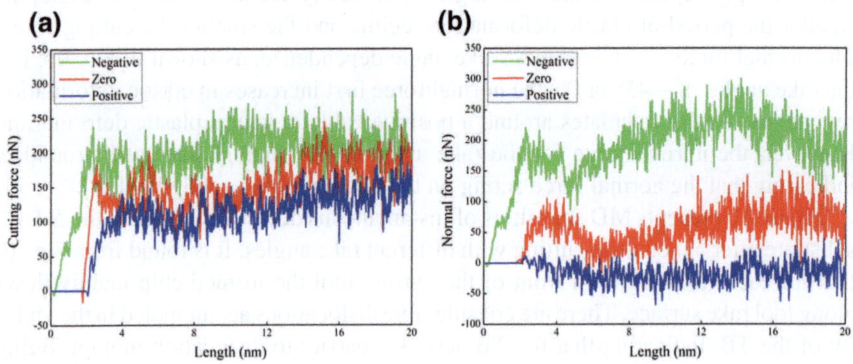

Fig. 9 Influence of rake angle on variations of machining force of NT Cu containing 26° inclined TBs. **a** Cutting force and **b** normal force [24]

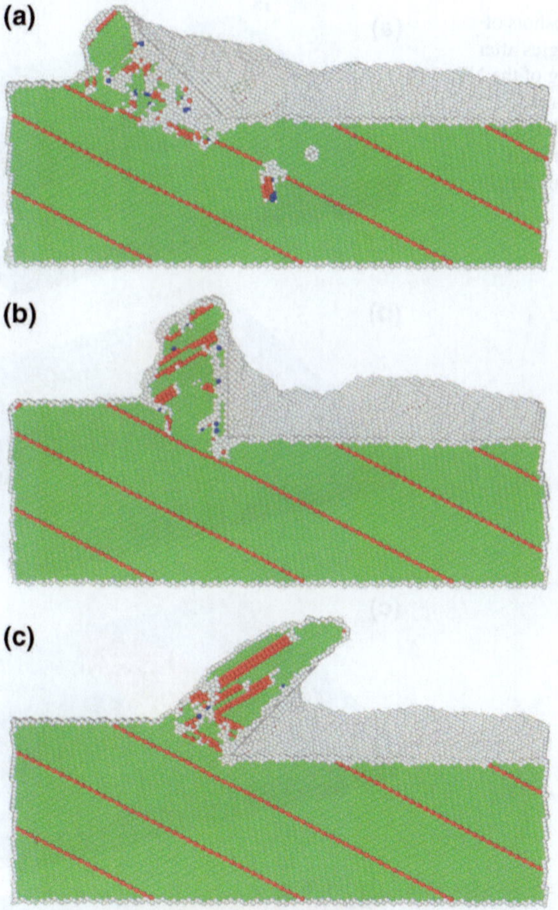

Fig. 10 MD snapshots of defect structures after nanometric cutting of the NT Cu with different rake angles—**a** −45°, **b** 0°, and **c** 45°. Atoms are colored according to their CNA values [24]

3.3 Effect of Inclination Angle

Figure 12 plots the cutting force–length curves and normal force–length curves in the cutting processes of NT Cu substrates containing aligned TBs with different inclination angles. The rake angle of diamond cutting tool is 45°. It is seen from Fig. 9a that the variation of the cutting force for each inclination angle has similar characteristics: It first becomes negative value due to the adhesion between diamond cutting tool and substrate when the cutting tool is approaching the substrate. After the cutting tool contacts with the substrate, the cutting force increases rapidly in the elastic deformation regime and then drops dramatically when the plasticity in the substrate initiates. Upon further cutting, the cutting force increases and finally

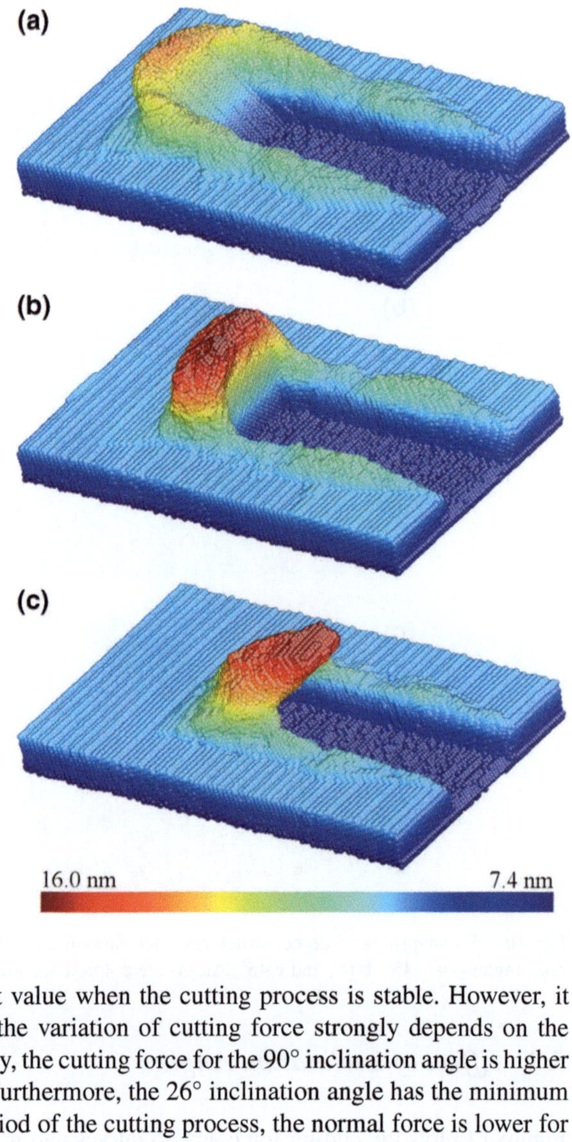

Fig. 11 MD snapshots of surface morphologies after nanometric cutting of the NT Cu with different rake angles—**a** −45°, **b** 0°, and **c** 45°. Atoms are colored according to atomic heights [24]

fluctuates around a constant value when the cutting process is stable. However, it is found from Fig. 9a that the variation of cutting force strongly depends on the inclination angle. Specifically, the cutting force for the 90° inclination angle is higher than the other four angles. Furthermore, the 26° inclination angle has the minimum cutting force. In the first period of the cutting process, the normal force is lower for the higher inclination angle. However, this trend is reversed in the later period of the cutting process.

Figure 13 presents cross-sectional views of MD snapshots of defect structures in the NT Cu substrate containing aligned TBs with different inclination angles after nanometric cutting processes. The defect structures in single crystal Cu substrate are also presented for comparison. It is found from Fig. 13 that for the single crystal Cu substrates, there are dislocations existed in both the formed chip volume and beneath the machined surface. Furthermore, there are multiple {111}<110> slip planes activated. In particular, there are mechanical TBs formed in the formed chip,

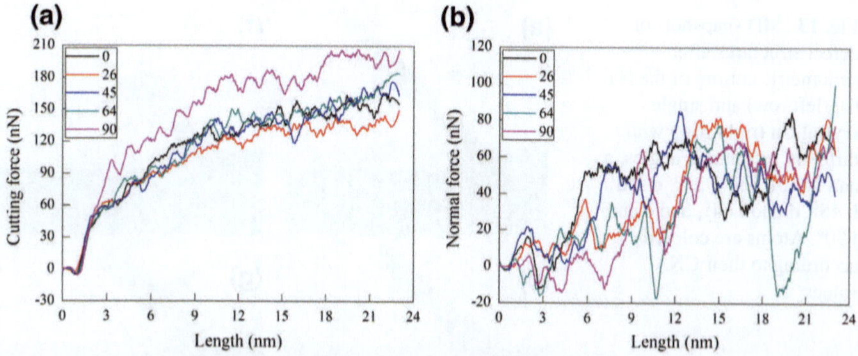

Fig. 12 Influence of inclination angle on the machining force of NT Cu. **a** Cutting force–length curves and **b** normal force–length curves

which have different inclination angles with respect to the rake face of the cutting tool. In contrast, dislocation density in the NT Cu substrates is significantly lower than that in the single crystal Cu substrates. Furthermore, the motion of dislocations in the NT Cu substrate is confined by the fine twin lamellas. In particular, there are considerable mechanical TBs that are parallel to the rake face of the cutting tool formed in the formed chip of NT Cu substrate. It is also found that the deformed TBs in the wake of the probe undergo purely recovery.

It is seen from Fig. 13 that the inclination angle has a significant influence on the microscopic deformation of NT Cu substrates. The dislocation density for the 0° inclination angle is significantly lower than that for the other four inclination angles. In the cutting process of the NT Cu with 26° inclination angle, the glide of lattice dislocations on the twin plane dominates the plastic deformation, accompanied with the nucleation of twinning partials caused by TB migration. Simultaneously, there is considerable chip containing inclined dislocations to TB formed in front of the cutting tool. Upon further cutting, there are also intersections of inclined dislocation with TB which leads to significant TB migration occurred, which subsequently acts as dislocation nucleation sites. For the 90° inclination angle, the blocking of dislocation motion by the perpendicular TBs to the free surface is the most pronounced. Figure 2e shows there are two dislocations intersecting with the same TB in the neighboring two twin lamellae. In contrast, dislocations mainly glide parallel to the TBs for the 64° inclination angle.

Figure 14 presents machined surface morphologies of the NT Cu substrate after nanometric cutting with different inclination angles. The formed chip is removed for each NT Cu substrate for better visualizing of machined surface quality. Atoms are colored according to their atomic heights. Figure 14 demonstrates that the inclination angle has a strong influence on the machined surface quality.

Figure 14 shows that the machined surface qualities for the inclination angle of 0° and 64° are better than the other three inclination angles. Furthermore, the comparison between single crystal and NT Cu substrates demonstrates that the machined surface

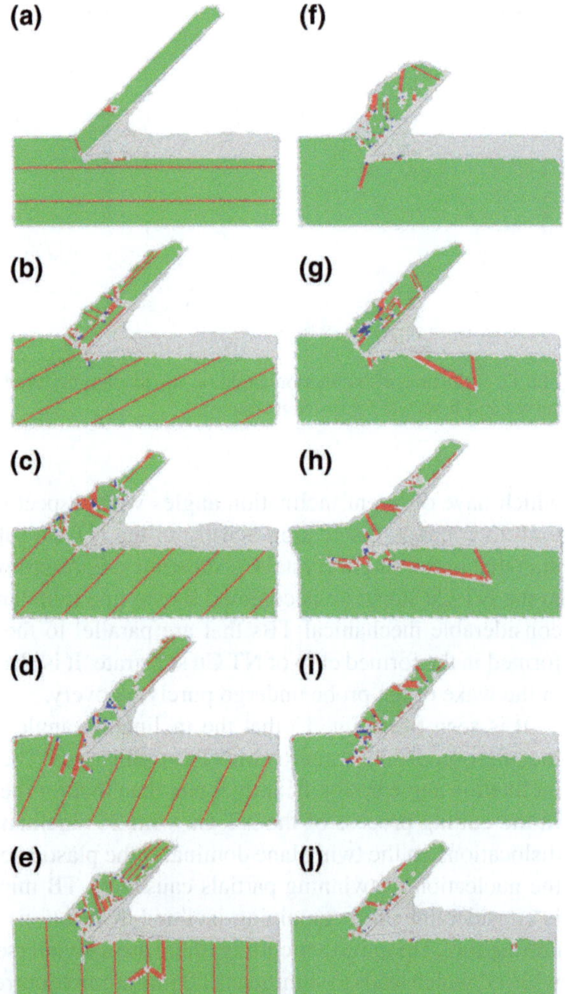

Fig. 13 MD snapshots of defect structures after nanometric cutting of the NT Cu (left row) and single crystal Cu (right row) with different inclination angles. **a** and **f** 0°, **b** and **g** 26°, **c** and **h** 45°, **d** and **i** 64°, and **e** and **j** 90°. Atoms are colored according to their CNA values

qualities for the inclination angle of 0° and 64° are better than their single crystal counterparts.

3.4 Effect of TB Spacing

Figure 15 plots the cutting force–length curves and normal force–length curves in the cutting processes of polycrystalline NT Cu substrates containing aligned TBs with different TB spacing. The rake angle of diamond cutting tool is 45°. It is seen from Fig. 15a that the variation of the cutting force for each TB spacing has similar

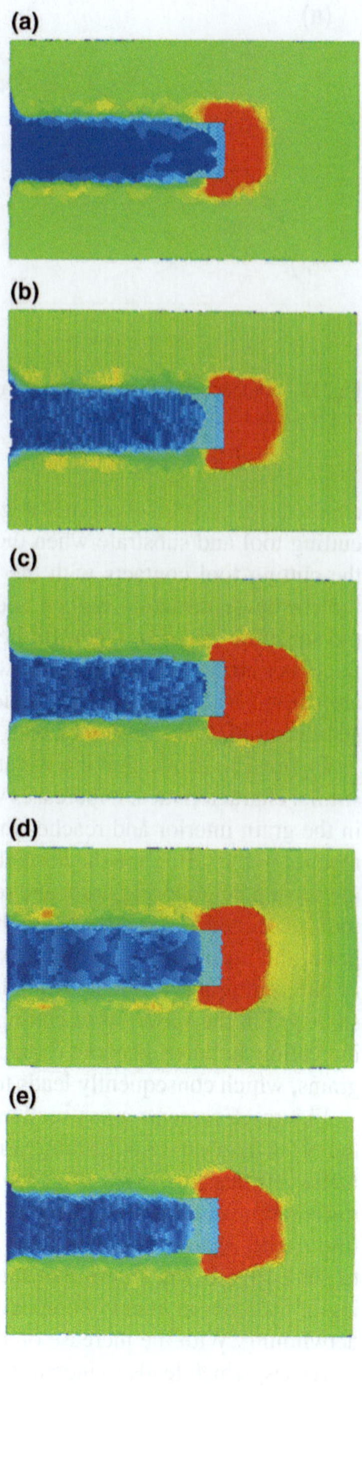

Fig. 14 MD snapshots of surface morphologies after nanometric cutting of the NT Cu with different inclination angles. **a** 0° **b** 26°, **c** 45°, **d** 64°, and **e** 90°. Atoms are colored according to their atomic heights

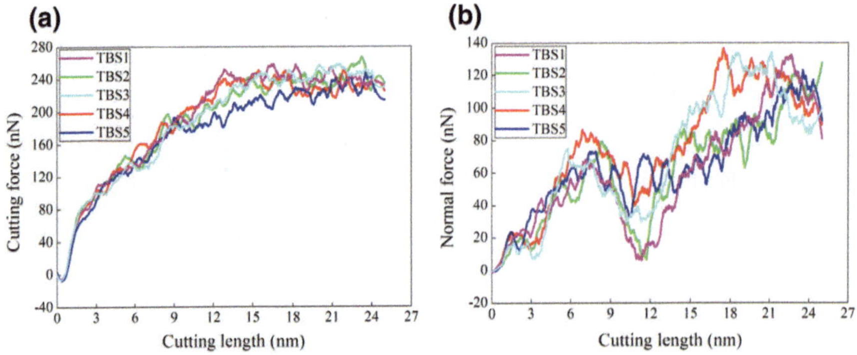

Fig. 15 Influence of TB spacing on the machining force of NT Cu. **a** Cutting force–length curves and **b** normal force–length curves

characteristics: It first becomes negative value due to the adhesion between diamond cutting tool and substrate when the cutting tool is approaching the substrate. After the cutting tool contacts with the substrate, the cutting force increases rapidly in the elastic deformation regime and then drops dramatically when the plasticity in the substrate initiates. Upon further cutting, the cutting force increases and finally fluctuates around a constant value when the cutting process is stable. The fluctuations are caused by successive dislocation nucleation events. Figure 15a shows that the stable cutting force decreases with increasing TB spacing.

Figure 15b shows that the variation of the normal force for each TB spacing has similar characteristics: It increases with strong fluctuations when cutting is performed in the grain interior and reaches the local maximum value when the cutting tool is reached at the GB. Then the normal force decreases when the cutting is performed in the second neighboring grain and reaches the local minimum value when the cutting tool is reached at the GB. When the cutting is performing in the third neighboring grain, the normal force increases with fluctuations. The variations of normal force in different grains originate from different crystal orientations. Furthermore, for the same grain, the variation of cutting force is also different for different TB spacing, indicating the heterogeneous characteristics of microscopic deformation in individual grains, which consequently leads to anisotropic cutting response.

Figure 16 presents cross-sectional views of MD snapshots of defect structures in the NT Cu substrate containing aligned TBs with different TB spacing after nanometric cutting processes. It is found from Fig. 16 that for small TB spacing, the microscopic deformation of the NT Cu is dominated by detwinning and TB migration, rather than dislocation slip. Furthermore, dislocations mainly glide parallel to TBs. Figure 16a shows that the formed chip for the NT Cu with the TB spacing of 0.63 nm is mainly composed of FCC atoms, which is a result of significant detwinning. With the increase of TB spacing, the space for dislocation motion also increases, which leads to increase of dislocation density. Figure 16 shows that there

Fig. 16 MD snapshots of defect structures after nanometric cutting of the NT Cu with different TB spacing. **a** 0.63 nm, **b** 1.25 nm, **c** 1.67 nm, **d** 2.92 nm, and **e** 3.75 nm. Atoms are colored according to their CNA values

Fig. 17 MD snapshots of surface morphologies after nanometric cutting of the NT Cu with different TB spacing. **a** 0.63 nm, **b** 1.25 nm, **c** 1.67 nm, **d** 2.92 nm, and **e** 3.75 nm. Atoms are colored according to their atomic heights

are dislocations both parallel and inclined to TBs observed for large TB spacing, and there are considerable dislocations formed in the chip volume.

Figure 17 presents machined surface morphologies of the NT Cu substrate with different TB spacing after nanometric cutting processes. Atoms are colored according to their atomic heights. Figure 17 demonstrates that the chip profile is similar to each TB spacing. However, the volume of surface pile up in the cutting zone is smaller for larger TB spacing, which is attributed to larger space for dislocation motion to accommodate plastic deformation. In addition, the volume of surface pile on both sides of the groove increases with increasing TB spacing, which is caused by more pronounced dislocation annihilation at the free surface. There are more dislocations gliding upward the free surface for larger TB spacing.

3.5 Effect of Grain Size

Figure 18 plots the cutting force–length curve and normal force–length curve in the cutting processes of polycrystalline NT Cu substrates with different grain sizes. The rake angle of diamond cutting tool is 45°. It is seen from Fig. 18a that the variation of the cutting force for each grain size has similar characteristics: It first becomes negative value due to the adhesion between diamond cutting tool and substrate when the cutting tool is approaching the substrate. After the cutting tool contacts with the substrate, the cutting force increases rapidly in the elastic deformation regime and then drops dramatically when the plasticity in the substrate initiates. Upon further cutting, the cutting force increases and finally fluctuates around a constant value when the cutting process is stable. The fluctuations are caused by successive dislocation nucleation events. While the number of grains is different for different grain sizes, the cutting lengths for transition points in the cutting force–length curves are also different for different grain sizes. It is seen from Fig. 18a that the cutting force is

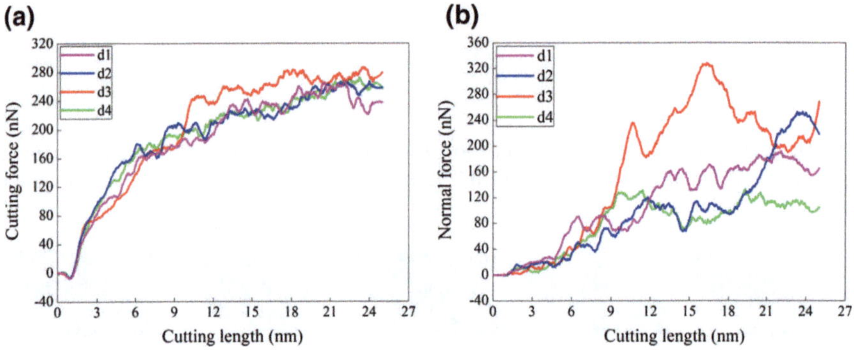

Fig. 18 Influence of grain size on the machining force of NT Cu. **a** Cutting force–length curves and **b** normal force–length curves

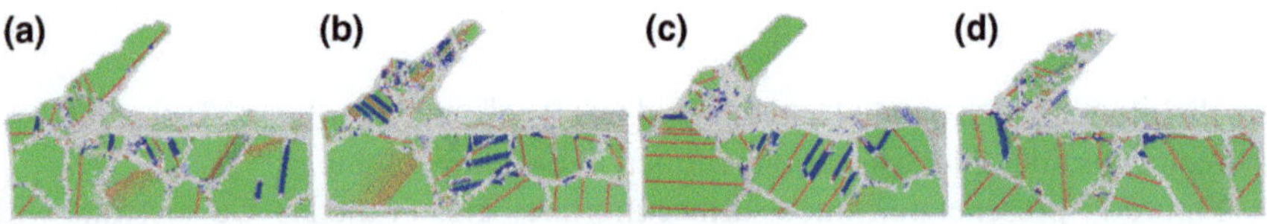

Fig. 19 MD snapshots of defect structures after nanometric cutting of the NT Cu with different grain sizes. **a** 6 nm, **b** 8 nm, **c** 10 nm, and **d** 14 nm. Atoms are colored according to their CNA values

the largest for the grain size of 10 nm (*d*3). Figure 18b shows that the variation of normal force with cutting length is different for each grain size. The normal force is the largest for the grain size of 10 nm (*d*3). It is indicated that there is a critical grain size of 10 nm for the largest cutting force and normal force.

Figure 19 presents cross-sectional views of MD snapshots of defect structures in the NT Cu substrate with different grain sizes after nanometric cutting processes. It is seen from Fig. 19 that the microscopic deformation behavior is different in different grain interiors in the same NT Cu substrate. Figure 19c indicates that lattice partial dislocations mainly glide parallel to twin planes in the centered grain. However, there are dislocations inclined to TBs observed in the right neighboring grain. In particular, there are considerable mechanical TBs formed in the NT Cu with small grain size. The newly formed TBs intersect with pre-existing TBs. The larger the grain size, the more dislocation activity, and the less pronounced the GB-associated mechanisms.

Figure 20 presents machined surface morphologies of the NT Cu substrate with different grain sizes after nanometric cutting processes. Atoms are colored according to their atomic heights. Figure 20 demonstrates that the chip profile is strongly dependent on grain size. With the increase of grain size, the uniformity along the rake face of the cutting tool decreases, and the number of dislocations in the chip volume increases. There is large space for dislocation motion in NT Cu with large grain size, and consequent dislocation–GB interactions lead to strong anisotropic microscopic deformation in individual grains, which deteriorates the long-range uniformity of

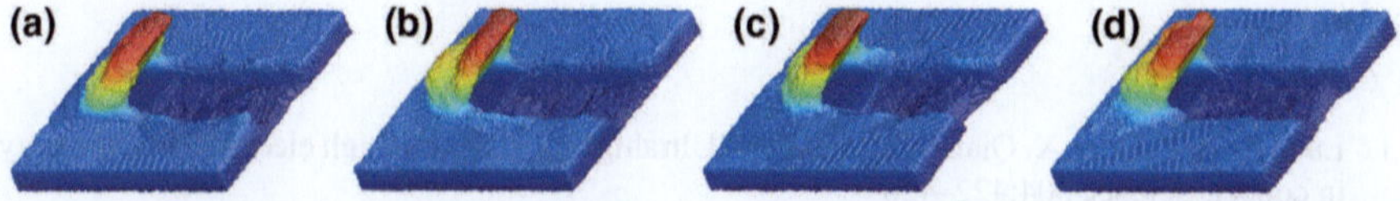

Fig. 20 MD snapshots of surface morphologies after nanometric cutting of the NT Cu with different grain sizes. **a** 6 nm, **b** 8 nm, **c** 10 nm, and **d** 14 nm. Atoms are colored according to their atomic heights

chip profile. Figure 20 also shows that the volume of surface pileup in front of the cutting tool increases with grain size, which is caused by the dislocation annihilation at the free surface. However, the volume of surface pileup on both sides of formed groove decreases with increasing grain size.

4 Summary

In summary, we perform MD simulations to investigate the nanometric cutting of NT Cu containing different internal microstructures using the diamond cutting tool. Simulation results indicate that in addition to dislocation slip that exclusively dominates the plastic deformation of single crystal Cu, dislocation–TB interactions and TBs-associated mechanisms are two important deformation modes for the NT Cu. Specifically, TBs act as both dislocation barriers and sources. Furthermore, TB migration accompanied with nucleation of twinning partials plays important role in the plastic deformation of the NT Cu. It is shown that the nanometric cutting of NT Cu exhibits strong dependence on both intrinsic and extrinsic parameters. It is found that both the dislocation density beneath machined surface and the machining force decrease with increasing rake angle due to the transition in stress state between tool surface and the material, and a compliant chip profile and fine machined surface quality of NT Cu can be achieved with a large rake angle of 45°. The strong competition between individual deformation modes is strongly dependent on TB inclination, TB spacing, and grain size, which leads to the strong anisotropic cutting response of NT Cu that originates from the heterogeneous localized deformation.

Acknowledgements The authors greatly acknowledge financial support from the Science Challenge Project (No. TZ2018006-0201-02), the Fundamental Research Funds for the Central Universities, the National Natural Science Foundation of China (51405106) and the State Key Laboratory of Precision Measuring Technology and Instruments (Tianjin University) (PIL1405).

References

1. Lu L, Shen Y, Chen X, Qian L, Lu K (2004) Ultrahigh strength and high electrical conductivity in copper. Science 304:422–426
2. Li XY, Wei YJ, Lu K, Gao HJ (2010) Dislocation nucleation governed softening and maximum strength in nano-twinned metals. Nature 464:877–880
3. Yan YD, Hu ZJ, Zhao XS, Sun T, Dong S, Li XD (2010) Top-down nanomechanical machining of three-dimensional nanostructures by atomic force microscopy. Small 6:724–728
4. Sun J, Luo X, Chang W, Ritchie JM, Chien J, Lee A (2012) Fabrication of periodic nanostructures by single-point diamond turning with focused ion beam built tool chips. J Micromech Microeng 22:115014
5. Wu ZX, Zhang YW, Srolovitz DJ (2009) Dislocation–twin interaction mechanisms for ultrahigh strength and ductility in nanotwinned metals. Acta Mater 57:4508
6. Cao AJ, Wei YG (2007) Molecular dynamics simulation of plastic deformation of nanotwinned copper. J Appl Phys 102:083511
7. Tsuru T, Kaji Y, Matsunaka D, Shibutani Y (2010) Incipient plasticity of twin and stable/unstable grain boundaries during nanoindentation in copper. Phys Rev B 82:024101
8. Anderoglu O, Misra A, Wang J, Hoagland RG, Hirth JP, Zhang X (2010) Plastic flow stability of nanotwinned Cu foils. Int J Plast 26:875
9. Kulkarni Y, Asaro RJ (2009) Are some nanotwinned fcc metals optimal for strength, ductility and grain stability? Acta Mater 57:4835
10. Qu SX, Zhou HF (2010) Hardening by twin boundary during nanoindentation in nanocrystals. Nanotechnology 21:335704
11. Wang B, Zhang ZY, Cui JF, Jiang N, Lyu JL, Chen GX, Wang J, Liu ZD, Yu JH, Lin CT, Ye F, Guo DM (2017) In situ TEM study of interaction between dislocations and a single nanotwin under nanoindentation. ACS Appl Mater Interfaces 9(35):29451–29456. https://doi.org/10.1021/acsami.7b11103
12. Stukowski A, Albe K, Farkas D (2010) Nanotwinned fcc metals: strengthening versus softening mechanisms. Phys Rev B 82:224103
13. Brown JA, Ghoniem NM (2010) Reversible–irreversible plasticity transition in twinned copper nanopillars. Acta Mater 58:886–894
14. Wei YJ (2011) Anisotropic size effect in strength in coherent nanowires with tilted twins. Phys Rev B 84:014107
15. Jang DC, Li XY, Gao HJ, Greer JR (2012) Deformation mechanisms in nanotwinned metal nanopillars. Nat Nanotechnol 7:594–601
16. Zhang JJ, Hartmaier A, Wei YJ, Yan YD, Sun T (2013) Mechanisms of anisotropic friction in nanotwinned Cu revealed by atomistic simulations. Modell Simul Mater Sci Eng 21:065001
17. Lu L, Chen X, Huang X, Lu K (2009) Revealing the maximum strength in nanotwinned copper. Science 323:607–610
18. Honeycutt JD, Andersen HC (1987) Molecular dynamics study of melting and freezing of small Lennard-Jones clusters. J Phys Chem 91:4950–4963
19. Zhang JJ, Wei YJ, Sun T, Hartmaier A, Yan YD, Li XD (2012) Twin boundary spacing-dependent friction in nanotwinned copper. Phys Rev B 85:054109
20. Mishin Y, Mehl MJ, Papaconstantopoulos DA, Voter AF, Kress JD (2001) Phys Rev B 63:224106
21. Yan YD, Sun T, Dong S, Luo XC, Liang YC (2006) Molecular dynamics simulation of processing using AFM pin tool. Appl Surf Sci 252:7523–7531
22. Plimpton S (1995) Fast parallel algorithms for short-range molecular dynamics. J Comput Phys 117:1–19
23. Stukowski A (2010) Visualization and analysis of atomistic simulation data with OVITO–the open visualization tool. Modell Simul Mater Sci Eng 18:015012
24. Zhang JJ, Geng L, Yan YD, Sun T (2015) Effect of tool geometry in nanometric cutting of nanotwinned Cu: a molecular dynamics study. Int J Nanomanuf 11:138–149

Investigation into Plastic Deformation and Machining-Induced Subsurface Damage of High-Entropy Alloys

Jia Li and Qihong Fang

Abstract High-entropy alloys (HEAs), which contain more than five principal elements with equal or near equal atomic percent, exhibit high wear resistant, high strength, and great plasticity. However, the plastic deformation mechanism and the machining-induced subsurface damage of HEAs at nanoscale are not yet fully understood, to limit their widely practical utility. Based on the experiment, AlCrFeCuNi HEA of atomic model is built through a melting and quick quenching method. In this work, we study the mechanical behaviors of AlCrFeCuNi HEA under uniaxial tensile loading and scratching processes by molecular dynamics (MD) simulations, in terms of the scratching force, atomic strain, atomic displacement, microstructural evolution, and dislocation density. The results show that the HEA obtained from MD simulations not only has high strength, but also exhibits good plasticity which is qualitatively consistent with the experiment. The dislocation gliding, dislocation pinning, and twinning subjected to the severe atomic lattice distortion and solid solution effects are still the main mechanism of plastic deformation in HEA. In addition, the larger tangential and normal forces and higher friction coefficient take place in HEA due to its outstanding strength and hardness, and high adhesion over the pure metal materials. Furthermore, the excellent comprehensive scratching properties of the bulk HEA are associated with the combined effects of multiple strengthening mechanisms, such as dislocation strengthening, deformation twinning strengthening as well as solute strengthening. This atomistic mechanism provides a fundamental understanding of plastic deformation and scratching behavior in HEA.

J. Li · Q. Fang (✉)
State Key Laboratory of Advanced Design and Manufacturing for Vehicle Body, Hunan University, Changsha 410082, People's Republic of China
e-mail: fangqh1327@hnu.edu.cn

J. Zhang et al. (eds.), *Simulation and Experiments of Material-Oriented Ultra-Precision Machining*, Springer Tracts in Mechanical Engineering,
https://doi.org/10.1007/978-981-13-3335-4_2

1 Introduction

The development of new materials drives the progress of human society, which experienced the Stone Age, the Iron Age, the Steam Engine Age, and the Information Age. To widen the application range of metals, we need obtain alloy materials with superior combinations of strength and toughness, for daily life, national defense, and outer space exploration [1, 2]. Recent years, "high-entropy alloys (HEAs)" or "multi-principal element alloys (MPEAs)," consisting of equal or near equal atomic multi-elements, have attracted a large number of scholars from different fields, due to their high wear resistant, high strength, high hardness, and good ductility [3–6]. Compared to the conventional alloy systems, HEAs have mainly four unique effects, such as high-entropy effects in thermodynamics, sluggish diffusion in kinetics, severe lattice distortion in structures, and cocktail effects in properties [7].

A large number of experimental studies have been investigated on the relationship between microstructure and mechanical properties in HEAs [8–12]. The tensile strength of the single-phase FCC CrMnFeCoNi HEAs can improve with reducing the temperature, due to the continuous steady strain hardening from a planar-slip dislocation to mechanical nanotwinning transition [3]. The fatigue life of $Al_{0.5}CoCrCuFeNi$ HEAs is greater than 10^7 cycles at stresses above 858 MPa, attributed to the multilevel defect densities introduced by fabrication and processing [13]. The dual-phase HEAs adjusted by the alloying contents exhibit the excellent strength–ductility combinations owing to transformation-induced plasticity [14]. In addition, the wear behavior of HEAs is important to their industrial applications, because the plastic deformation and work-hardening take place during the wear process [15]. The tribological properties of the CuCrFeTiZn HEAs show a good combination of plasticity and wear resistance, which are the result of the presence of FCC and BCC phases distributed homogenously [16]. Compared to the traditional alloy systems, the TiZrHfNb HEAs exhibit an improved wear resistance and the lower coefficient of friction [17]. Many of these experimental observations have deeply revealed the deformation behavior of single-phase and dual-phase HEAs.

To intuitively observe the microstructure evolution controlling the mechanical properties, large-scale MD simulations have played an essential part in exploring the microstructural evolution mechanism of metal and alloy materials [18, 19]. The rapidly solidified structure of NiAlCuCoTiVZnZr HEA and the thin film growth of AlCoCrCuFeNi HEA have been modeled using MD simulation, indicating that the degree of disorder increases with the number of elements [20, 21]. The mechanical behaviors of AlCrFeCuNi and $Al_{0.1}CoCrFeNi$ HEAs under uniaxial tension are presented via MD simulations, showing the dislocation gliding, dislocation pinning, and twinning due to the severe atomic lattice distortion and solid solution are still the main mechanisms of plastic deformation in HEAs [22, 23]. The elastic and plastic deformations in FeCrCuAlNi HEA and $Cu_{29}Zr_{32}Ti_{15}Al_5Ni_{19}$ high-entropy bulk metallic glass (HE-BMG) are reported during indentation process, revealing that the equal element addition can significantly improve the mechanical properties of HEA as compared to the conventional alloys [24, 25]. The nanoscratching mechanical response of AlCrCuFeNi HEA is studied, finding that the higher friction coefficient takes place in HEA due to its outstanding strength and hardness, and high adhesion and fracture toughness over the pure metal materials [26]. Moreover, the stacking fault energy (SFE) in HEA increases the probability to initiate dislocation and twinning, which is conducive to the formation of complex deformation modes [27]. The MD simulations aforementioned not only give valuable insights into the mechanical properties, but also the structure–property relationship in HEA. These works maybe facilitate the design and preparation of advanced HEA with outstanding mechanical properties.

In this chapter, MD modeling is utilized to investigate the effects of severe atomic lattice distortion on plastic deformation and machining-induced subsurface damage of single-crystal and nanocrystalline HEAs. Section 2 describes the MD methods and the models of the single-crystal and nanocrystalline HEAs. Section 3 discusses the plastic deformation behavior in the HEAs. Machining-induced subsurface damage is demonstrated in Sect. 4. Section 5 concludes this chapter.

2 Models and Methods

2.1 *Building Model*

Figure 1a shows that the MD model of nanoindentation and nanoscratching consists of the spherical indenter and the HEA sample. The MD simulation parameters could be obtained [22, 25, 26]. To build the atomic model of HEA, the atoms of Fe (Cr, Cu, Al) are randomly selected and replaced with Ni atoms until the desired composition is obtained for meeting the experimental measurement. Using the melting and quenching simulations, the HEA sample can be obtained. The flowchart in the FeCrCuAlNi HEA preparation is shown in Fig. 1b, and the equiatomic FeCrCuAlNi alloys form a single FCC phase. The atomic model in FeCrCuAlNi is initially relaxed at 300 K for 100 ps under periodic boundary conditions. The HEA sample can be gradually heated to 1500 K at the heating rate 0.004 K/fs and then maintains at 1500 K for 100 ps to allow HEA to melt. The HEA sample is quenched to 300 K with cooling rate 0.004 K/fs and finally maintains at 300 K for 100 ps. Thus, the HEA sample can be prepared for the MD simulation of nanoindentation and nanoscratching.

2.2 *Simulation Method*

Here, the HEA substrate contains three kinds of atoms: boundary atoms, thermostat atoms, and Newtonian atoms [24–26]. The bottom atoms in the HEA substrate are fixed to support the workpiece, and the adjacent-layer atoms maintain a constant temperature of 293 K to release heat energy from nanoindentation and nanoscratch-

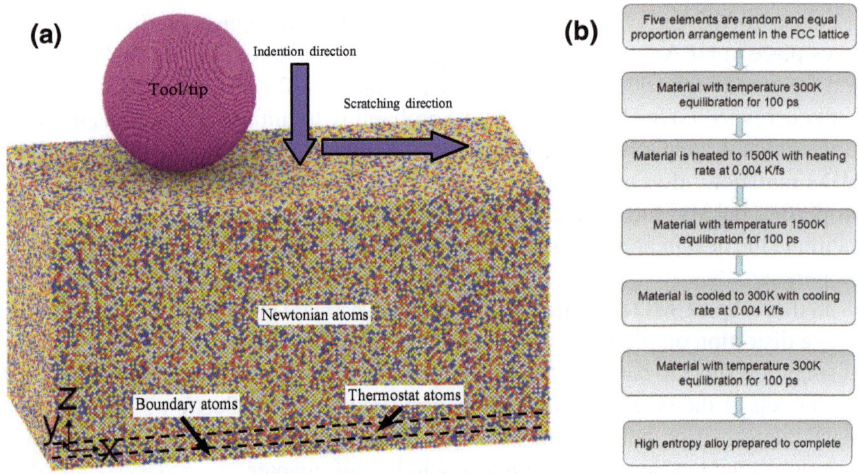

Fig. 1 **a** MD model of HEA FeCrCuAlNi under nanoindentation and nanoscratching. Cr, Fe, Ni, Cu, Al. **b** The flowchart of preparation for FeCrCuAlNi HEA. Reproduced from Ref. [22] with permission from The Royal Society of Chemistry. Reproduced from [26] with permission

ing [24–26]. The motion equations of Newtonian atoms meet Newton's second law. A time step of MD simulation is 1 fs. The indenter moves at a constant velocity of $10\ \mathrm{m\,s^{-1}}$ during the loading and unloading processes with the maximum penetration depth of 3 nm. Both the relaxation, nanoindentation and nanoscratching simulations are used in microcanonical ensemble (NVE). In this chapter, the open-source LAMMPS code performs the MD simulation [28], and the open-source software Ovito visualizes the microstructure evolutions generated by MD simulation [29]. Based on the common neighbor analysis (CNA), four types of atoms are identified in different colors, where green atoms stand for FCC phase, blue atoms for BCC phase, red atoms for stacking faults, and white atom for grain boundaries, other atoms or dislocation cores.

The mixing potentials, consisting of embedded atom method (EAM) potential and Morse potential, are used to the FeCrCuAlNi HEA during the tension, nanoindentation, and nanoscratching. The EAM potential is also used to the indentation of $Cu_{29}Zr_{32}Ti_{15}Al_5Ni_{19}$ HE-BMG and the plastic deformation of $Co_{25}Ni_{25}Fe_{25}Al_{7.5}Cu1_{17.5}$ HEA. The interactions of atoms in Fe–Ni–Cr, Al–Al, and Cu–Cu are applied by the EAM potential [20–22, 30, 31], which is expressed as

$$E = F_\alpha \sum_{j \neq i} \rho_i\left(R_{i,j}\right) + \frac{1}{2} \sum_{j \neq i} \phi_{\alpha,\beta}\left(R_{i,j}\right) \tag{1}$$

where the total energy E is the sum of the embedding energy F, and the short-range pair potential energy ϕ. ρ stands for the electron density. α and β indicate the element types of atoms i and j, respectively.

The interactions of the remained atoms are employed by the Morse potential, which can be given

$$U = D\{\exp[-2\alpha(r_{ij} - r_0)] - 2\exp[-\alpha(r_{ij} - r_0)]\} \quad (2)$$

where D is the cohesion energy, α a constant parameter, r_{ij} the distance between the two atoms, and r_0 the distance at equilibrium. Using the Lorentz–Berthelot mixing rule [32], the parameters of Morse potential in different materials are calculated. The parameters of D, r_0, and α for a mixed pair of atoms in the materials A and B meet the following formulas [32]

$$D_{A-B} = \sqrt{D_A D_B} \quad (3)$$

$$\alpha_{A-B} = (\alpha_A + \alpha_B)/2 \quad (4)$$

$$r_{0A-B} = \sqrt{\sigma_A \sigma_B} + \ln 2/\alpha_{A-B} \quad (5)$$

$$\sigma_{A,B} = r_{0A,B} - \ln 2/\alpha_{A,B} \quad (6)$$

where D_{A-B} is the fitted dimer energy, α_{A-B} is the lattice constant, and r_{0A-B} is the equilibrium distance for materials A and B. The ratio between the chosen cutoff radius and the equilibrium distance is less than 4%, to improve the efficiency of calculation. Here, the cutoff radius used for the remained atom is chosen to be 9.0 Å. The detailed parameters can be obtained in Ref. [22].

A purely virtual spherical repulsive rigid indenter is used in the indentation simulation [25, 30]:

$$F(r) = K\theta(R - r)(R - r)^2 \quad (7)$$

where R is the spherical indenter radius, and r is the distance between the center of the spherical indenter and workpiece. A specified force constant K is 3 eV/Å, and $\theta(R - r)$ is the standard step function. For the nanoscratching simulation, the interaction between indenter and workpiece is used to the Morse potential [26].

3 Plastic Deformation Behavior

3.1 Single-Crystal High-Entropy Alloy

3.1.1 Elastic Deformation Behavior

To verify the validity of the present mixing potentials, Fig. 2a plots the cohesive energy distribution with the increasing atomic density at different crystal structure AlCrFeCuNi HEAs. The FCC HEA has the minimum cohesive energy, indicating the stable phase of FCC HEA [31, 32]. This trend shows good agreement with the

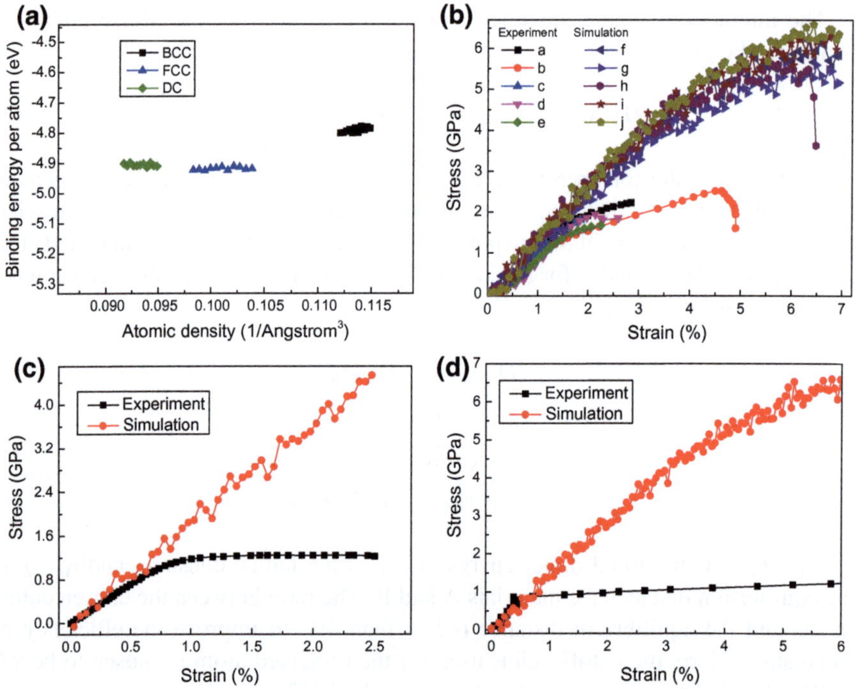

Fig. 2 **a** Cohesive energy of different crystal structure HEAs as a function of the atomic density. **b** Stress versus strain curves of $AlCrFeCuNi_x$ (x from 0.6 to 1.4) for various Ni content. $x = 0.6$ in curve a, $x = 0.8$ in curve b, $x = 1.0$ in curve c, $x = 1.2$ in curve d, and $x = 1.4$ in curve e from experiment [33], and $x = 0.6$ for curve f, $x = 0.8$ for curve g, $x = 1.0$ for curve h, $x = 1.2$ for curve i, and $x = 1.4$ for curve j from MD simulation. **c** Tensile stress–strain curves of $Al_{0.5}CrCuFeNi_2$ from experiment [34] and MD simulation. **d** Compressive stress–strain curves of $AlCrCuFeNi_2$ in experiment [35] and MD simulation. Reproduced from Ref. [22] with permission from The Royal Society of Chemistry

experiments [33–35], where the $AlCrFeCuNi_x$ HEA is consisted of FCC and BCC phases. Hence, the FCC AlCrFeCuNi HEA is selected as reasonable study object.

The stress–strain relationship in $AlCrFeCuNi_x$ (x from 0.6 to 1.4) HEA with a variety of Ni-solute concentration at room temperature is shown in Fig. 2b. Here, the curves a–e represent the experimental data [33], and the curves f–j mean the MD simulation data. Figure 2c, d presents the comparison of $Al_{0.5}CrCuFeNi_2$ [34] and $AlCrCuFeNi_2$ [35] HEAs from experiment and simulation. From Fig. 2b–d, in the elastic stage, the results obtained by MD simulation are consistent with the experiments [34, 35]. However, the nanosized $AlCrFeCuNi_x$ HEA exhibits high strength and good plasticity, which deviates from the experiments. This trend is owing to smaller and stronger mechanical properties in nanoscale metal materials [36]. In the classical mechanics, Young's modulus, one of the important characterizations of mechanical properties of materials, is defined as

$$E = \sigma/\varepsilon \tag{8}$$

where σ and ε are the stress and strain. Based on linear fitting of Eq. (8), Young's modulus is about 125 GPa in $AlCrFeCuNi_x$ HEA, agreeing with 118 GPa from the experimental result [33].

3.1.2 Plastic Deformation Behavior

The stress versus strain curve in $AlCrFeCuNi_{1.4}$ HEA under tensile loading is depicted in Fig. 3. This curve can be calculated from average value of the 10 groups of MD simulation data, to eliminate the random inherent in the isothermal–isobaric (NPT) ensemble. In the initial deformation stage, the stress linearly increases with the increasing strain, similar to the mechanical behavior in the pure metals subjected to loading. The HEAs show the ideal elastic–plastic behavior [33–35], indicating the homogeneous plastic flow. Figure 3a–e examines the evolution of microstructure for $AlCrFeCuNi_{1.4}$ HEA. After the yielding, the HEA shape obviously changes in Fig. 3b–d, where the plastic deformation relies on the nucleation and motion of dislocations.

The local crystalline classification is visualized in $AlCrFeCuNi_{1.4}$ HEA [37]. Here, a single layer composed by HCP atoms stands for a twinning boundary (TB), two adjacent HCP layers are an intrinsic stacking fault (ISF), and two HCP layers with a FCC layer between them are an extrinsic stacking fault (ESF). With the increasing strain, the ISF embryos and deformation twinning are observed in Fig. 3g–j. Besides, at larger plastic deformation stage, deformation twinning grows (Figs. 3f–j and 4c), to enhance the plasticity of $AlCrFeCuNi_{1.4}$ HEA. Meanwhile, the secondary ISF and ESF occur (Fig. 3b–d). As a result, the secondary partial dislocations are blocked by the primary ISF. Furthermore, dislocations interact with the TB (Fig. 4c–e), and detwinning takes place (Fig. 4b–i).

Figure 5 shows a comparison of microstructure evolution in $AlCrFeCuNi_{1.4}$ HEA and pure metals. A large number of ISFs occur, and more slip systems are activated in $AlCrFeCuNi_{1.4}$ HEA, leading to a good plastic behavior. In HEA, deformation twinning occurs at the strain of 14.3% (Fig. 4b, c), and the reversible detwinning happens with the further loading (Fig. 4b–i). At the strain of 25%, there is the presence of twin formation in Cu, Ni, and Cr (Figs. 4a and 5). The $AlCrFeCuNi_{1.4}$ HEA can form a single FCC phase with low SFE [38]. For the low SFE HEA, the partial dislocations with the large SF width come from the dissociation of the perfect dislocation, which is difficult to cause the occurrence of cross-slip and climb [38]. Figure 6 demonstrates the shear localization of HEA under uniaxial loading, where the red atom means high shear strain. As well known, the essentially smooth deformation changes to the highly localized deformation with the increasing strain, and this process results in the shear band (SB) formation. As shown in Fig. 6a–e, SBs initiate at the free surface and subsequently propagate across the substrate, resulting in a sudden drop of flow stress (Fig. 3). With the increasing strain, SBs and shear transformation zones extend in Fig. 6b–e, which would eventually cause the fracture

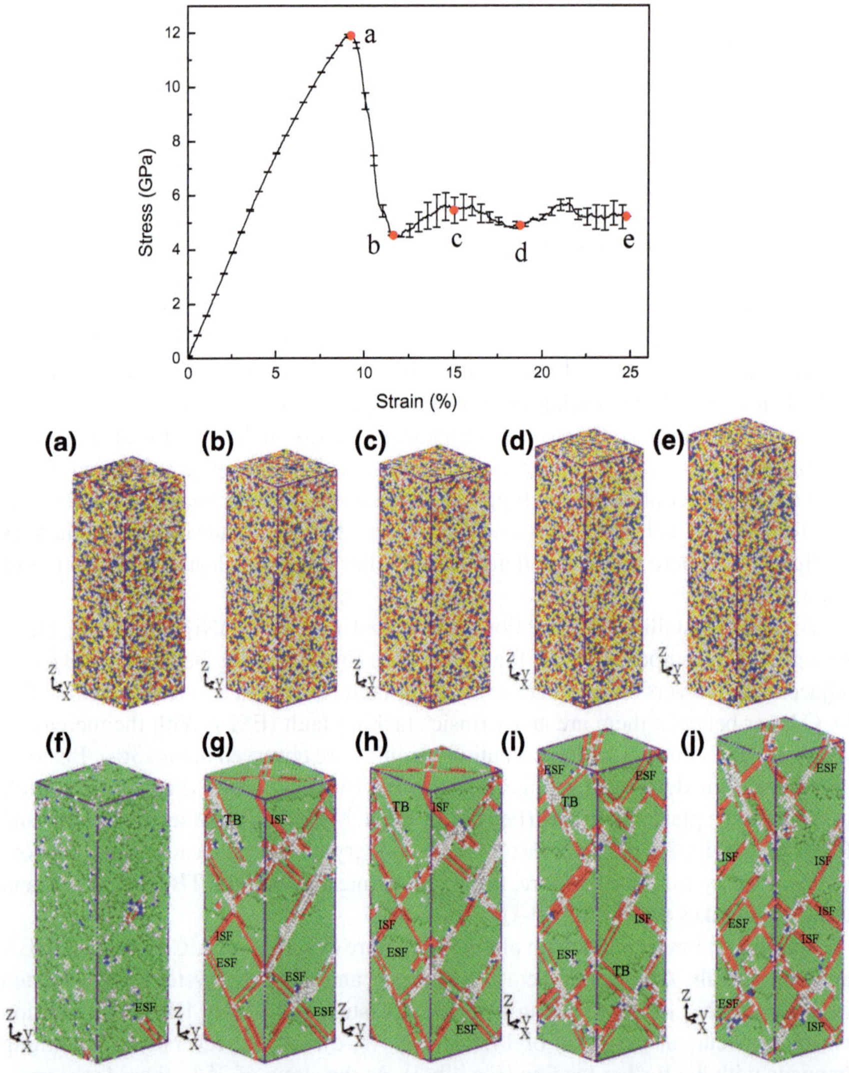

Fig. 3 Tensile stress–strain curve, and atomic microstructure along *z*-direction in $AlCrFeCuNi_{1.4}$ HEA (**a**–**e**). Microstructure of $AlCrFeCuNi_{1.4}$ at the strain: **f** 9.1%, **g** 11.7%, **h** 15.2%, **i** 18.8%, and **j** 25%. Reproduced from Ref. [22] with permission from The Royal Society of Chemistry

of substrate. The SB formation depends on the stress concentration, reducing the strength of HEA. Hence, the dislocation gliding controls the plastic deformation of HEA, due to the dislocation and solute atoms interaction to inhibit the dislocation movement and activate more slip systems.

Using dislocation extraction algorithm (DXA) [39], the interactions between the dislocations are observed clearly, as shown in Fig. 6f–j. As a result, after yielding the

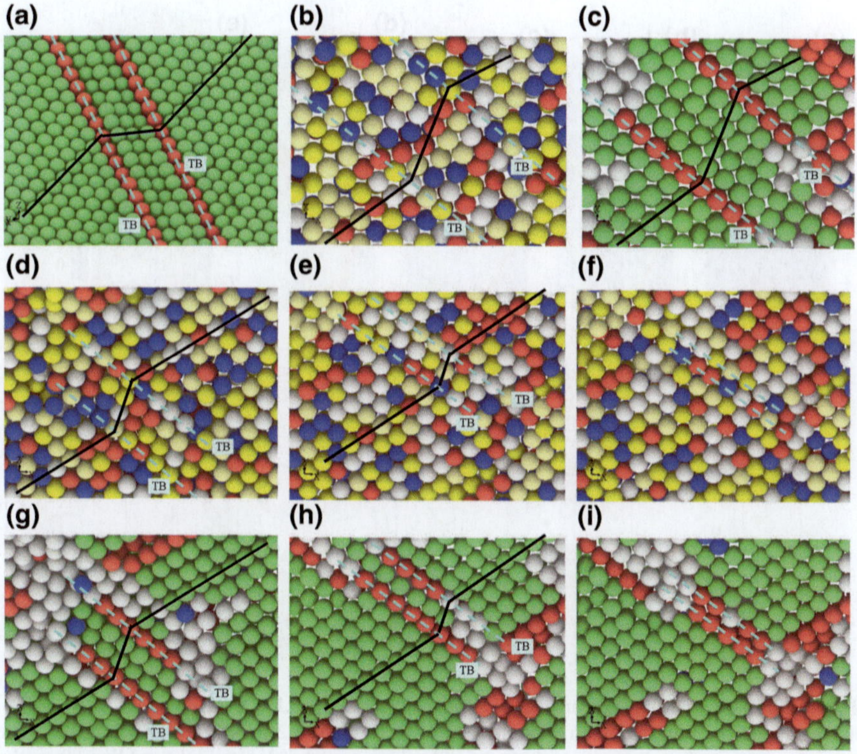

Fig. 4 Deformation twinning in Cr at strain 25% (**a**). Deformation twinning and detwinning in HEA at strain 14.3% (**b, c**), 21.2% (**d, g**), 23.3% (**e, h**), and 24.3% (**f, i**). Reproduced from Ref. [22] with permission from The Royal Society of Chemistry

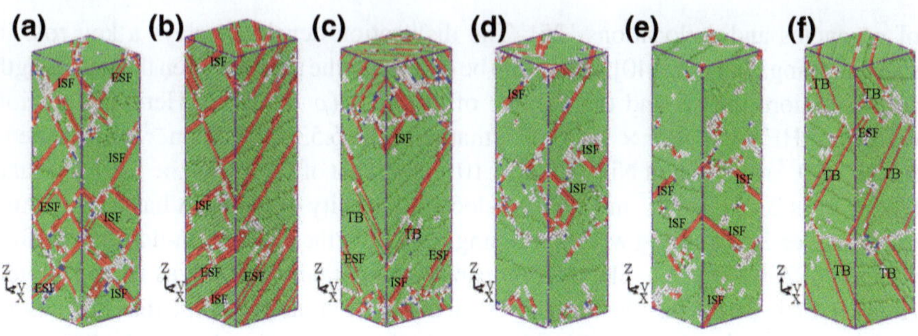

Fig. 5 Microstructure in $AlCrFeCuNi_{1.4}$ HEA (**a**), Al (**b**), Cu (**c**), Fe (**d**), Ni (**e**), and Cr (**f**) at the strain of 25%. Reproduced from Ref. [22] with permission from The Royal Society of Chemistry

partial dislocations emit in slip planes of (1 1 1), (1 −1 1), and (−1 1 1) (Fig. 6f), to reduce the stress concentration. When the formed dislocations encounter the emitted dislocations, they either annihilate or interact with each other, causing the formation

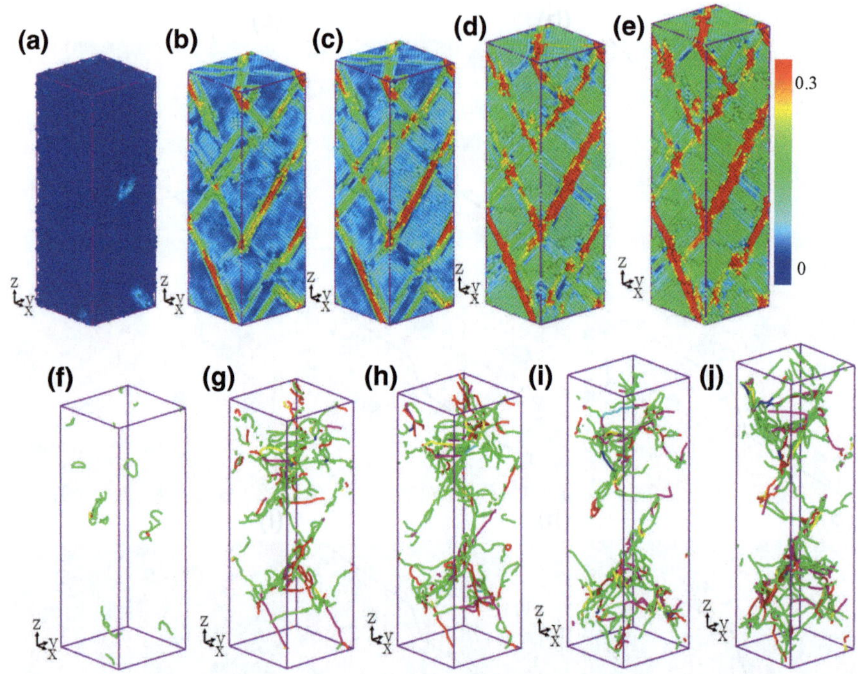

Fig. 6 Shear strain of $AlCrFeCuNi_{1.4}$ HEA at tensile strain: **a** 9.1%, **b** 11.7%, **c** 15.2%, **d** 18.8%, and **e** 25%. Dislocation evolution at strain: **f** 9.1%, **g** 11.7%, **h** 15.2%, **i** 18.8%, and **j** 25%. The blue line represents perfect dislocation, green line Shockley partial dislocation, the red line other, sky blue line Frank partial dislocation, pink line stair-rod dislocation, and yellow line Hirth dislocation. Reproduced from Ref. [22] with permission from The Royal Society of Chemistry

of vacancies and dislocations [25]. The dislocation density ρ plays a key role in strengthening materials [40], which can be defined as the ratio between the total length of dislocation lines L and the volume of crystal $V(\rho = L/V)$. Here, dislocation density of HEA is 8.469×10^{17} m^{-2}, that of Cu is 6.539×10^{17} m^{-2}, that of Fe is 5.720×10^{17} m^{-2}, that of Ni is 4.763×10^{17} m^{-2}, that of Cr is 3.122×10^{17} m^{-2}, and that of Al is 2.902×10^{17} m^{-2}. The dislocation density of the HEA has a maximum value and leads to a strong work hardening, revealing the dislocation-dominated high strength of $AlCrFeCuNi_{1.4}$ HEA. Compared with pure metals such as Al and Cr, the $AlCrFeCuNi_{1.4}$ HEA has the massive Hirth dislocations and stair-rod dislocations, to further enhance the strength of HEA owing to the formation of Lomer–Cottrel lock.

3.2 *Nanocrystalline High-Entropy Alloy*

3.2.1 Mechanical and Physical Properties

The deformation of the $Co_{25}Ni_{25}Fe_{25}Al_{7.5}Cu_{17.5}$ (at.%) HEA nanocrystalline is depicted in Fig. 7a, b [27]. Here, the ISF γ_{ISF} is 25.6 mJ m^{-2} with the error 4.6 mJ m^{-2} in Fig. 7c, which agrees with the previous experiments, such as 19.7 mJ m^{-2} in $Ni_{23}Fe_{20}Cr_{17}Co_{20}Mn_{20}$ [41], 49 mJ m^{-2} in FeCoNiCrCu [42], and 25 mJ m^{-2} in FeMnNiCoCr [38]. The atomic lattice distortion strongly affects SFE [36–38, 43], which can influence the dislocation nucleation and motion. The random element distribution and atomic size can affect the degree of local atomic lattice distortion, causing the SFE fluctuation [41] (Fig. 7c). Indeed, the plastic deformation of the low SFE steels is controlled by the phase transformation. Therefore, the phase transformation could take place in the low SFE HEAs, which is similar to the behavior of the low SFE steels. The stress–strain curves of nanocrystalline HEA show that the yield strength of HEA increases with the strain rate in the range of 5×10^6 to 2×10^8 s^{-1} (Fig. 7d). Compared to the previous experimental value of 1.795 GPa, the yield strength at strain rate 5×10^6 s^{-1} is 1.866 GPa with 4% error in nanocrystalline HEA [44]. Hence, the yield strength of HEA depends on the strain rates, which agrees with the result of nanosecond-laser shock-induced phase transformations [45].

3.2.2 Phase Transformation Process

The primary plastic deformation process in the nanocrystalline HEA depends on the dislocation reaction and grain-boundary-deformation behavior (Fig. 8a, b) [44, 45], such as the nucleation and emission of partial dislocations from grain boundary and the formation of HCP structures. When the dislocation behavior is suppressed due to the high stress from the nucleated SF, the phase transformation from FCC to BCC surprisingly occurs (Fig. 8c–e). The nucleus of BCC arises from the SF-SF intersections or grain boundaries (Fig. 8c–e) [46, 47], grows along the habit plane (Fig. 8d–e), and further extends to the surroundings. Based on the results of Olson and Cohen [46], two intersecting shear systems produce the nucleation sites and embryos of BCC phase in low SFE austenites. Accordingly, phase transformation of HEA depends on its low SFE, which is similar to the martensitic phase transformation in low SFE steels [48]. Subsequently, the BCC phase merges with the other BCC phase and then forms the larger BCC phase to release the high strain (Fig. 8e). Figure 8f further confirms that more than 20% of HEA has been transformed to the BCC phase. A schematic illustration for the formation and growth of BCC phase induced by the SF-SF intersection is shown in Fig. 8g.

The FCC to BCC transformation in Fig. 9a satisfies the Nishiyama–Wasserman (N–W) relationship, namely (111)fcc||(110)bcc and [1–10] fcc||[−111]bcc orientation [49, 50]. Figure 9b, c shows the nucleated BCC phase at the strain of 5%, and the crystallographic relationship of the original FCC structure and the generated

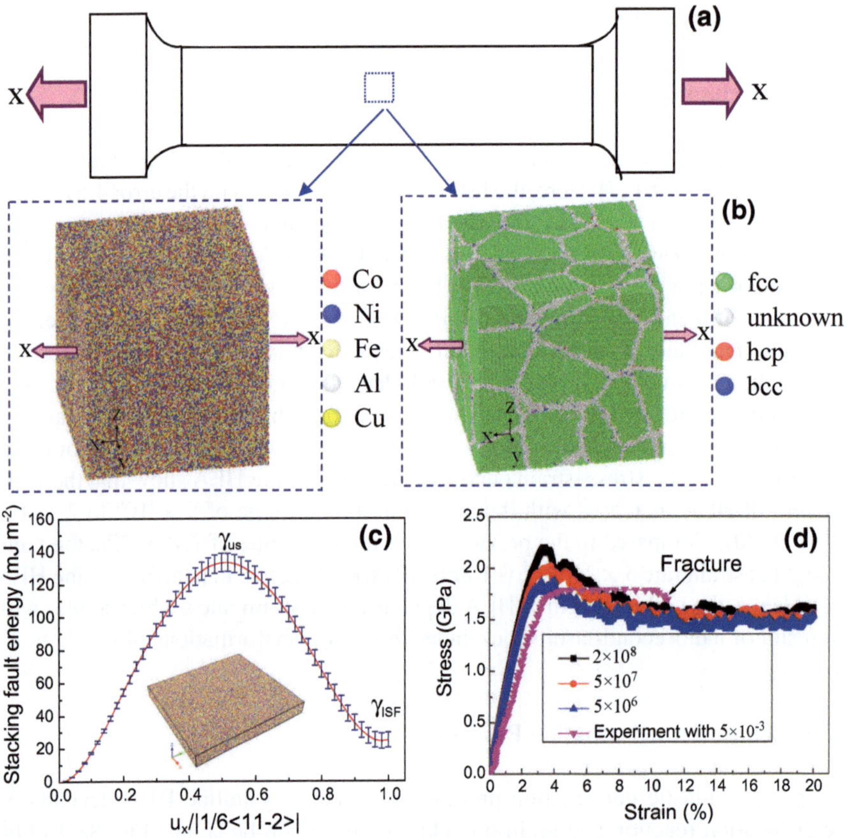

Fig. 7 Atomistic model of $Co_{25}Ni_{25}Fe_{25}Al_{7.5}Cu_{17.5}$ HEA with average grain size of 10 nm according to atomic type (**a**) and CNA (**b**). Energy barriers of displacement pathway ⟨11-2⟩ direction (**c**), where error bar is computed by the element random distribution, and γ_{us} is the unstable SFE. Stress–strain of nanocrystalline HEA subjected to compressive loading from MD simulations and experiment [44] (**d**). Reproduced from [27] with permission

BCC structure. In addition, most of the atoms deviate from the perfect FCC/BCC lattice sites, and their average positions along the projected direction still form a FCC/BCC lattice (Fig. 9c). Hence, the experimentally observed crystal structure in $Co_{25}Ni_{25}Fe_{25}Al_{7.5}Cu_{17.5}$ HEA is only a result of a wide range average of local atomic structure, which is neither long-range nor short-range order in HEA system [44].

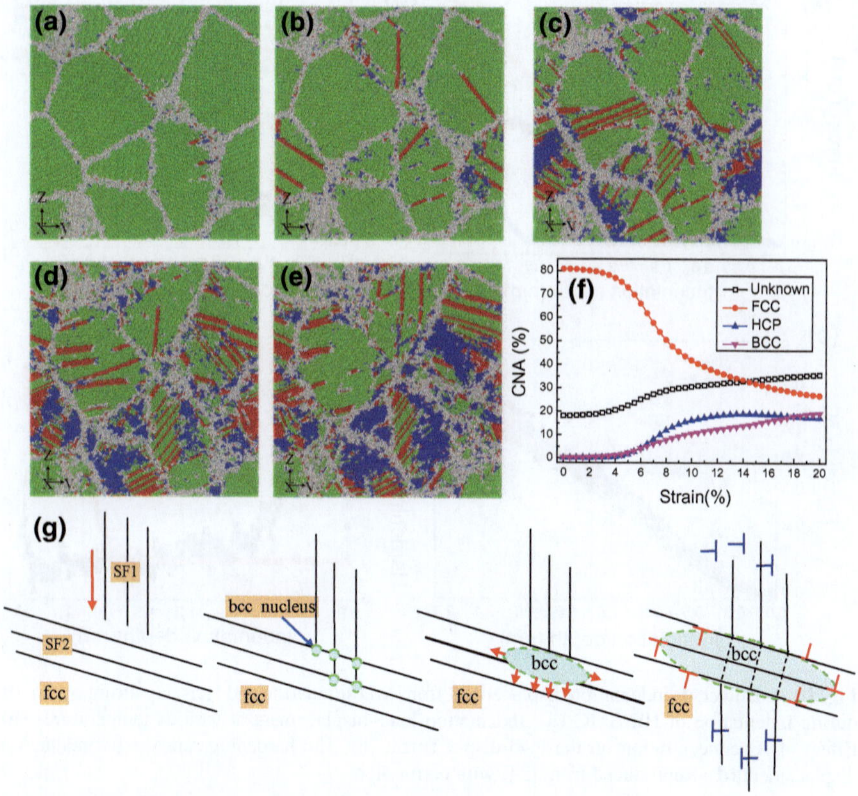

Fig. 8 Microstructural evolution at strains: **a** 0%, **b** 3%, **c** 8%, **d** 17%, and **e** 20%. The fractional phase content with strain (**f**). Illustration of the BCC phase formation and growth induced by the SF-SF intersection (**g**). Reproduced from [27] with permission

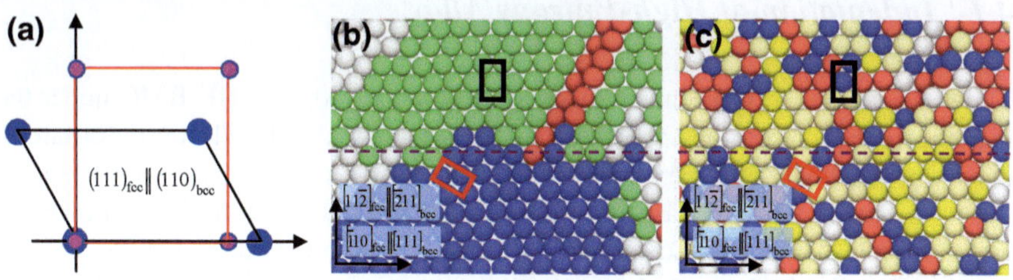

Fig. 9 **a** Illustration of the (111)fcc to (110)bcc phase transformation to meet the N–W orientations. Large and small circles show the atoms of FCC and BCC lattices. Phase-transforming system at strain 5% (**b**) and (**c**). The black square means FCC lattice, and the red rectangle shows BCC lattice. Reproduced from [27] with permission

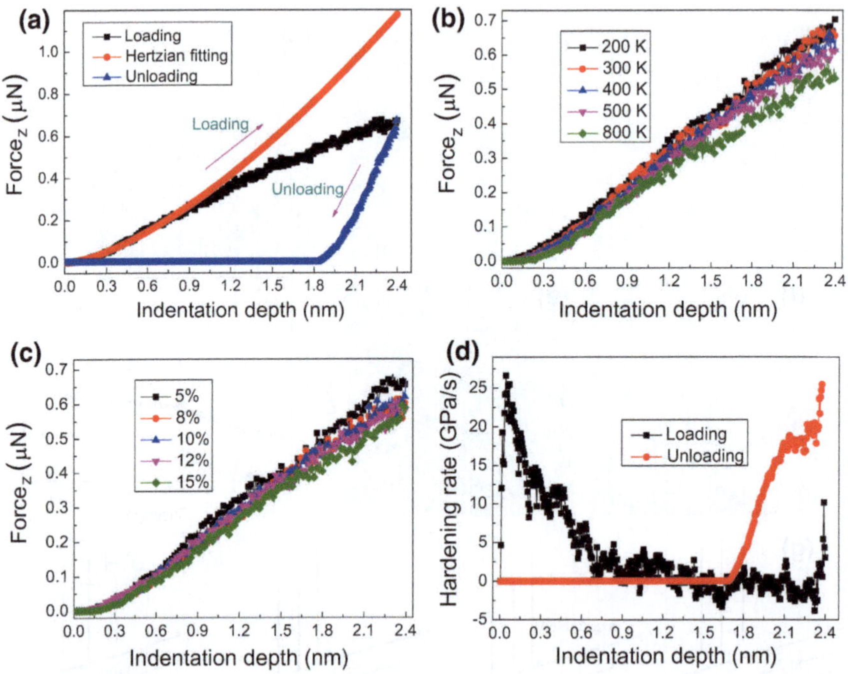

Fig. 10 Load versus indentation displacement from MD simulation and Hertzian fitting by Eq. (9) during indentation of HE-BMG (**a**). Indentation load–displacement at various temperatures (**b**). Effect of Al concentration on the load-displacement (**c**). The hardening rate versus indentation displacement (**d**). Reproduced from [24] with permission

4 Machining-Induced Subsurface Damage

4.1 Indentation of High-Entropy Alloy

The classical load–displacement curve in $Cu_{29}Zr_{32}Ti_{15}Al_5Ni_{19}$ HE-BMG under the nanoindentation process is presented in Fig. 10a. Firstly, the elastic deformation occurs in HE-BMG and then transforms to the plastic deformation, which is similar to the deformation behavior induced by the indentation in traditional metallic glass [51–53]. Moreover, a displacement burst occurs in the curve. On the basis of the Hertzian elastic theory [54–56], the correlation of the indented load P and the penetrated depth h during the elastic deformation can be expressed as follows

$$P = 4E_* R^{1/2} h^{3/2} / 3 \tag{9}$$

where R indicates the radius of indenter, and E_* represents the reduced Young's modulus which can be derived from the equation

$$\frac{1}{E_*} = \frac{1 - \nu_i^2}{E_i} + \frac{1 - \nu_s^2}{E_s} \tag{10}$$

where ν_i and E_i represent Poisson's ratio and Young's modulus of indenter. ν_s and E_s are Poisson's ratio and Young's modulus of HE-BMG. In the present MD simulation, owing to the adoption of virtual rigid indenter, Eq. (10) can be simplified as [57]

$$E_s = \left(1 - \nu_s^2\right) E_* \tag{11}$$

Figure 10a shows the Hertzian fitting from Eq. (9). Compared with the indentation experiment of 105.4 GPa, the reduced modulus is 106.1 GPa with a standard deviation of 0.4 GPa and has about 0.6% deviation [58]. Recently, Poisson's ratio acquired from experiment of HE-BMG is 0.348 for $Ti_{20}Zr_{20}Hf_{20}Be_{20}Cu_{20}$ [59], 0.351 for $Ti_{20}Zr_{20}Hf_{20}Be_{20}Cu_{20}Ni_{10}$ [60], and 0.354 for $Ti_{16.7}Zr_{16.7}Hf_{16.7}Ni_{16.7}Cu_{16.7}Be_{16.7}$ [61]. Here, Poisson's ratio of $Cu_{29}Zr_{32}Ti_{15}Al_5Ni_{19}$ HE-BMG is set as 0.35. Thus, Young's modulus of HE-BMG $Cu_{29}Zr_{32}Ti_{15}Al_5Ni_{19}$ calculated from Eq. (11) is 93.1 GPa. Furthermore, Young's modulus of BMG can be calculated based on the mixtures rule, as shown in [62]

$$\frac{1}{E} = \sum_{i=1}^{n} \frac{f_i}{E_i} \tag{12}$$

where E_i is Young's modulus of the element i, and f_i represents the atomic percentage of the element i. Young's modulus of HE-BMG calculated from Eq. (12) is 101 GPa. Compared to the result of experiment and MD simulation, the corresponding Young's modulus of HE-BMG on the basis of the rule of mixtures is larger. Therefore, the MD simulation is more accurate to predict the mechanical properties of HE-BMG compared to the mixtures rule.

The correlation between the indentation load and displacement at various temperatures is shown in Fig. 10b. As the temperature increases, the elastic modulus of HE-BMG reduces. In addition, the maximum indentation force also decreases at the same indentation depth, showing the high-temperature softening behavior of HE-BMG. This trend agrees with the previous experiments [51–53]. The Al element plays a critical role in the strength and ductility of HEA [8, 63, 64]. Here, the indentation force–displacement in HE-BMG with various Al content is studied in Fig. 10c. The indentation force decreases with the increase of Al element, resulting in the reduction of Young's modulus in HE-BMG owing to the low stiffness of pure Al [62].

Hardness H is the characteristic of material to resist hard objects pushing into the material surface and can be expressed as

$$H = F/A_c \tag{13}$$

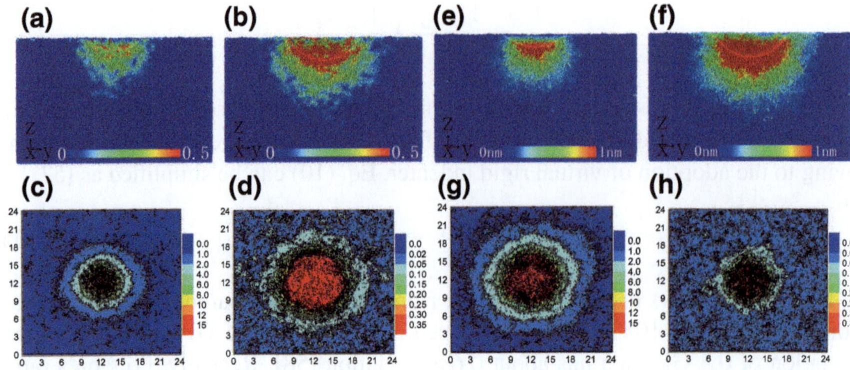

Fig. 11 Shear strain (**a–d**), and atomic displacement (**e–h**), at the indentation depths of (**a**, **c**, **e**, **g**) 1.2, and (**b**, **d**, **f**, **h**) 2.4 nm. Reproduced from [24] with permission

where the load is F, and the area of pressing surface A_c is $2\pi Rh$ [30, 57]. Using Eq. (13), the hardness in $Cu_{29}Zr_{32}Ti_{15}Al_5Ni_{19}$ HE-BMG is 8.8 GPa from MD simulation and is in good agreement with 7.45 GPa from the indentation experiment [58]. The strong size effect causes the about 18% deviation compared to the experiment.

The ratio between hardness and elasticity modulus (H/E ratio) can give contribution to the wear resistance [65, 66]. Higher H/E ratio leads to lower friction and higher wear resistance [65, 66]. Here, H/E ratio in $Cu_{29}Zr_{32}Ti_{15}Al_5Ni_{19}$ HE-BMG is 0.094 from MD simulation. The hardness hardening rate during indentation is shown in Fig. 10d. As the indentation depth increases, the hardening rate rapidly goes up and then decreases slowly at loading process. However, the hardening rate decreases at unloading process.

Figure 11a–d shows the shear strain distribution below the indenter at different depths of indentation. The nucleation of shear area is occurred at the contact zone between workpiece and indenter, and extends toward the inside of HE-BMG. In the $Cu_{29}Zr_{32}Ti_{15}Al_5Ni_{19}$ workpiece, the disordered structure leads to the asymmetric distribution for shear region, although the spherical indenter is symmetric, due to the highly degree of disordered and HE effect. Figure 11c, d shows that the gradient-strain distribution continues to decline from the indenter–workpiece contacting center to the boundary of workpiece below the indenter, indicating the plastic anisotropy. That is to say, the region of the highest shear strain is located around indentation at the 1.2 and 2.4 nm indention depth, qualitatively according with the Hertzian elastic theory [54].

During the process of nanoindentation, the atomic displacement distribution in HE-BMG is shown in Fig. 11e–h. An analysis of local displacement is implemented at selected depth of indentation (Fig. 11e, f), indicating the seriously deformed behavior in a local cluster [22, 25]. In indentation region, the displacement of atomic could still keep relative symmetry with the increase of indenter, showing that the displacement of HE-BMG is isotropic. This trend agrees with the result in Fig. 11g, h, where the contours of atomic displacement show a clear symmetry.

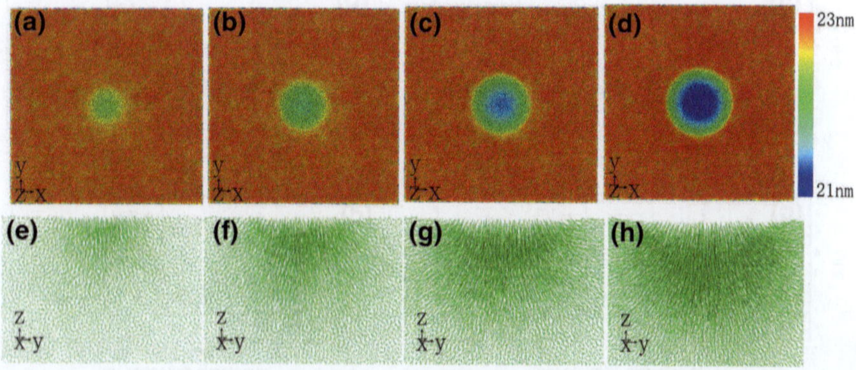

Fig. 12 Surface morphology at the indentation depth: **a** 0.6, **b** 1.2, **c** 1.8, and **d** 2.4 nm. The cross-section snapshots of atomic displacement vectors at the indentation depth: **e** 0.6, **f** 1.2, **g** 1.8, and **h** 2.4 nm. Reproduced from [24] with permission

To investigate the behavior of plastic deformation, the surface morphology at the different indentation depths is performed in the surrounding indentation region (Fig. 12a–d). When the indentation depth changes from 1.8 to 2.4 nm, the brittle deformation of HE-BMG transforms to the ductile model. At the indentation depth of 0.6 nm, a brittle deformation of HE-BMG agrees with the result of previous experiment and simulation [62]. When the indentation depth is larger than 2.4 nm, an appearance of many atom pileups suggests a typical model of ductile deformation. The shear stress flow from the indentation process could drive the surface atom groups to move. This phenomenon affects the brittleness and toughness of HE-BMG [51, 67].

Figure 12e–h shows the atomic moving vector in $Cu_{29}Zr_{32}Ti_{15}Al_5Ni_{19}$ HE-BMG, where the length and direction of arrow are the moving distance and moving direction of atoms. As can be seen, the significant curvilinear motion occurs around indenter, instead of below the indenter, owing to the shear stress flow. The surface atoms could move along the opposite direction of indention as the indentation continues to penetrate (Fig. 12h). Hence, the brittle to ductile transition relies on the indentation depth in the HE-BMG, which can be achieved by the integral sliding in the local atom groups at different layers [51, 67] (Fig. 12e–h). In other words, the homogeneous and inhomogeneous stress flow results in the plastic deformation of HE-BMG attributed to that the activation energy for the occurrence of shear deformation decreases by the severe distortion.

Figure 13a–e presents the partial RDF, including the Zr–Zr, Zr–Ti, Zr–Ni, Zr–Al, Zr–Cu, Ti–Ti, Ti–Ni, Ti–Al, Ti–Cu, Ni–Ni, Ni–Al, Ni–Cu, Al–Al, Al–Cu, and Cu–Cu pairs. The long-range order could reflect the crystal structure. RDF is seen as the key link between macroscopic thermodynamic properties and intermolecular interactions [22, 25]. The first RDF peak sharply changes among the fifteen partial RDFs, indi-

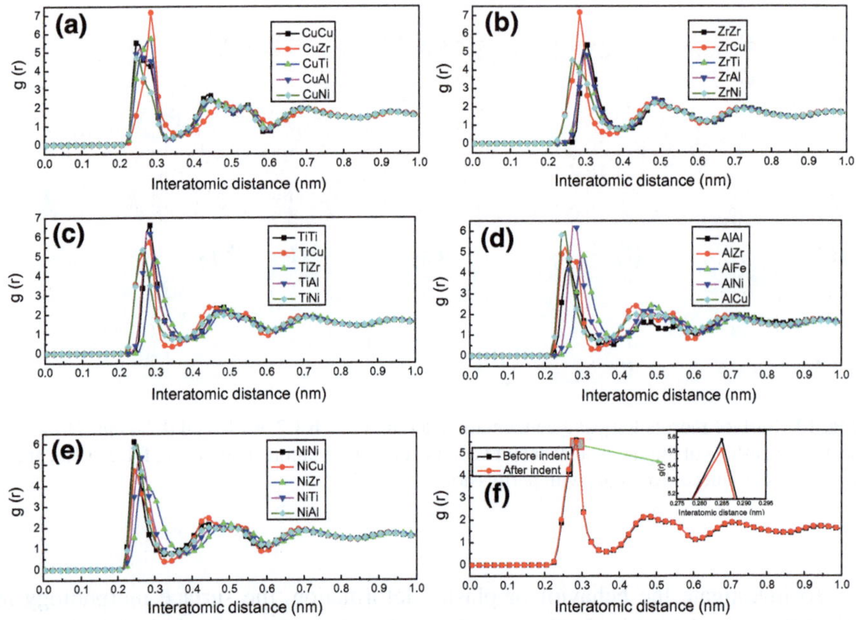

Fig. 13 After indentation the partial RDF versus interatomic distance in (**a–e**) and RDF versus interatomic distance before and after indentation in (**f**). Reproduced from [24] with permission

cating the amorphous property of HE-BMG. The first partial RDF peak of Al is obviously shifted to the left/right owing to the difference in atomic radius (Fig. 13d). It can be obtained to the atomic size for different elements [26]. Hence, the difference of various element radii plays a major role in formation ability of amorphous HE-BMG [68]. The previous study reveals the formation ability of amorphous HE-BMG which relies on the element concentration and the value of the critical pressure melts quenching [69, 70]. The RDF of HE-BMG indicates that the HE-BMG is the amorphous structure at room temperature (Fig. 13f). Figure 13f also shows that the slight reduction is observed in the first peak after indentation. Thus, it is expected that the RDF analysis can give the possible relationship between the structural modification and the formation ability of the amorphous HE-BMG.

4.2 Scratching of High-Entropy Alloy

Figure 14a, b shows the relationship of force and scratching distance in the AlCrCuFeNi HEA and pure metals. Here, the tangential force and the normal force are defined as the total force from the atoms of scratching tip [71–73]. Since a constant material is removed by the scratching tip for the stable scratching stage, the tangential force needs to be a stable value (Fig. 14a). Specially, the tangential force of AlCr-

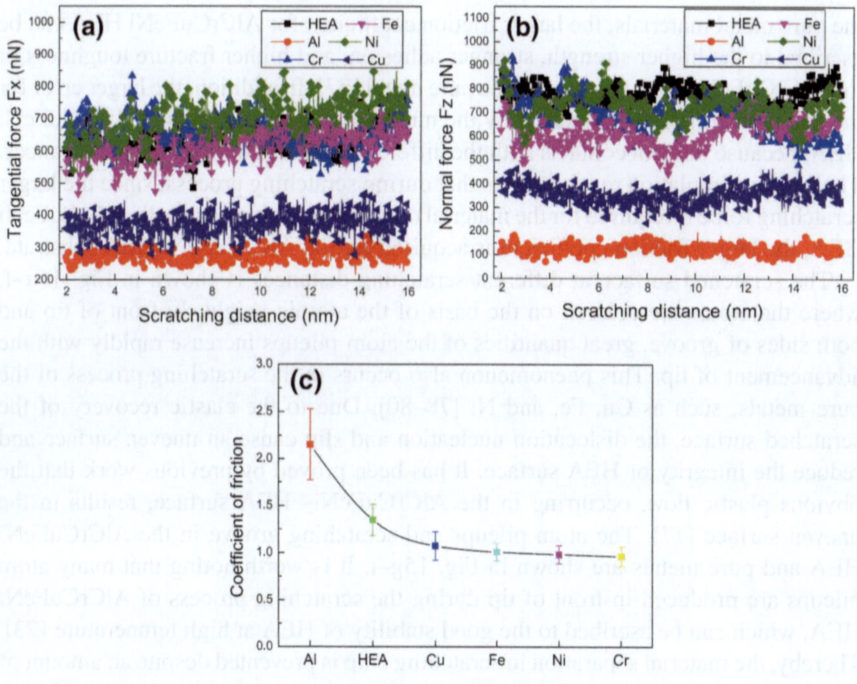

Fig. 14 Scratching force and scratching distance in HEA and pure metals (**a**, **b**). The friction coefficient versus different materials (**c**). Error bars indicate mean deviation for the stable scratching process. Reproduced from [26] with permission

CuFeNi HEA overtakes the force of pure Cu and Al, attributed to that AlCrCuFeNi HEA has the higher strength and adhesion as compared with pure metals of Cu and Al [26, 42, 74–76]. With the advancement of the scratching tip, the normal force fluctuates slightly and achieves the maximum value in AlCrCuFeNi HEA (Fig. 14b). The stable normal force is required for the scratching of pure metal workpiece with the constant scratching depth. However, the drastic concussion of the normal force in AlCrCuFeNi HEA depends on the difference of the strength, yield, and fracture toughness in different elements [26, 74, 77]. In order to cut a constant material, the AlCrCuFeNi HEA with the higher adhesion and fracture toughness [78] requires a consistently larger normal force than that of pure metal.

Figure 14c presents the friction coefficients for pure metal materials and AlCrCuFeNi HEA during the scratching process. The ratio of tangential force to normal force is called as the friction coefficient, namely $k = F_x/F_z$. Figure 14c illustrates that the properties affect the friction coefficient for the metal materials. It is seen that except for Al with the strong adhesion, the friction coefficients of the pure metal materials are smaller than that of AlCrCuFeNi HEA, which presents the good adhesion for AlCrCuFeNi HEA. Furthermore, the previous result on the worn surface of $AlCrCuFeNi_2$ HEA reveals the features of adhesion mechanism [77]. Compared with the pure metal materials, the larger friction coefficient for AlCrCuFeNi HEA can be ascribed to the higher strength, stronger adhesion, and higher fracture toughness of the AlCrCuFeNi HEA than that of the pure metal [77]. In addition, the larger error for the friction coefficient is induced by the multicomponent elements of AlCrCuFeNi HEA, because the tool contacts with the different types of atoms as the tip advances. The current simulation result verifies that during scratching process, since the larger scratching force is required for the material removal of the high strength AlCrCuFeNi HEA, the larger friction coefficient is acquired for the AlCrCuFeNi HEA substrate.

The scratched surface at different scratching distances is shown in Fig. 15a–f, where the atoms are colored on the basis of the atomic height. In front of tip and both sides of groove, great quantities of the atom pileups increase rapidly with the advancement of tip. This phenomenon also occurs in the scratching process of the pure metals, such as Cu, Fe, and Ni [78–80]. Due to the elastic recovery of the scratched surface, the dislocation nucleation and slip cause an uneven surface and reduce the integrity of HEA surface. It has been proved by previous work that the obvious plastic flow, occurring in the $AlCrCuFeNi_2$ HEA surface, results in the uneven surface [77]. The atom pileups and scratching groove in the AlCrCuFeNi HEA and pure metals are shown in Fig. 15g–l. It is worth noting that many atom pileups are produced in front of tip during the scratching process of AlCrCuFeNi HEA, which can be ascribed to the good stability of HEA at high temperature [73]. Thereby, the material separation in scratching chip is prevented despite an amount of scratching heat. In addition, it can be seen in Fig. 15g–l that great quantities of atom pileups are obviously observed on both sides of groove in the pure metals. Since the mechanical properties of scratched materials affect the shape and size of groove, the atoms in the surface of soft Al workpiece can be easily peeled off and pile up on both sides of the groove (Fig. 15g–l).

The subsurface damaged structures are shown in Fig. 16 for the pure metal and AlCrCuFeNi HEA workpieces. The special region, including dislocation, void, SF, and TB, is called as the subsurface damaged structure. When the tip scratches through the HEA workpiece, the serious lattice distortion is induced by the scratching. As a result, many voids are nucleated from the subsurface. Besides, it is seen from Fig. 16f that lots of SFs appear in the AlCrCuFeNi HEA, which is owing to the low SFE of HEA confirmed by the experiment and simulation [81]. Furthermore, since the more slip systems are activated and dislocations slip along the {111} planes in AlCrFeCuNi HEA, severer plastic deformation and work hardening occur in AlCrFeCuNi HEA workpiece [82] (Fig. 16). Thus, compared with the traditional crystal metals, the AlCrCuFeNi HEA has a better plasticity, which is consistent with the previous experiments [3].

The non-uniform distributions in strain and strain rate along depth of the AlCrCuFeNi HEA workpiece are induced by the scratching process [83] (Fig. 17a), which is also observed in Cu [84, 85]. With the increasing depth, the atomic strain decreases quickly at the scratching distance of 16 nm (Fig. 17b). The higher shear strain concentrates in the subsurface region; thus, the strain hardening varies along the depth direction. The ductility of HEA material is enhanced remarkably by the intensive slip bands, which is in good agreement with the planar-slip dislocation to control

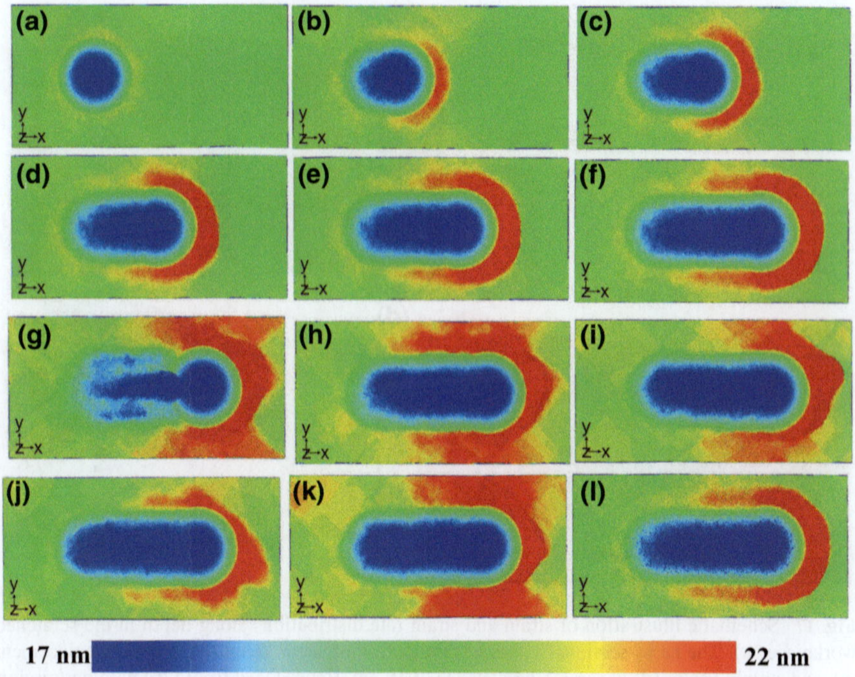

Fig. 15 Variation of surface morphology at the scratching distances: 0 (**a**), 3 (**b**), 6 (**c**), 9 (**d**), 12 (**e**), and 15 nm (**f**). Variation of surface morphology for different materials: Al (**g**), Cr (**h**), Cu (**I**), Fe (**j**), Ni (**k**), and AlCrCuFeNi (**l**). Reproduced from [26] with permission

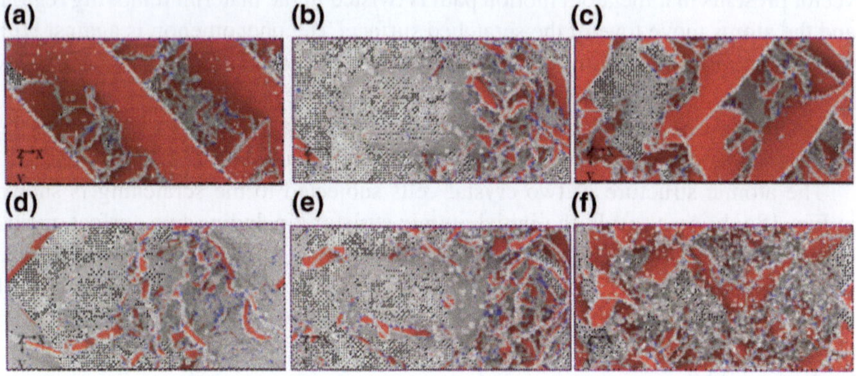

Fig. 16 Variation of subsurface damaged structure: Al (**a**), Cr (**b**), Cu (**c**), Fe (**d**), Ni (**e**), and AlCrCuFeNi HEA (**f**). FCC atoms are excluded for a clearer visualization of damaged structures. Reproduced from [26] with permission

the plastic deformation [3]. Figure 17c shows the displacement localization, and the overlapping zigzag shape is formed by the nonlinearly move of atoms in AlCrCuFeNi

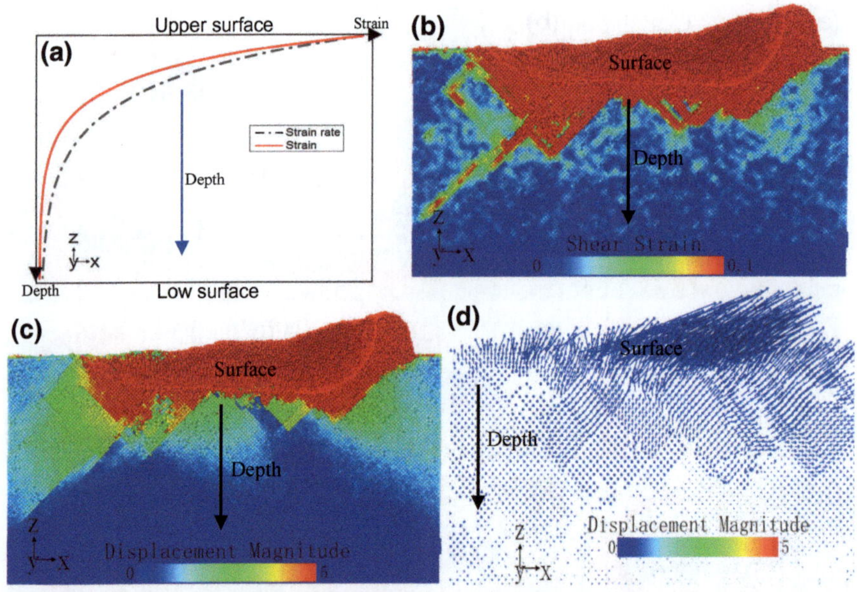

Fig. 17 Schematic illustration of strain and strain rate distributions along depth in the scratched workpiece (**a**). The cross-sectional views of distribution of shear strain (**b**), atomic displacement (**c**), and atomic vector (**d**) at scratching distance of 16 nm. Reproduced from [26] with permission

HEA. Hence, the dislocation slipping relies on the atomic displacement. The atomic vector presents that the atom motion path is twisted in the material removing region, and the atoms move toward the scratched surface. The phenomenon is against with the conventional machining view that all atoms except the chip atoms move away from the scratched surface owing to the compressive stress induced by the scratching. It may be explained by the fact that the severer atomic lattice distortion produces the strong stress to promote the atom motion in the AlCrCuFeNi HEA (Fig. 17d).

The atomic structure of two crystal cells subjected to the scratching is shown in Fig. 18a, b. As a result, the initial atomic structure including two perfect crystal cells has severe lattice distortion owing to the significant difference of atomic size [86, 87] between Al and other four kind elements [26] (Fig. 18a). The variation of atomic structure after the scratching process is demonstrated in Fig. 18b, indicating the occurrence of serious lattice deformation due to the plastic deformation induced by the scratching process.

Here, the critical twinning stress from scratching τ_p can be expressed as [88]

$$\tau_p = 2\alpha\mu b_{\mathrm{p}}/\lambda + \gamma_{\mathrm{sf}}/b_{\mathrm{p}} \tag{14}$$

where α is a constant, μ is the shear modulus, b_{p} is the partial Burgers vector, and λ is the twin thickness between adjacent TB, and γ_{sf} is the SFE. In Fig. 18c, the average twinning thickness is 0.62 nm. On the basis of results [26, 78, 88], the

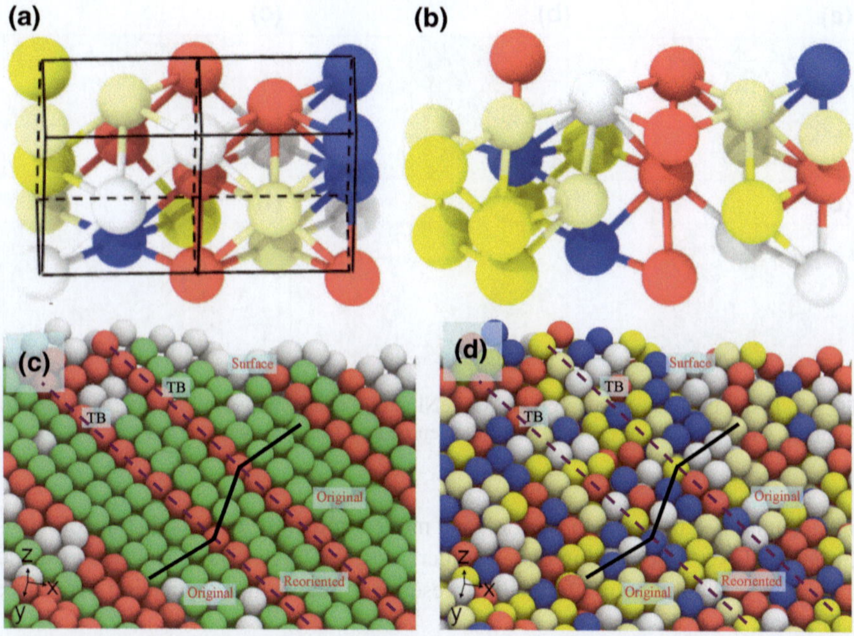

Fig. 18 Schematic illustration of two crystal cells before (**a**) and after (**b**) scratching in AlCrCuFeNi HEA. Deformation twinning at scratching distance 16 nm (**c**, **d**). Reproduced from [26] with permission

critical twinning stress is approximately 14.6 GPa for the deformation twinning. As a matter of fact, the local stress fields originating from the Suzuki interaction and lattice distortion would significantly reduce the critical twinning stress for the deformation twinning. Currently, the deformation twinning occurred in CrMnFeCoNi HEA is a typical {111} twinning structure. It is conducive to the damage tolerance due to that the formation of deformation twinning can consume more energy to obtain a more stable structure compared with the dislocation slipping. Moreover, TBs can strongly reduce the mobility of dislocation and lead to the dislocation pileup along TBs to strengthen HEA. It is very according to the experiments in $CoNiFeCrAl_{0.6}Ti_{0.4}$ containing nanoscale twins [86].

The dislocation substructures are observed with increase of scratching distance (Fig. 19), indicating that the interaction between dislocations is still one part of the significant contribution for the work hardening in the AlCrCuFeNi HEA. As a result, dislocation numbers increase rapidly at the scratching distance range from 0 to 3 nm, but decrease gradually at the scratching distance of 6–15 nm. It demonstrates that the number of dislocation is significantly dependent upon the scratching distance. Once the stress induced by the scratching exceeds the critical stress, the dislocations begin nucleating and emitting from the {111} slip planes to release the high stress below the tool. As the tool continues to go forward, the original dislocations meet with the emitted dislocations and then interact with each other. Shockley partial dislocations

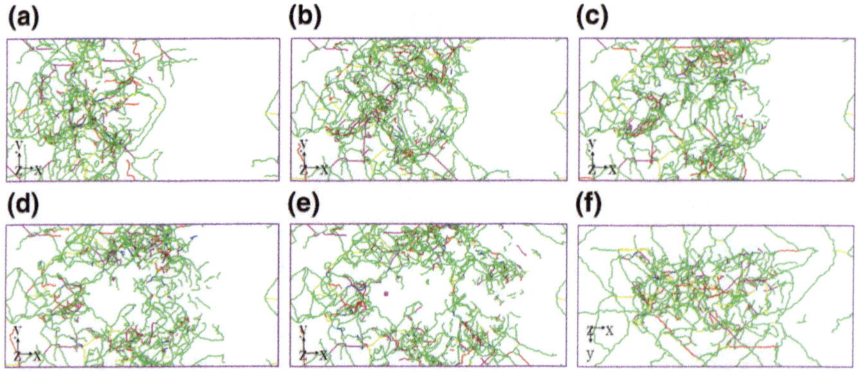

Fig. 19 Dislocation substructure of AlCrCuFeNi at the scratching distance: 0 nm (**a**), 3 nm (**b**), 6 nm (**c**), 9 nm (**d**), 12 nm (**e**), and 15 nm (**f**). Reproduced from [26] with permission

are accumulated underneath the tool, and move through the entire substrate of AlCrCuFeNi HEA. As scratching distance increases, the large dislocation cluster could disappear gradually owing to the TB presence to reduce the dislocation mean free path.

The evolution of different kind dislocations in AlCrCuFeNi HEA is shown in Fig. 19a–f. It is well established that the ductility of the metal materials decides on the capacity of dislocation motion, and the dislocation motion difficulty in metal materials leads to the high strength [26]. Many Shockley partial dislocations are observed in pure Cu and AlCrCuFeNi HEA, in good agreement with the previous data [81]. In addition, the dislocations act as the effective obstacles to hinder the motion of other dislocations, forming the complex dislocation nets which would result in the work hardening in HEA workpiece (Fig. 19). After the scratching, the dislocation density of AlCrCuFeNi HEA is 1.1×10^{17} m^{-2}, that of Cr is 7.04×10^{16} m^{-2}, that of Al is 1.28×10^{17} m^{-2}, that of Cu is 1.49×10^{17} m^{-2}, that of Fe is 9.7×10^{16} m^{-2}, and that of Ni is 9.5×10^{16} m^{-2}. Compared with the pure metals of Cr, Fe, and Ni, AlCrCuFeNi HEA has the higher dislocation density. It results in not only a strong work hardening owing to the interaction of dislocation and solute atom, but also a good plasticity owing to many dislocations nucleated from the activated slip systems. Hence, dislocation still plays a critical role in the deformation and strengthening in AlCrCuFeNi HEA. Compared with the pure metals such as Ni, Fe, and Cr, the massive Hirth dislocations and stair-rod dislocations as well as the formation of Lomer–Cottrel lock enhance the strength of AlCrCuFeNi HEA. The dislocation contribution to the strain hardening is evaluated in terms of the dislocation mean free path approach [80]. In addition, the serious lattice distortion would activate more slipping systems, due to hindering the dislocation movement.

The number of dislocation substructure is shown in Fig. 20. A large number of Shockley partial dislocations occur in the HEA and pure metal. Compared to pure metals, the quantity of perfect dislocation in HEA is minimal due to the low SFE of HEA. A comparison of the HEA and pure metals shows there is no obvious difference

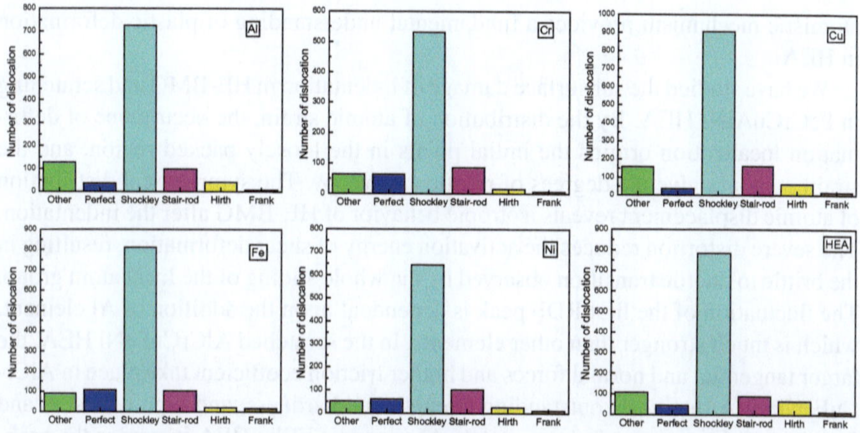

Fig. 20 Number of various type dislocation substructures in Al, Cr, Cu, Fe, Ni, and AlCrCuFeNi HEA after scratching. Reproduced from [26] with permission

in the number of other type dislocation substructure. The dislocation substructure domains the plastic deformation behavior below the tip, depending on the low SFE effect on the evolution of microstructure during the scratching process.

5 Conclusions

In this chapter, we present the atomistic modeling of the plastic deformation and subsurface damage during the uniaxial deformation, indentation, and scratching of HEAs. The main results are summarized as follows.

We have studied the mechanical behaviors of AlCrFeCuNi HEA under uniaxial tensile loading. The elastic properties and stress–strain relations from MD simulations are in excellent agreement with recent experimental results. Above all, the $AlCrFeCuNi_{1.4}$ HEA not only has high strength, but also exhibits good plasticity, qualitatively consistent with the experiment. Similar to the mechanical properties of single-crystal metals, stress fluctuation during plastic deformation of HEA is always accompanied with the generation and motion of dislocation and SF with the increase of strain. In addition, the dislocation–dislocation interaction, dislocation–solid solution interaction, deformation twinning and detwinning occur after the yield point. Furthermore, the dislocation gliding, dislocation pinning due to the severe lattice distortion and solid solution, and twinning are still the main mechanisms of plastic deformation in the $AlCrFeCuNi_{1.4}$ HEA. The kinetics of strain-induced phase transformation from FCC→BCC in nanocrystalline HEA are investigated by MD simulations. Interestingly, owing to the structural dissipation and strain relaxation by the volume expansion, the phase transformation agreeing with N–W relationship could convert the stress state from tensile to compressive in single-crystal HEA. This atomistic mechanism provides a fundamental understanding of plastic deformation in HEA.

We have studied the subsurface damage of indentation in HE-BMG and scratching in FeCrCuAlNi HEA. By the distribution of atomic strain, the occurrence of deformation localization origins the initial points in the loosely packed region, and the strain contours give the degrees of plastic anisotropy. The symmetrical distribution of atomic displacement reveals isotropic behavior of HE-BMG after the indentation. The severe distortion reduces the activation energy of shear deformation, resulting in the brittle to ductile transition observed by the whole sliding of the local atom group. The fluctuation of the first RDF peak is dependent upon the addition of Al element, which is much stronger than other elements. In the scratched AlCrCuFeNi HEA, the larger tangential and normal forces and higher friction coefficient take place in AlCrCuFeNi HEA due to its outstanding strength and hardness, and high adhesion and fracture toughness over the pure metals. Moreover, SFE in HEA increases the probability to initiate dislocation and twinning, which is conducive to the formation of complex deformation modes. Compared to the pure metals, the segregation potency of solutes into TB is raised due to the decreasing segregation energy of TB, resulting in the stronger solute effects on improving twinning properties for HEA. The higher dislocation density and the more activated slipping planes lead to the outstanding plasticity of AlCrCuFeNi HEA. The solute atoms as barriers to hinder the motion of dislocation and the severe lattice distortion to suppress the free slipping of dislocation are significantly stronger obstacles to strengthen HEA. This work provides a basis for further understanding and tailoring SFE in mechanical properties and deformation mechanism of HEA, which maybe facilitate the design and preparation of new HEA with high performance. Further, this effective strategy is used to accelerate the discovery of excellent mechanical properties of HEA.

To obtain the best comprehensive performance (including strength, ductility, hardness and wear resistance) in HEA, in future we should generate large amounts of data using high-throughput experiment and simulation, and then build the microstructure–property database, finally predict the performance of HEA based on machine learning. Under the condition of less and even free experiment and simulation, the ultimate goal realizes to accurately predict the properties of new HEA system through machine learning combined with the design–process–microstructure–property–application database.

Acknowledgements The authors would like to deeply appreciate the support from the NNSFC (11572118, 11772122 and 51871092), the Fundamental Research Funds for the Central Universities (531107051151), and the National Key Research and Development Program of China (2016YFB0700300).

References

1. Ritchie RO (2011) The conflicts between strength and toughness. Nat Mater 10:817
2. Zheng X, Smith W, Jackson J, Moran B, Cui H, Chen D, Spadaccini CM (2016) Multiscale metallic metamaterials. Nat Mater 15:1100
3. Gludovatz B, Hohenwarter A, Catoor D, Chang EH, George EP, Ritchie RO (2014) A fracture-resistant high-entropy alloy for cryogenic applications. Science 345:1153–1158
4. Zhang Y, Zuo TT, Tang Z, Gao MC, Dahmen KA, Liaw PK, Lu ZP (2014) Microstructures and properties of high-entropy alloys. Prog Mater Sci 61:1–93
5. Yeh JW, Chen SK, Lin SJ, Gan JY, Chin TS, Shun TT, Chang SY (2004) Nanostructured high-entropy alloys with multiple principal elements: novel alloy design concepts and outcomes. Adv Eng Mater 6:299–303
6. Schuh B, Mendez-Martin F, Völker B, George EP, Clemens H, Pippan R, Hohenwarter A (2015) Mechanical properties, microstructure and thermal stability of a nanocrystalline CoCrFeMnNi high-entropy alloy after severe plastic deformation. Acta Mater 96:258–268
7. Yeh JW (2006) Recent progress in high-entropy alloys. Ann Chim Sci Mater 31:633–648
8. He JY, Liu WH, Wang H, Wu Y, Liu XJ, Nieh TG, Lu ZP (2014) Effects of Al addition on structural evolution and tensile properties of the FeCoNiCrMn high-entropy alloy system. Acta Mater 62:105–113
9. He JY, Wang H, Huang HL, Xu XD, Chen MW, Wu Y, Lu ZP (2016) A precipitation-hardened high-entropy alloy with outstanding tensile properties. Acta Mater 102:187–196
10. Gao X, Lu Y, Zhang B, Liang N, Wu G, Sha G, Zhao Y (2017) Microstructural origins of high strength and high ductility in an $AlCoCrFeNi_{2.1}$ eutectic high-entropy alloy. Acta Mater 141:59–66
11. Lu Y, Gao X, Jiang L, Chen Z, Wang T, Jie J, Zhao Y (2017) Directly cast bulk eutectic and near-eutectic high entropy alloys with balanced strength and ductility in a wide temperature range. Acta Mater 124:143–150
12. Tracy CL, Park S, Rittman DR, Zinkle SJ, Bei H, Lang M, Mao WL (2017) High pressure synthesis of a hexagonal close-packed phase of the high-entropy alloy CrMnFeCoNi. Nat. Commun 8:15634
13. Hemphill MA, Yuan T, Wang GY, Yeh JW, Tsai CW, Chuang A, Liaw PK (2012) Fatigue behavior of $Al_{0.5}CoCrCuFeNi$ high entropy alloys. Acta Mater 60:5723–5734
14. Li Z, Pradeep KG, Deng Y, Raabe D, Tasan CC (2016) Metastable high-entropy dual-phase alloys overcome the strength–ductility trade-off. Nature 534:227
15. Wu JM, Lin SJ, Yeh JW, Chen SK, Huang YS, Chen HC (2006) Adhesive wear behavior of $Al_xCoCrCuFeNi$ high-entropy alloys as a function of aluminum content. Wear 261:513–519
16. Yadav S, Sarkar S, Aggarwal A, Kumar A, Biswas K (2018) Wear and mechanical properties of novel $(CuCrFeTiZn)_{100-x}Pb_x$ high entropy alloy composite via mechanical alloying and spark plasma sintering. Wear 410:93–109
17. Ye YX, Liu CZ, Wang H, Nieh TG (2018) Friction and wear behavior of a single-phase equiatomic TiZrHfNb high-entropy alloy studied using a nanoscratch technique. Acta Mater 147:78–89
18. Zepeda-Ruiz LA, Stukowski A, Oppelstrup T, Bulatov VV (2017) Probing the limits of metal plasticity with molecular dynamics simulations. Nature 550:492
19. Wu Z, Ahmad R, Yin B, Sandlöbes S, Curtin WA (2018) Mechanistic origin and prediction of enhanced ductility in magnesium alloys. Science 359:447–452
20. Kao SW, Yeh JW, Chin TS (2008) Rapidly solidified structure of alloys with up to eight equal-molar elements—a simulation by molecular dynamics. J Phys Cond Matter 20:145214
21. Xie L, Brault P, Thomann AL, Yang X, Zhang Y, Shang G (2016) Molecular dynamics simulation of Al–Co–Cr–Cu–Fe–Ni high entropy alloy thin film growth. Intermetallics 68:78–86
22. Li J, Fang Q, Liu B, Liu Y, Liu YW (2016) Mechanical behaviors of AlCrFeCuNi high-entropy alloys under uniaxial tension via molecular dynamics simulation. RSC Adv. 6:76409–76419
23. Sharma A, Balasubramanian G (2017) Dislocation dynamics in $Al_{0.1}CoCrFeNi$ high-entropy alloy under tensile loading. Intermetallics 91:31–34

24. Fang QH, Yi M, Li J, Liu B, Huang Z (2018) Deformation behaviors of $Cu_{29}Zr_{32}Ti_{15}Al_5Ni_{19}$ high entropy bulk metallic glass during nanoindentation. Appl Sur Sci 443:122–130
25. Li J, Fang QH, Liu B, Liu Y, Liu YW (2016) Atomic-scale analysis of nanoindentation behavior of high-entropy alloy. J Micro Mol Phys 1:1650001
26. Wang Z, Li J, Fang QH, Liu B, Zhang LC (2017) Investigation into nanoscratching mechanical response of AlCrCuFeNi high-entropy alloys using atomic simulations. Appl Sur Sci 416:470–481
27. Li J, Fang QH, Liu B, Liu Y (2018) Transformation induced softening and plasticity in high entropy alloys. Acta Mater 147:35–41
28. Plimpton S (1995) Fast parallel algorithms for short-range molecular dynamics. J Comput Phys 117:1–19
29. Stukowski A (2009) Visualization and analysis of atomistic simulation data with OVITO–the open visualization tool. Model Simul Mater Sci Eng 18:15012
30. Zhu PZ, Fang FZ (2012) Molecular dynamics simulations of nanoindentation of monocrystalline germanium. Appl Phys A 108:415–421
31. Mishin Y, Mehl MJ, Papaconstantopoulos DA, Voter AF, Kress JD (2001) Structural stability and lattice defects in copper: Ab initio, tight-binding, and embedded-atom calculations. Phys Rev B 63:224106
32. Cohen AJ, Gordon RG (1975) Theory of the lattice energy, equilibrium structure, elastic constants, and pressure-induced phase transitions in alkali-halide crystals. Phys Rev B 12:3228
33. Pi JH, Pan Y, Zhang H, Zhan L (2012) Microstructure and properties of $AlCrFeCuNi_x$, $0.6\leq x\leq 1.4$, high-entropy alloys. Mater Sci Eng A 534:228–233
34. Ma SG, Qiao JW, Wang ZH, Yang HJ, Zhang Y (2015) Microstructural features and tensile behaviors of the $Al_{0.5}CrCuFeNi_2$ high-entropy alloys by cold rolling and subsequent annealing. Mater Des 88:1057–1062
35. Ma SG, Jiao ZM, Qiao JW, Yang HJ, Zhang Y, Wang ZH (2016) Strain rate effects on the dynamic mechanical properties of the $AlCrCuFeNi_2$ high-entropy alloy. Mater Sci Eng, A 649:35–38
36. Li XY, Gao HJ (2016) Mechanical metamaterials: smaller and stronger. Nat Mater 15:373–374
37. Honeycutt JD, Andersen HC (1987) Molecular dynamics study of melting and freezing of small Lennard-Jones clusters. J Phys Chem 91:4950–4963
38. Zaddach AJ, Niu C, Koch CC, Irving DL (2013) Mechanical properties and stacking fault energies of NiFeCrCoMn high-entropy alloy. JOM 65:1780–1789
39. Stukowski A, Albe K (2010) Extracting dislocations and non-dislocation crystal defects from atomistic simulation data. Model Simul Mater Sci 18:085001
40. Zhao WS, Tao NR, Guo JY, Lu QH, Lu K (2005) High density nano-scale twins in Cu induced by dynamic plastic deformation. Scripta Mater 53:745–749
41. Liu S, Wei YJ (2017) The Gaussian distribution of lattice size and atomic level heterogeneity in high entropy alloys. Ext Mech Lett 11:84–88
42. Deng Y, Tasan CC, Pradeep KG, Springer H, Kostka A, Raabe D (2015) Design of a twinning-induced plasticity high entropy alloy. Acta Mater 94:124–133
43. Zhang YH, Zhuang Y, Hu A, Kai JJ, Liu CT (2017) The origin of negative stacking fault energies and nano-twin formation in face-centered cubic high entropy alloys. Scripta Mater 130:96–99
44. Fu Z, Chen W, Wen H, Zhang D, Chen Z, Zheng B, Lavernia EJ (2016) Microstructure and strengthening mechanisms in an FCC structured single-phase nanocrystalline $Co_{25}Ni_{25}Fe_{25}Al_{7.5}Cu_{17.5}$ high-entropy alloy. Acta Mater. 107:59–71
45. Yaakobi B, Boehly TR, Meyerhofer DD, Collins TJB, Remington BA, Allen PG, Eggert JH (2005) EXAFS measurement of iron bcc-to-hcp phase transformation in nanosecond-laser shocks. Phys Rev Lett 95:075501
46. Olson GB, Cohen MJ (1972) A mechanism for the strain-induced nucleation of martensitic transformations. J Less Common Metals 28:107–118
47. Kadau K, Germann TC, Lomdahl PS, Holian BL (2005) Atomistic simulations of shock-induced transformations and their orientation dependence in bcc Fe single crystals. Phys Rev B 72:064120

48. Zhang HW, Hei ZK, Liu G, Lu J, Lu K (2003) Formation of nanostructured surface layer on AISI 304 stainless steel by means of surface mechanical attrition treatment. Acta Mater 51:1871–1881
49. Wang B, Urbassek HM (2013) Molecular dynamics study of the α–γ phase transition in Fe induced by shear deformation. Acta Mater 61:5979–5987
50. Diao J, Gall K, Dunn ML (2003) Surface-stress-induced phase transformation in metal nanowires. Nat Mater 2:656–660
51. Telford M (2004) The case for bulk metallic glass. Mater Today 7:36–43
52. Greer AL, Ma E (2007) Bulk metallic glasses: at the cutting edge of metals research. MRS Bull 32:611–619
53. Wang WH, Dong C, Shek CH (2004) Bulk metallic glasses. Mater Sci Eng, R 44:45–89
54. Hertz H (1896) Miscellaneous papers. Macmillan, New York
55. Zhang JJ, Sun T, Hartmaier A, Yan YD (2012) Atomistic simulation of the influence of nanomachining-induced deformation on subsequent nanoindentation. Comp Mater Sci 59:14–21
56. Zhang JJ, Sun T, Yan Y, He Y, Liang Y, Dong S (2011) Atomistic investigation of probe-based nanomachining on Cu twin boundaries. J Comp The Nano 8:2344–2349
57. Qiu C, Zhu P, Fang F, Yuan D, Shen X (2014) Study of nanoindentation behavior of amorphous alloy using molecular dynamics. Appl Sur Sci 305:101–110
58. Pi J, Wang Z, He X, Bai Y, Zhen R (2016) Nanoindentation mechanical properties of glassy $Cu_{29}Zr_{32}Ti_{15}Al_5Ni_{19}$. J Alloys Comp 657:726–732
59. Zhao SF, Shao Y, Liu X, Chen N, Ding HY, Yao KF (2015) Pseudo-quinary $Ti_{20}Zr_{20}Hf_{20}Be_{20}(Cu_{20-x}Ni_x)$ high entropy bulk metallic glasses with large glass forming ability. Mater Des 87:625–631
60. Gong P, Jin J, Deng L, Wang S, Gu J, Yao K, Wang X (2017) Room temperature nanoindentation creep behavior of TiZrHfBeCu(Ni) high entropy bulk metallic glasses. Mater Sci Eng, A 688:174–179
61. Ding HY, Shao Y, Gong P, Li JF, Yao KF (2014) A senary TiZrHfCuNiBe high entropy bulk metallic glass with large glass-forming ability. Mater Lett 125:151–153
62. Wang WH (2012) The elastic properties, elastic models and elastic perspectives of metallic glasses. Prog Mater Sci 57:487–656
63. Rao JC, Diao HY, Ocelík V, Vainchtein D, Zhang C, Kuo C, Liaw PK (2017) Secondary phases in $Al_xCoCrFeNi$ high-entropy alloys: An in-situ TEM heating study and thermodynamic appraisal. Acta Mater 131:206–220
64. Borkar T, Gwalani B, Choudhuri D, Mikler CV, Yannetta CJ, Chen X, Banerjee R (2016) A combinatorial assessment of $Al_xCrCuFeNi_2$ ($0<x<1.5$) complex concentrated alloys: microstructure, microhardness, and magnetic properties. Acta Mater 116:63–76
65. Nishikawa M, Soyama H (2011) Two-step method to evaluate equibiaxial residual stress of metal surface based on micro-indentation tests. Mater Des 32:3240–3247
66. Cheng JB, Liang XB, Xu BS (2014) Effect of Nb addition on the structure and mechanical behaviors of CoCrCuFeNi high-entropy alloy coatings. Surf Coat Tech 240:184–190
67. Wang WH (2014) High-entropy metallic glasses. JOM 66:2067–2077
68. Guo S, Liu CT (2011) Phase stability in high entropy alloys: formation of solid-solution phase or amorphous phase. Prog Nat Sci 21:433–446
69. Azumo S, Nagayama K (2006) Amorphous formation and magnetic properties of Nd–Fe—Co–Al alloys by gas flow type levitation process. Mater Trans 47:2842–2845
70. Hu ZQ, Ding BZ, Zhang HF, Li DJ, Yao B, Liu HZ, Wang AM (2001) Formation of non-equilibrium alloys by high pressure melt quenching. Sci Technol Adv Mat 2:41–48
71. Li J, Fang QH, Liu YW, Zhang LC (2015) Scratching of copper with rough surfaces conducted by diamond tip simulated using molecular dynamics. Int J Adv Manuf Tech 77:1057–1070
72. Li J, Fang QH, Liu YW, Zhang LC (2014) A molecular dynamics investigation into the mechanisms of subsurface damage and material removal of monocrystalline copper subjected to nanoscale high speed grinding. Appl Surf Sci 303:331–343

73. Zhang JJ, Begau C, Geng L, Hartmaier A (2015) Atomistic investigation of wear mechanisms of a copper bi-crystal. Wear 332:941–948
74. Diao H, Xie X, Sun F, Dahmen KA, Liaw PK (2016) Mechanical properties of high-entropy alloys. Springer International Publishing, Berlin
75. Wang Y, Yang Y, Yang H, Zhang M, Qiao J (2017) Effect of nitriding on the tribological properties of $Al_{1.3}CoCuFeNi_2$ high-entropy alloy. J Alloy Comp 725:365–372
76. Reihanian M, Ebrahimi R, Tsuji N, Moshksar MM (2008) Analysis of the mechanical properties and deformation behavior of nanostructured commercially pure Al processed by equal channel angular pressing (ECAP). Mater Sci Eng, A 473:189–194
77. Li W, Liaw PK, Gao Y (2018) Fracture resistance of high entropy alloys: a review. Intermetallics 99:69–83
78. Liu Y, Ma S, Gao MC, Zhang C, Zhang T, Yang H, Qiao J (2016) Tribological properties of $AlCrCuFeNi_2$ high-entropy alloy in different conditions. Metal Mater Trans A 47:3312–3321
79. Zhang JJ, Wei YJ, Sun T, Hartmaier A, Yan YD, Li X (2012) Twin boundary spacing-dependent friction in nanotwinned copper. Phys Rev B 85:054109
80. Gao Y, Urbassek HM (2016) Scratching of nanocrystalline metals: a molecular dynamics study of Fe. Appl Surf Sci 389:688–695
81. Gao Y, Lu C, Huynh NN, Michal G, Zhu HT, Tieu AK (2009) Molecular dynamics simulation of effect of indenter shape on nanoscratch of Ni. Wear 267:1998–2002
82. Tsai CW, Chen YL, Tsai MH, Yeh JW, Shun TT, Chen SK (2009) Deformation and annealing behaviors of high-entropy alloy $Al_{0.5}CoCrCuFeNi$. J Alloys Compd 486:427–435
83. Zhang F, Meng B, Geng Y, Zhang Y (2016) Study on the machined depth when nanoscratching on 6H-SiC using Berkovich indenter: Modelling and experimental study. Appl Sur Sci 368:449–455
84. Lu K, Lu J (2004) Nanostructured surface layer on metallic materials induced by surface mechanical attrition treatment. Mater Sci Eng, A 375:38–45
85. Fang TH, Li WL, Tao NR, Lu K (2011) Revealing extraordinary intrinsic tensile plasticity in gradient nano-grained copper. Science 331:1587–1590
86. Fu Z, Chen W, Fang S, Zhang D, Xiao H, Zhu D (2013) Alloying behavior and deformation twinning in a $CoNiFeCrAl_{0.6}Ti_{0.4}$ high entropy alloy processed by spark plasma sintering. J Alloys Compd 553:316–323
87. Gao MC, Yeh JW, Liaw PK, Zhang Y (2016) High-entropy alloys: fundamentals and applications. Springer, Berlin
88. Chen MW, Ma E, Hemker KJ, Sheng HW, Wang YM, Cheng XM (2003) Deformation twinning in nanocrystalline aluminum. Science 300:1275

Investigation into the Realization of a Single Atomic Layer Removal in Nanoscale Mechanical Machining of Single Crystalline Copper

Pengzhe Zhu and Jianyong Li

Abstract It is widely believed that the minimum depth of material removal of single crystalline workpieces is one single atomic layer in nanoscale mechanical machining. However, direct evidence for this is still lacking. In this work, the minimum depth of material removal of single crystalline copper in nanoscale mechanical machining is investigated through nanoscratching using molecular dynamics simulations. We demonstrate that the minimum depth of material removal of copper workpiece can achieve a single atomic layer under certain machining conditions in nanoscale machining process. It is found that the minimum depth of material removal is closely associated with the crystal orientation and scratching direction of copper workpiece. Our results also demonstrate that even when the depth of material removal is a single atomic layer of copper workpiece under certain machining conditions, the workpiece material is not removed in a layer-by-layer fashion, which rejects the hypothesis that single crystalline metal materials can be continuously and stably removed one layer of atoms after another in nanoscale mechanical machining. These understandings not only shed light on the material removal mechanism in nanoscale mechanical machining but also provide insights into the control and optimization of nanoscale machining process.

1 Introduction

In the last two decades, miniaturization toward nanoscale has been the trend of technological development in a variety of industries such as mechanical, optical, and electronic applications. The ever-increasing demand for devices and systems with nanoscale size and/or nanoscale accuracy [such as microelectromechanical systems (MEMS) and nanoelectromechanical systems (NEMS)] has motivated the development of nanoscale manufacturing technologies [1–3]. Among the nanoscale manu-

P. Zhu (✉) · J. Li
School of Mechanical, Electronic and Control Engineering,
Beijing Jiaotong University, Beijing, China
e-mail: pzzhu@bjtu.edu.cn

J. Zhang et al. (eds.), *Simulation and Experiments of Material-Oriented Ultra-Precision Machining*, Springer Tracts in Mechanical Engineering,
https://doi.org/10.1007/978-981-13-3335-4_3

facturing technologies, nanoscale mechanical machining processes such as cutting, grinding, and tip-based nanomanufacturing have been attracting much attention since they can machine a wide range of materials and produce three-dimensional complex nanoscale devices with high accuracy controllably and stably [4–6]. However, as the nanoscale mechanical machining involves only the removal of a few atoms or layers of atoms, the nanoscale material removal mechanisms cannot be accurately described by the conventional macroscale machining theory based on continuum mechanics due to the discrete nature of materials at the nanoscale [7–10]. At this stage, the material removal mechanisms in the nanoscale mechanical machining processes are still not fully understood. Many researchers assume that the nanoscale material removal takes places via chip formation as in macrosale machining process [4, 5]. However, Komanduri et al. demonstrated that most of the material removal in the nanoscale machining process occurs by plowing instead of chip formation in front of the tool as in macroscale and microscale machining [10]. In addition, Fang et al. proposed a new cutting model that material removal in the nanoscale machining process is based on extrusion rather than shearing in macroscale cutting process [7, 8]. In a word, owing to the complexity of nanoscale machining process, there are still many fundamental questions concerning the material removal that need to be urgently addressed. One of the most important questions is: What is the limit of nanoscale mechanical machining? That is, what is the minimum depth of material removal in nanoscale mechanical machining? The minimum depth of material removal can be defined as the minimum depth above which the workpiece atoms can be removed stably from the workpiece surface under perfect machining conditions. As the minimum depth of material removal is a measure of the extreme accuracy achievable in nanoscale machining, the investigation of the limit of nanoscale mechanical machining can not only contribute to the understanding of nanoscale material removal mechanisms, but also offer insights into the control and optimization of nanoscale machining process [4, 11, 12].

However, despite its importance, little attention has been paid to the study of the minimum depth of material removal in nanoscale machining process, although many researchers have been focusing on the investigation of minimum depth (thickness) of cut under which no stable chip formation in front of the tool would occur in microscale machining during the past decades [13–17]. A pioneering study employing molecular dynamics (MD) simulations and experiments by Shimada et al. showed that the minimum thickness of cut is about 1 nm [11, 12]. It is obvious that the minimum depth of material removal could be expected to be less than 1 nm. Another significant study using MD simulations found that the minimum wear depth of Cu(111) surface is equal to the critical indentation depth (0.65–0.87 nm) associated with the first force drop in the force–depth curve of nanoindentation [18]. However, the friction model ignores atomic-scale surface roughness of probe produced by discrete atoms and adhesion interaction between the probe and Cu(111) surface [18], both of which have considerable effect on contact area and stress distribution in the nanoscale contact and should be taken into consideration [19–22]. Another inspiring simulation work presented a monoatomic layer removal mechanism in chemical mechanical polishing (CMP) by modeling the nanoscratching process of single crystalline silicon using a spherical silica tool [23]. It is assumed that in CMP process the material was removed one atomic layer after another. But in their work, the material removal is neither continuous nor stable. In fact, it is very natural to hypothesize that the minimum depth of material removal is one atomic layer of workpiece for single crystalline materials such as copper and silicon. The demonstration of atom-by-atom manipulation [24–26] and recently proposed atom-by-atom wear mechanism [27–29] make this hypothesis even more convincing. However, so far, although a great many experiments and simulations have been widely carried out to investigate the nanoscale machining mechanisms [4–6, 30–38], a study verifying the hypothesis that the minimum depth of material removal is one single atomic layer of workpiece in nanoscale mechanical machining process is still lacking. Furthermore, the view that single crystalline metal materials can be continuously and stably removed one layer of atoms after another in nanoscale mechanical machining has never been demonstrated.

This work intends to fill this gap. As it is rather difficult to directly observe the nanoscale machining process by in situ experiments with atomic-scale resolution, MD simulation that can capture atomic details provides a powerful tool to gain deeper insights into the minimum depth of material removal and fundamental machining mechanisms of nanoscale machining process. Therefore, in this study, MD simulations of the nanoscratching process of single crystalline copper using diamond tool are performed to explore the minimum depth of material removal in the nanoscale machining process. Copper is chosen as the workpiece since it is a common and typical metal and has numerous applications in various fields such as integrated circuits, MEMS, and NEMS [39, 40]. In this work, we demonstrate that the minimum depth of material removal of copper workpiece can achieve a single atomic layer under certain machining conditions in nanoscale machining process, but the workpiece material is not removed in a layer-by-layer fashion. Our results reject the hypothesis that single crystalline metal materials can be continuously and stably removed one layer of atoms after another in nanoscale mechanical machining.

2 Simulation Models and Methods

In this study, we use MD simulations to investigate the nanoscratching process of single crystalline copper. The MD simulations are conducted using the large-scale atomic/molecular massively parallel simulator (LAMMPS) [41]. The simulation model consists of a hemispherical rigid diamond tool and a single crystalline copper workpiece (see Fig. 1). Since single crystalline copper exhibits significant anisotropic effects in deformation behavior under nanoindentation [42] and nanomachining [9, 43], we performed MD simulations of nanoscratching on three different surfaces of Cu(001), (110), and (111) planes. For Cu(001) surface, the coordinate systems are taken as x-[100], y-[010], and z-[001] and the size of workpiece is $21.69 \times 14.46 \times 2.89\ \text{nm}^3$. For Cu(110) surface, the coordinate systems are taken as x-[001], y-[$1\bar{1}0$],

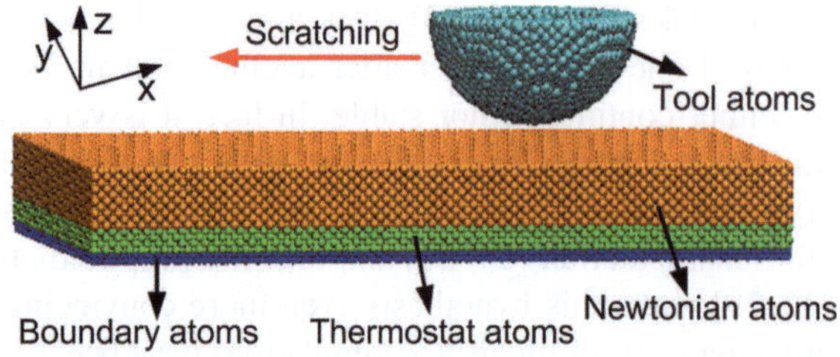

Fig. 1 Simulation model of nanoscratching process

and z-[110] and the size is $21.69 \times 14.57 \times 2.81$ nm^3. And for Cu(111) surface, the coordinate systems are taken as x-[$1\bar{1}0$], y-[$11\bar{2}$], and z-[111] and the size is $21.73 \times 14.61 \times 2.92$ nm^3.

The Cu workpiece is composed of three kinds of atoms: boundary atoms, thermostat atoms, and Newtonian atoms, as shown in Fig. 1. The two layers of atoms at the bottom of the workpiece are kept fixed in space as boundary atoms to prevent the workpiece from translating during the nanoscratching process. The next four layers of atoms adjacent to the boundary atoms are thermostat atoms which are kept at a constant temperature of 300 K by the velocity scaling method [44] to mimic the heat dissipation in real machining process. The remaining workpiece atoms are Newtonian atoms that are free of constraints and move freely according to the interatomic forces. The equations of motion are integrated with a velocity Verlet algorithm with a time step of 1 fs. Periodic boundary conditions are imposed in the x and y directions.

Interatomic forces within Cu are derived from an embedded atom method (EAM) potential [45]. The EAM potential has been very successful in modeling the elastic properties, defect formation energies, and fracture mechanisms of various metals [45, 46]. It has also been successfully applied to describe the surface properties of metals such as surface energies and surface reconstructions [45–48]. As the tool is treated as a rigid body, the interactions between tool atoms are ignored. The interaction between Cu and tool is modeled by the widely used Morse potential [44]:

$$V(r) = D(\mathrm{e}^{-2\alpha(r-r_0)} - 2\mathrm{e}^{-\alpha(r-r_0)}) \quad (1)$$

where $V(r)$ is a pair potential energy function; D is the cohesion energy; α is the elastic modulus; r and r_0 are the instantaneous and equilibrium distance between two atoms, respectively. The cutoff radius of the Morse potential is chosen as 9.0 Å, which ensures that the calculations will not consume large amounts of computational time calculating the forces that are near zero. The standard Morse potential parameters [35–38, 49] are given as $D = D^* = 0.087$ eV, $\alpha = 5.14$ Å^{-1}, and $r_0 = 2.05$ Å. It should be noted that the interaction strength of material bonds generally increases with the increase of cohesion energy D [50, 51]. It is well-known that adhesion plays an essential role in nanoscale machining process. Hence, in order to examine the effects of adhesion between tool and workpiece on the minimum depth of material removal, we also performed MD simulations of nanoscratching for different values of D [50,

51]. If not noted, we adopt the standard Morse potential parameters for the interaction between Cu and tool (standard adhesion).

In the simulations, hemispherical rigid diamond tools with 3.0, 4.0, and 5.0 nm radius created from perfect diamond atomic lattices are initially positioned above the single crystal copper workpiece surface. After the system is fully relaxed, the tool moves downwards to the Cu workpiece with a constant velocity until a desired scratching depth and then slides on the surface along the negative x direction with a constant scratching velocity. It should be pointed out that the indentation and scratching velocities reported in the references are on the order of 1–200 m/s for most MD simulations to save computational cost [32–38, 49, 52]. If not noted, the indentation velocity and scratching velocity are 100 m/s in this work. It should be mentioned that the indentation and scratching velocities in the simulations are very high due to the small time step (femtosecond) used in MD simulation to capture atomic motion. In real applications, a typical cutting velocity is in the range of 1–10 m/s for nanoscale cutting process [32] and up to 50 m/s for grinding [53]. The maximum cutting speed in ultra-high-speed grinding exceeds 250 m/s already [53]. However, a typical velocity in the AFM-based scratching process is only in the range of nm/s to μm/s [54]. To study the effect of velocity, we have modeled the machining process for different scratching velocities of 20, 50, 100, and 200 m/s. It is obvious that the velocities in the simulations are comparable to the operating conditions in some cutting and grinding experiments. However, it is not the case for the AFM-based scratching process. Until now, bridging the gap in velocities between MD and AFM experiments still remains a big challenge. But it has been demonstrated that although the velocity in MD simulations exceeds the AFM experimental velocity by several orders of magnitude, the MD simulations can still produce consistent results with the experiments and the model predictions are still meaningful [54, 55].

The scratching depth h is defined as the distance from the top surface of the workpiece to the bottom of the tool along z direction. If the bottom of the tool is above the top surface of workpiece, the scratching depth is negative. In this case, the material removal of workpiece may still occur due to the attractive interaction between tool and workpiece.

3 Effect of Machining Parameters on Material Removal Process

We first model the sliding process of a hemispherical diamond tool with a radius of 4 nm over a Cu(001) surface. Figure 2 shows the material removal behavior of a Cu(001) workpiece for several typical scratching depths at the scratching velocity of 100 m/s. It should be noted that the discussions presented here are based not only on the MD simulation snapshots, but also on the observation of the animations of the scratching process. In order to identify the machining conditions under which a single atomic layer removal of workpiece can be realized, we also quantitatively

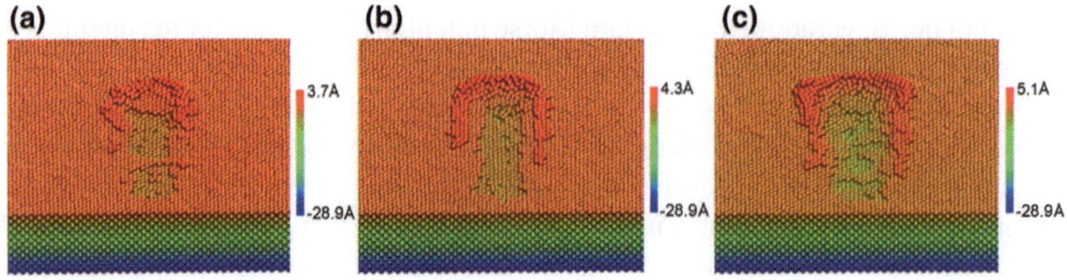

Fig. 2 Deformation behavior of a Cu(001) workpiece for scratching depths of **a** −0.18 nm, **b** −0.08 nm, and **c** 0 nm at the scratching distance of 8 nm. Atoms are colored according to their heights in z direction

analyze the variation of the number of atomic layers removed with the simulation parameters, as shown in Table 1. During the scratching process, some workpiece atoms are removed and piled up in front of and on both sides of the tool. In the real machining process, the removed material may be taken away by the cutting fluids [56] or further removed through successive machining by cross-feed of the tool relative to the workpiece [10]. When the scratching depth is small ($h = -0.18$ nm), part of the topmost atomic layer of workpiece along the scratch path is not removed and material removal also occurs in the second topmost atomic layer of workpiece (see Fig. 2a and Table 1). For a bigger scratching depth ($h = -0.08$ nm), the material removal also involves the two topmost atomic layers of workpiece. However, the topmost atomic layer along the scratch path has been completely removed, and more atoms in the second topmost atomic layer of workpiece are removed (see Fig. 2b). As the scratching depth increases to 0 nm, more atoms in the second topmost atomic layer of workpiece are removed and the material removal expands to the third topmost atomic layer (see Fig. 2c). It is obvious that a continuous and stable monolayer removal of a Cu(001) workpiece cannot be realized just by changing the scratching depth.

The scratching forces for different scratching depths during the scratching process of a Cu(001) workpiece are shown in Fig. 3. The scratching forces are obtained by summing the atomic forces of workpiece atoms on tool atoms. The fluctuation of forces during the scratching process is mainly due to the thermal motion of workpiece atoms and the formation and movement of dislocations in the workpiece which leads to the release of the accumulated strain energy and causes the temporary drop of the friction force and normal force [9, 35]. It can be seen from Fig. 3 that for the scratching depth of −0.18 nm, the friction force is bigger than the normal force. However, the normal force increases much more rapidly than the friction force as the scratching depth increases from −0.18 to 0 nm. Therefore, the friction coefficient decreases from larger than 1.0 (1.25) to smaller than 1.0 (0.49). It is interesting to point out that for the scratching depth of −0.18 nm, at the initial stage of scratching, the normal force is negative (attractive interaction). A comparison of Figs. 2a and 3a shows that the attractive interaction between the tool and workpiece can also cause material removal. With the accumulation of removed workpiece atoms in front of the tool during the scratching process, the normal force becomes positive. As much

Table 1 Variation of the number of atomic layers removed with the simulation parameters

Tool radius (nm)	Scratching depth (nm)	Scratching velocity (m/s)	Interaction strength (D/D^*)	Number of layers removed (n)
4	−0.18	100	1	$0<n<2$
4	−0.08	100	1	$1<n<2$
4	0	100	1	$1<n<3$
4	−0.18	20	1	$0<n<2$
4	−0.18	50	1	$0<n<2$
4	−0.18	200	1	$0<n<2$
3	−0.13	100	1	$0<n<2$
5	−0.21	100	1	$0<n<3$
4	0	100	2	$1<n<3$
4	0	100	0.5	$1<n<2$
4	0	100	0.1	$n=1$

Here, "$0<n<2(3)$" means the material removal involves two (three) topmost atomic layers of workpiece, but each of the two (three) atomic layers along the scratch path is not completely removed. "$1<n<2(3)$" implies that the topmost atomic layer of the workpiece along the scratch path is completely removed, and the second topmost (and the third) atomic layer(s) along the scratch path is (are) not completely removed

work has been carried out to investigate such effects as scratching depth, scratching velocity, tool size, and crystal orientation on the scratching forces [9, 10, 18, 35], the variation of scratching forces during the scratching process will not be the main focus of this work.

Next, we investigate the basic factors that play an important role in the material removal process and check whether a single atomic layer removal can be realized by varying the machining conditions. We study the deformation behavior of a Cu(001) workpiece for different scratching velocities of 20, 50, and 200 m/s (see the case with a scratching depth of −0.18 nm in Fig. 4 and Table 1). It is observed that while the topmost atomic layer of workpiece along the scratch path is not completely removed, a small number of atoms in the second topmost atomic layer of workpiece are inevitably removed for all cases, which indicates that a single atomic layer removal cannot be realized just by changing the scratching velocity.

We further explore the effect of tool size on the possibility of realization of a single atomic layer removal in the scratching process. It is found that whether the tool radius is bigger or smaller, the topmost atomic layer of workpiece along the scratch path is not completely removed while a small number of atoms in the second topmost atomic layer of workpiece are inevitably removed during the scratching process (see Fig. 5). For the tool radius of 5 nm, the material removal also occurs in the third topmost atomic layer of workpiece (see Fig. 5a and Table 1). This shows that a single atomic layer removal in the scratching process cannot be realized just by changing the tool size.

As adhesion between the tool and workpiece plays a crucial role in the material removal process [50, 51], it is natural to expect that changing adhesion interaction

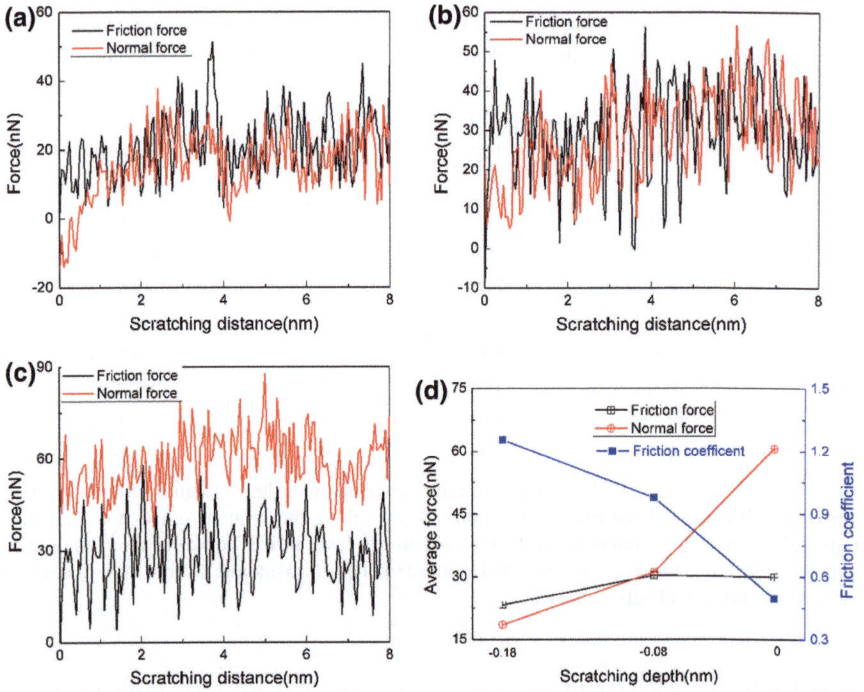

Fig. 3 Force–distance curves with scratching depths of **a** −0.18 nm, **b** −0.08 nm, and **c** 0 nm for a Cu(001) workpiece. The average forces calculated for the scratching distance from 2 to 8 nm where scratching is in a steady state and corresponding friction coefficients are shown in (**d**)

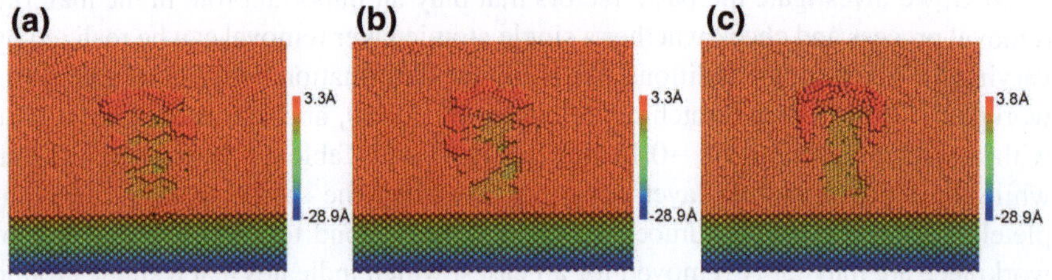

Fig. 4 Deformation behavior of a Cu(001) workpiece for scratching velocity of **a** 20 m/s, **b** 50 m/s, and **c** 200 m/s at the scratching depth of −0.18 nm and scratching distance of 8 nm. The radius of tool is 4 nm. Atoms are colored according to their heights in z direction

between the tool and workpiece may provide a possible new approach to realizing a single atomic layer removal in scratching process. Hence, we performed MD simulations of nanoscratching process for different values of cohesion energy D between the tool and workpiece. It should be noted that smaller cohesive energy D corresponds to weaker adhesion interaction between tool and workpiece [50, 51]. We take the case with a scratching depth of 0 nm as an example (see Fig. 6). For a relatively strong adhesion ($D = 2D^*$), material removal occurs in the three topmost atomic

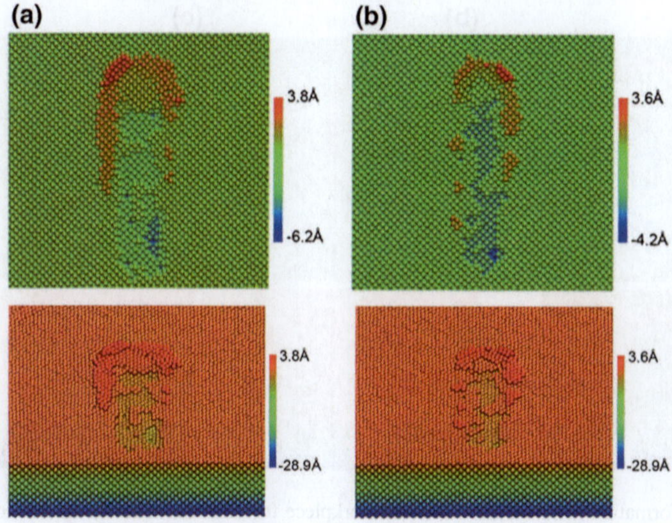

Fig. 5 Deformation behavior of a Cu(001) workpiece. **a** The radius of tool is 5 nm, and scratching depth is −0.21 nm. **b** The radius of tool is 3 nm, and scratching depth is −0.13 nm. Atoms are colored according to their heights in z direction

layers of the workpiece (see Fig. 6a and Table 1), while for an intermediate adhesion ($D = 0.5D^*$), material removal takes place in the two topmost atomic layers of workpiece (see Fig. 6b). It is of interest to find that for a weak adhesion between tool and workpiece ($D = 0.1D^*$), the material removal depth is only a single atomic layer of workpiece (see Fig. 6c). Furthermore, a perfectly ordered crystalline surface is observed behind the tool in the machined groove along the scratch path, and material removal behavior is continuous and stable (Fig. 6c). This shows that the limit of nanoscale mechanical machining is a single atomic layer of workpiece. The results suggest that reduction in adhesion between tool and workpiece plays a critical role in the realization of a single atomic layer removal of workpiece, which is consistent with the finding that a weaker adhesion will result in less subsurface deformation and damage (wear) [51, 57, 58]. In practice, the adhesion interaction between the tool and workpiece can be changed by proper lubrication [50, 51], adopting the widely used monolayer lubricants such as self-assembled monolayer films to provide boundary lubrication (in cases where use of liquid lubricants is contraindicated under nanoconfinement) [22] or adsorption of gas molecules on the workpiece and tool surfaces from environment [59, 60].

The results above clearly show that the scratching velocity and tool size have little effect on the realization of a single atomic layer material removal in the nanoscale machining process, while reduction in adhesion between the tool and workpiece plays an important role. This finding sheds light on the control and optimization of nanoscale machining process.

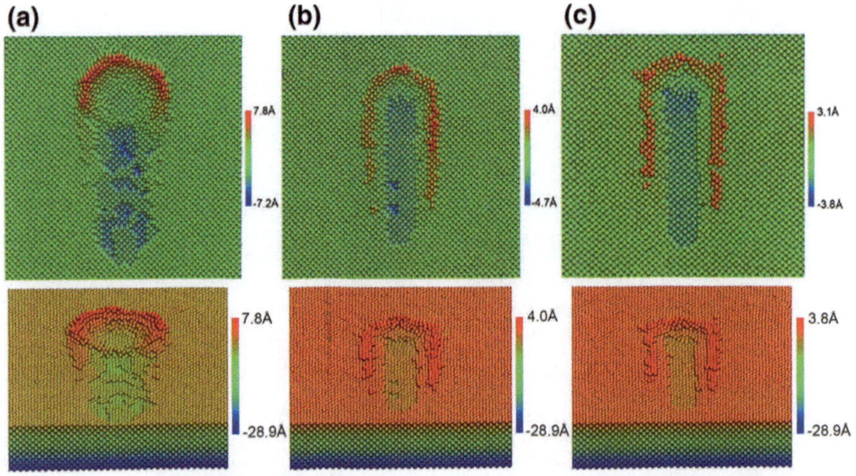

Fig. 6 Deformation behavior of a Cu(001) workpiece for different adhesion interaction strength between tool and workpiece. **a** $D = 2D^*$, **b** $D = 0.5D^*$, and **c** $D = 0.1D^*$. The radius of tool is 4 nm. The scratching depth is 0 nm, and scratching velocity is 100 m/s. Atoms are colored according to their heights in z direction

In order to gain more insights into the observed single atomic layer material removal for a weak adhesion between tool and workpiece, we further investigate the nanosratching process for different scratching depths under the condition of a weak adhesion (see Fig. 7a–e). For a small scratching depth ($h = -0.08$ nm), the material removal involves only the topmost atomic layer of workpiece along the scratch path, but the material removal process is neither continuous nor stable (see Fig. 7a). When the scratching depth is relatively bigger ($h = -0.02$ and 0 nm), the material removal depth is only a single atomic layer of workpiece, and a perfectly ordered crystalline surface is observed in the machined groove along the scratch path (see Fig. 7b, c). When the scratching depth further increases and the tool penetrates slightly into the workpiece ($h = 0.02$ and 0.1 nm), the material removal involves two topmost atomic layers of workpiece along the scratch path (see Fig. 7d, e). It is reasonable to expect that more atoms are removed for a bigger scratching depth.

However, it is still unknown whether the machined workpiece surface formed by a single atomic layer removal mechanism persists after retraction and relaxation. To address this issue, we remove the tool from the workpiece at a scratching distance of 8 nm and then fully relax the positions of workpiece atoms. We take the case with a scratching depth of 0 nm for an example (see Fig. 7f). It can be seen that the machined workpiece surface persists after retraction and relaxation, which validates the speculation that the minimum depth of material removal can achieve a single atomic layer of workpiece under perfect machining conditions in nanoscale mechanical machining process. But unfortunately, it is still unclear whether material removal occurs only in the topmost atomic layer of workpiece or not in Fig. 7b, c. To investigate this problem, the atoms of workpiece after nanoscratching are colored

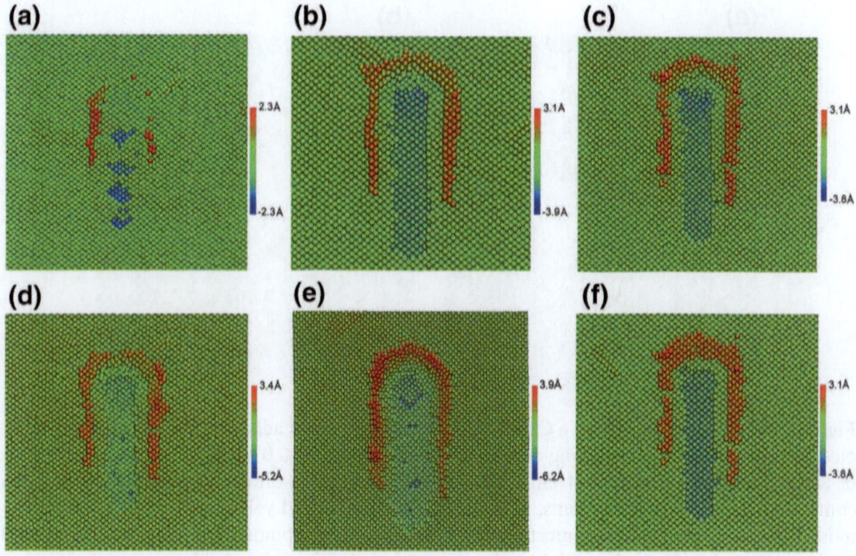

Fig. 7 Deformation behavior of a Cu(001) workpiece at scratching depths of **a** −0.08 nm, **b** −0.02 nm, **c** 0.0 nm, **d** 0.02 nm, and **e** 0.1 nm for a weak adhesion between tool and workpiece ($D = 0.1D^*$) at the scratching distance of 8 nm. The atomic configuration of workpiece after retraction and relaxation with a scratching depth of 0 nm is shown in (**f**). The radius of tool is 4 nm. Atoms are colored according to their heights in z direction

in different atomic layers of initial configuration as shown in Fig. 8. Moreover, we quantitatively summarize the number of atoms that move between the three topmost atomic layers of initial configuration after nanoscratching; see Table 2. A comparison of Figs. 7b, c, 8 and Table 2 shows that after the scratching process, nine atoms in the second topmost (second) layer of the Cu(001) workpiece are removed and some atoms (43 and 40 atoms for scratching depth of −0.02 and 0 nm, respectively) are displaced and incorporated into the initially topmost (first) atomic layer along the scratch path. At the same time, some atoms (81 and 74 atoms for the scratching depth of −0.02 and 0 nm, respectively) in the initially topmost layer of workpiece are also displaced and incorporated into the second topmost layer of the Cu(001) workpiece along the scratch path. In addition, some atoms (12 and 14 atoms for scratching depth of −0.02 nm and 0 nm, respectively) in the third topmost (third) atomic layer are also displaced and incorporated into the second topmost layer of the Cu(001) workpiece along the scratch path. Specifically for the scratching depth of −0.02 nm, one atom in the third topmost (third) atomic layer is even displaced and incorporated into the topmost layer of the Cu(001) workpiece. It should be pointed out that some atoms in the second topmost layer of Cu(001) workpiece are inevitably displaced and implanted into the third topmost layer during the scratching process for the scratching depth of −0.02 and 0 nm. It is interesting that a perfectly ordered crystalline surface is formed in the machined groove along the scratch path. It can be

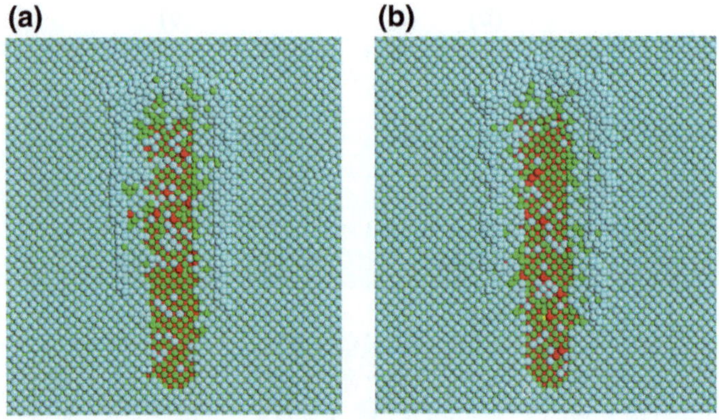

Fig. 8 Deformation behavior of a Cu(001) workpiece for a weak adhesion between tool and workpiece($D = 0.1D^*$) at the scratching depths of a −0.02 nm and **b** 0.0 nm with a scratching distance of 8 nm. The radius of tool is 4 nm. The atoms are colored in different atomic layers of initial configuration. The light blue atoms, green atoms, red atoms, and yellow atoms represent the topmost (first) layer, the second topmost (second) layer, the third topmost (third) layer, and the fourth topmost layer of atoms in the initial configuration of workpiece, respectively

Table 2 Number of atoms that move between the three topmost atomic layers of initial configuration of Cu(001) workpiece after nanosratching

Scratching depth (nm)	From first layer to second layer	Removed in second layer	From second layer to first layer	From third layer to second layer	From third layer to first layer
−0.02	81	9	43	12	1
0	74	9	40	14	0

found that though the topmost (single) atomic layer of the Cu(001) workpiece along the scratch path appears to be completely removed under certain conditions (Fig. 7b, c), the material removal actually involves two topmost atomic layers of workpiece (Fig. 8 and Table 2). Furthermore, some atoms in the initially topmost layer of workpiece are inevitably displaced and implanted into the second topmost atomic layer of workpiece along the scratch path. Therefore, even though the minimum depth of material removal can achieve a single atomic layer of the Cu(001) workpiece under certain conditions in nanoscale mechanical machining process, the workpiece is not removed in a layer-by-layer fashion. This rejects the prevailing hypothesis that single crystalline metal materials can be continuously and stably removed one layer of atoms after another in nanoscale mechanical machining.

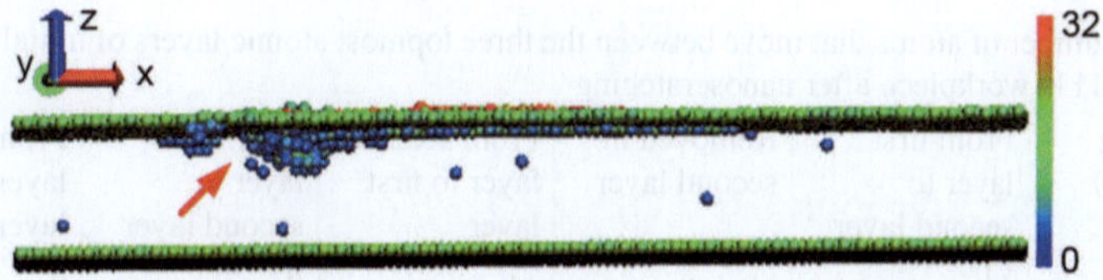

Fig. 9 Deformation behavior of a Cu(001) workpiece for a weak adhesion between tool and workpiece ($D = 0.1D^*$) at a scratching depth of 0 nm with a scratching distance of 8 nm. The radius of tool is 4 nm. Atoms are colored according to their CSP values

4 Effect of Crystal Orientation and Scratching Direction on the Material Removal Process

The material removal behaviors of workpiece are closely related to the deformation mechanism of workpiece. To further investigate the deformation mechanism of the workpiece, we clarify the defect structures of workpiece during the scratching process using the centrosymmetry parameter (CSP) [61]. Figure 9 shows the typical defect structures of workpiece for a weak adhesion between tool and workpiece ($D=0.1D^*$). Atoms are colored according to their CSP values. Note that the atoms inside the workpiece with CSP smaller than 3 are removed as they are assumed to be in perfect face-centered cubic configuration including elastic deformation [9, 62]. Red arrow is used to outstanding dislocations. Plastic deformation takes place in the workpiece as dislocations (or dislocation embryos) are generated within the workpiece in the scratching process.

The fact that single crystalline copper exhibits significant anisotropic effects during nanoindentation [42] and nanoscratching [9, 43] encourages us to further study the effect of crystal orientation and scratching direction on the material removal behaviors of workpiece in the scratching process. We carried out MD simulations for two more combinations of crystal orientation and scratching direction, i.e. (111)[$\bar{1}$10] and (110)[001]. Figures 10 and 11 show the deformation behavior of the Cu(111) workpiece. Table 3 presents the number of atoms that move between the three topmost atomic layers of initial configuration of the Cu(111) workpiece after nanoscratching. When the scratching depth is small ($h = -0.12$ nm or smaller), the workpiece undergoes a pure elastic deformation and no material removal occurs (Figs. 10a and 11a). When the scratching depth is bigger and in the range of −0.11 to 0 nm, the minimum depth of material removal is a single atomic layer of workpiece and a perfectly ordered crystalline surface is observed in the machined groove along the scratch path (See Fig. 10b–e). When the scratching depth further increases and the tool penetrates slightly into the workpiece ($h=0.02$ nm), the material removal involves two topmost atomic layers of workpiece along the scratch path (see Fig. 10f). However, similar to the case of Cu(001) surface, the finding that the minimum depth of material removal is a single atomic layer of Cu(111) workpiece does not mean that the material removal is in a layer-by-layer fashion (one layer of atoms after another). A

Table 3 Number of atoms that move between the three topmost atomic layers of initial configuration of Cu(111) workpiece after nanoscratching

Scratching depth (nm)	From first layer to second layer	Removed in second layer	From second layer to first layer	From third layer to second layer	From third layer to first layer
−0.11	11	0	11	0	0
−0.08	13	3	8	0	0
−0.03	30	6	26	0	0
0	38	4	35	2	0
0.02	57	36	63	4	0

comparison of Figs. 10b–e, 11b–e and Table 3 shows that during the scratching process some atoms in the topmost (first) layer of the Cu(111) workpiece are inevitably implanted into the second topmost (second) atomic layer and some atoms in the second topmost layer of the Cu(111) workpiece are also displaced and incorporated into the initially topmost layer along the scratch path. It can also be found that for the scratching depths between −0.08 and 0.0 nm, some atoms in the second topmost layer of the Cu(111) workpiece are removed as well along the scratch path although the material removal only occurs in the topmost layer of Cu(111) workpiece for the scratching depth of −0.11 nm (see Figs. 10c–e, 11c–e and Table 3). Specifically, for the scratching depth of 0.0 nm, two atoms initially in the third topmost layer of workpiece are incorporated into the second topmost layer of the Cu(111) workpiece along the scratch path(see Figs. 10e, 11e and Table 3). With a bigger scratching depth ($h=0.02$ nm), four atoms initially in the third topmost layer of workpiece are incorporated into the second topmost layer and a great number of atoms (36 atoms) originally in the second topmost layer of workpiece are removed along the scratch path (see Figs. 10f, 11f and Table 3).

Figures 12 and 13 show the deformation behavior of the Cu(111) workpiece for a weak adhesion between tool and workpiece ($D=0.1D^*$). Table 4 shows the number of atoms that move between the three topmost atomic layers of initial configuration of Cu(111) workpiece after nanosratching for a weak adhesion between tool and workpiece ($D=0.1D^*$). It can be found that when the scratching depth is small ($h=-0.08$ nm), the workpiece undergoes a pure elastic deformation and no material removal occurs (Figs. 12a and 13a). Compared with the situation for a standard adhesion between tool and workpiece, when the scratching depth is in a wider range of −0.03 to 0.12 nm, the minimum depth of material removal can achieve a single atomic layer of the Cu(111) workpiece and a perfectly ordered crystalline surface can be observed in the machined groove along the scratch path (see Fig. 12b–e). When the scratching depth further increases to 0.15 nm, the material removal involves two topmost atomic layers of workpiece along the scratch path (see Fig. 12f).

Similar to the situation for a standard adhesion, the material removal of the Cu(111) workpiece is not in a layer-by-layer fashion either for a weak adhesion. A comparison of Figs. 12b–e, 13b–e and Table 4 shows that during the scratch-

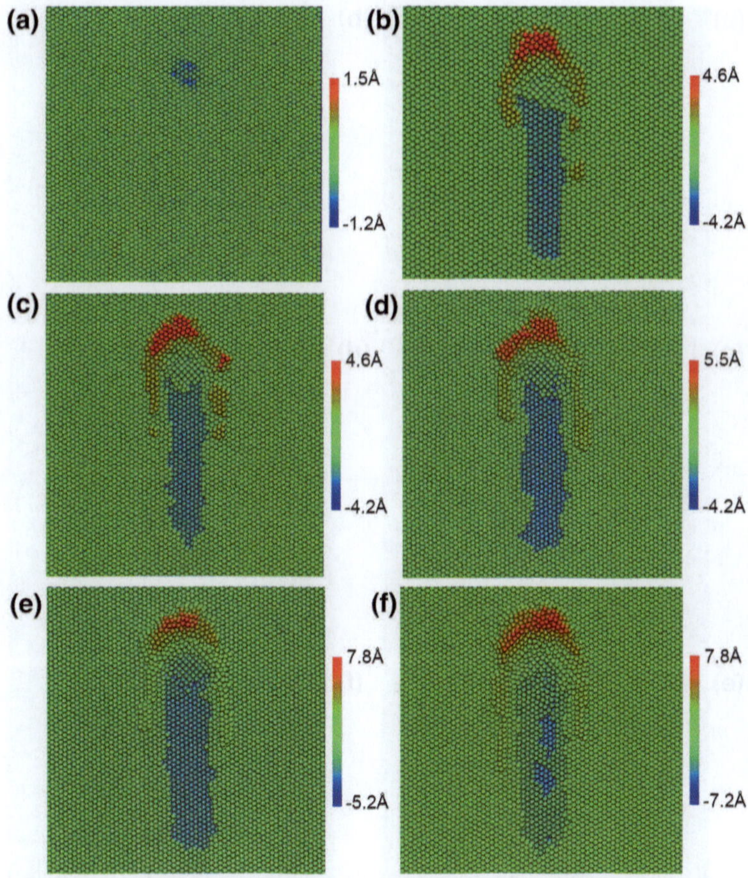

Fig. 10 Deformation behavior of a Cu(111) workpiece at scratching depths of **a** −0.12 nm, **b** −0.11 nm, **c** −0.08 nm, **d** −0.03 nm, **e** 0 nm, and **f** 0.02 nm for a scratching distance of 8 nm. The radius of tool is 4 nm. Atoms are colored according to their heights in z direction

ing process some atoms in the topmost (first) layer of the Cu(111) workpiece are inevitably implanted into the second topmost (second) atomic layer and some atoms in the second topmost layer of the Cu(111) workpiece are also displaced and incorporated into the initially topmost layer along the scratch path. It can also be found that for the scratching depths between 0.02 and 0.12 nm, some atoms in the second topmost layer of the Cu(111) workpiece are removed as well along the scratch path although the material removal only occurs in the topmost layer of Cu(111) workpiece for the scratching depth of −0.03 nm (see Figs. 12c–e, 13c–e and Table 4). Specifically, for the scratching depth of 0.12 nm, one atom (seven atoms) initially in the third topmost layer of workpiece is (are) incorporated into the (second) topmost layer of the Cu(111) workpiece along the scratch path. Besides, 2 atoms (25 atoms) initially in the third topmost layer of workpiece are incorporated into the (second)

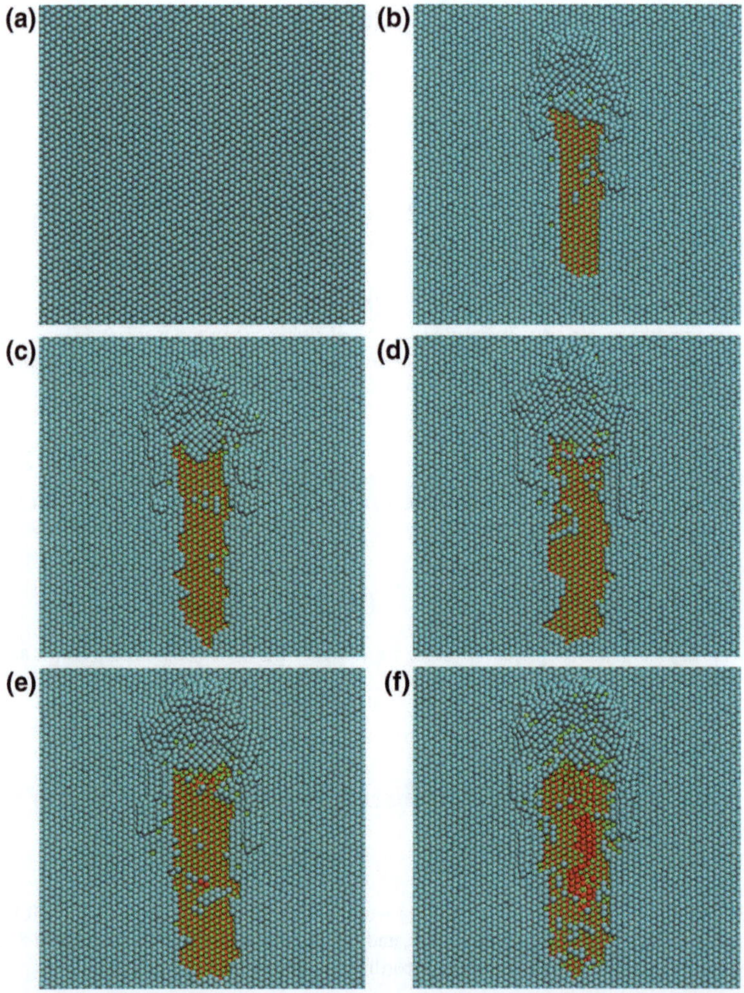

Fig. 11 Deformation behavior of a Cu(111) workpiece at scratching depths of **a** −0.12 nm, **b** −0.11 nm, **c** −0.08 nm, **d** −0.03 nm, **e** 0 nm, and **f** 0.02 nm for a scratching distance of 8 nm. The radius of tool is 4 nm. The atoms are colored in different atomic layers of initial configuration. The cyan atoms, green atoms, red atoms, and yellow atoms represent the topmost (first) layer, the second topmost (second) layer, the third topmost (third) layer, and the fourth topmost layer of atoms in the initial configuration of workpiece, respectively

topmost layer and many atoms (36 atoms) originally in the second topmost layer of workpiece are removed along the scratch path (see Figs. 12f, 13f and Table 4).

Figure 14 shows the deformation behavior of the Cu(110) workpiece for a standard adhesion (a–c) and a weak adhesion (d–f) between the tool and workpiece. We also summarize the corresponding variation of the number of atomic layers removed with the interaction strength and scratching depth, as shown in Table 5. It is found

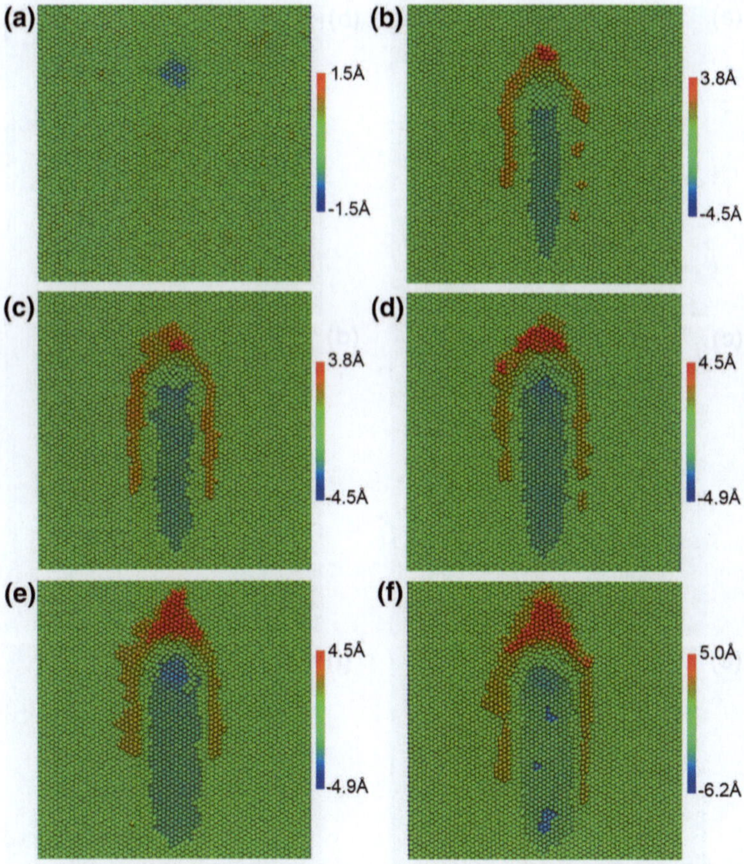

Fig. 12 Deformation behavior of a Cu(111) workpiece at scratching depths of **a** −0.08 nm, **b** −0.03 nm, **c** 0.02 nm, **d** 0.07 nm, **e** 0.12 nm, and **f** 0.15 nm for a weak adhesion between tool and workpiece($D = 0.1D^{*}$) with a scratching distance of 8 nm. The radius of tool is 4 nm. Atoms are colored according to their heights in z direction

that whether for a standard adhesion or weak adhesion, when the scratching depth is rather small (see Fig. 14a, d), though material removal only takes places in the topmost atomic layer of the Cu(110) workpiece, the material removal is neither continuous nor stable and part of the topmost atomic layer along the scratch path is not removed. However, when the scratching depth increases, the material removal involves two or three topmost atomic layers of workpiece along the scratch path (see Fig. 14b, c, e, f and Table 5). Therefore, in contrast to the case with Cu(001) and Cu(111) workpiece, a single atomic layer removal cannot be achieved for the Cu(110) workpiece in nanoscale mechanical machining process.

So what is the underlying mechanism of the different deformation behaviors for different crystal orientations and scratching directions of single crystal copper workpiece? In fact, the crystal orientation and scratching direction dependence of material

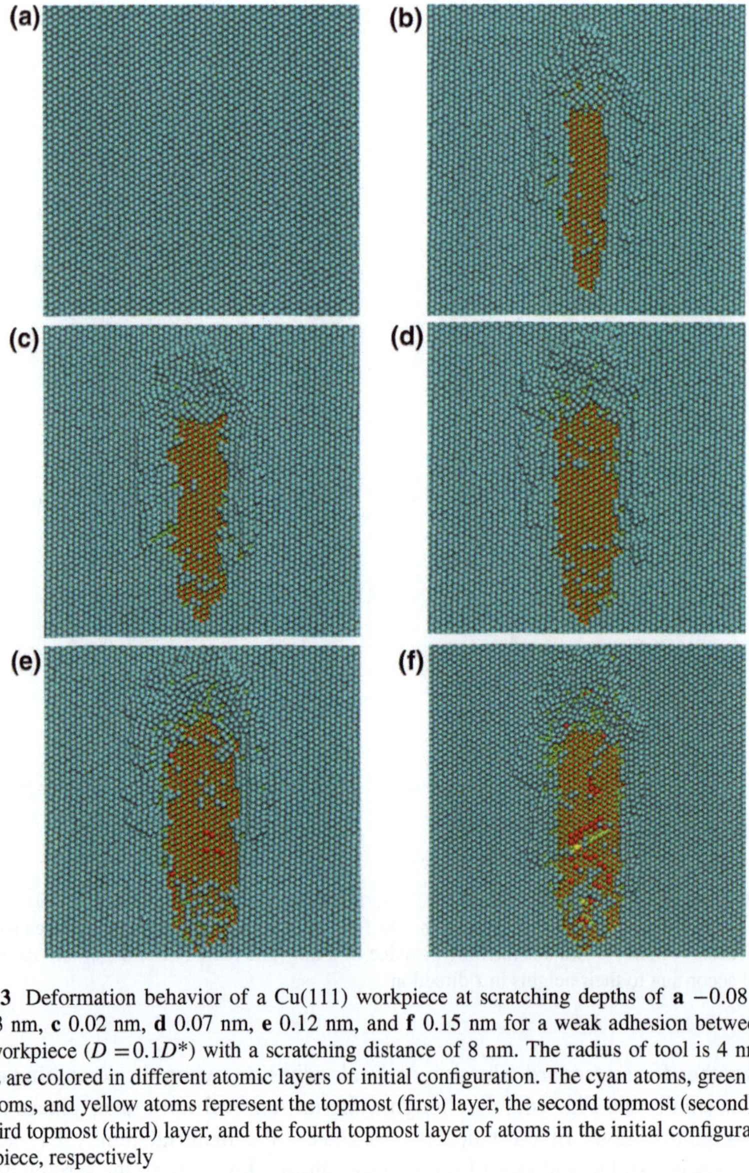

Fig. 13 Deformation behavior of a Cu(111) workpiece at scratching depths of **a** −0.08 nm, **b** −0.03 nm, **c** 0.02 nm, **d** 0.07 nm, **e** 0.12 nm, and **f** 0.15 nm for a weak adhesion between tool and workpiece ($D = 0.1D^*$) with a scratching distance of 8 nm. The radius of tool is 4 nm. The atoms are colored in different atomic layers of initial configuration. The cyan atoms, green atoms, red atoms, and yellow atoms represent the topmost (first) layer, the second topmost (second) layer, the third topmost (third) layer, and the fourth topmost layer of atoms in the initial configuration of workpiece, respectively

removal behavior are due to the difference in the atomic structure of workpiece [43, 63]. Table 6 shows the atomic density (number of atoms per unit area) and the distance between the planes with respect to the orientation of single crystal copper. Figure 15 shows the typical defect structures of the Cu(111) workpiece (a) and the Cu(110) workpiece (b). It should be noted that both the atomic density and interplanar distance of copper, a face-centered cubic (FCC) crystal, are maximum for (111) planes

Table 4 Number of atoms that move between the three topmost atomic layers of initial configuration of Cu(111) workpiece after nanosratching for a weak adhesion between tool and workpiece ($D = 0.1D^*$) with a scratching distance of 8 nm

Scratching depth (nm)	From first layer to second layer	Removed in second layer	From second layer to first layer	From third layer to second layer	From third layer to first layer
−0.03	15	0	12	0	0
0.02	23	2	17	0	0
0.07	44	6	27	0	0
0.12	74	16	37	7	1
0.15	65	36	65	25	2

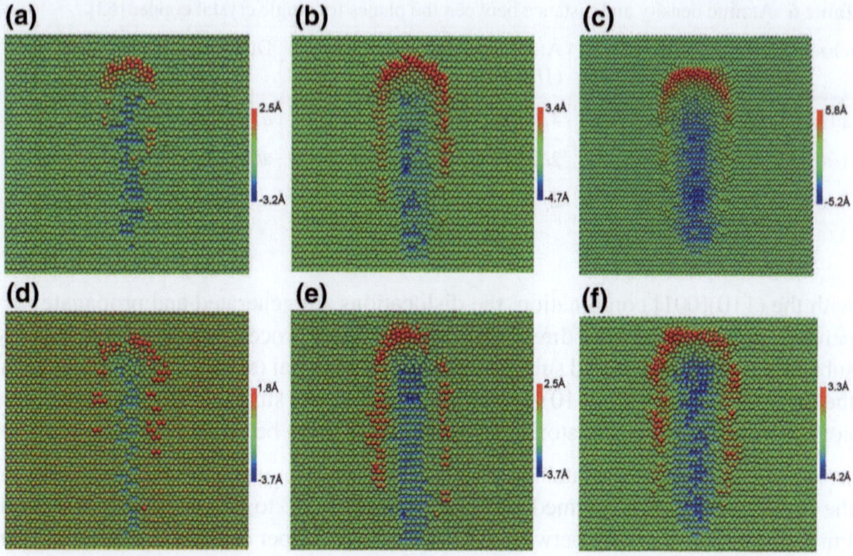

Fig. 14 Deformation behavior of a Cu(110) workpiece for a standard adhesion between tool and workpiece at scratching depths of **a** −0.16 nm, **b** −0.08 nm, and **c** 0.02 nm with a scratching distance of 8 nm. Deformation behavior of Cu(110) workpiece for a weak adhesion between tool and workpiece ($D = 0.1D^*$) at scratching depths of **d** −0.06 nm, **e** 0.0 nm, and **f** 0.07 nm with a scratching distance of 8 nm. The radius of tool is 4 nm. Atoms are colored according to their heights in z direction

and minimum for (110) planes. For FCC crystals, slip occurs on the densest planes ({111}) and along the close packed directions (<110> family). As a result, for the (111)[$\bar{1}$10] combination, the dislocations will be generated and propagate parallel to the scratching direction on the (111) planes in the scratching process (see Fig. 15a), enabling easy removal of a single atomic layer of Cu(111) [43, 63]. Besides, the distance between the (111) planes of copper is the largest (2.087 Å, see Table 6). Therefore, it is reasonable that realizing a single atomic layer material removal is rather easy for Cu(111) workpiece under perfect machining conditions. However,

Table 5 Variation of the number of atomic layers removed with the simulation parameters for a Cu(110) workpiece

Interaction strength (D/D^*)	Scratching depth (nm)	Number of layers removed (n)
1	−0.16	$0<n<1$
1	−0.08	$1<n<3$
1	0.02	$1<n<3$
0.1	−0.06	$0<n<1$
0.1	0	$1<n<2$
0.1	0.07	$1<n<2$

The meaning of n is the same as that in Table 1

Table 6 Atomic density and distance between the planes for single crystal copper [63]

Orientation	Atomic density per unit area (1/Å^2)	Distance between planes (Å)
{111}	$4/\sqrt{3}a^2$(0.177)	$a/\sqrt{3}$(2.087)
{001}	$2/a^2$(0.153)	$a/2$(1.807)
{110}	$2/\sqrt{2}a^2$(0.108)	$a/2\sqrt{2}$(1.278)

with the (110)[001] combination, the dislocations are generated and propagate perpendicular to the scratching direction in the scratching process, which can lead to easy subsurface deformation and subsurface material removal (see Fig. 15b). In addition, the distance between the (110) planes of copper is very small (1.278 Å). Hence, it is not surprising that a single atomic layer removal cannot be achieved for the Cu(110) workpiece in the [001] scratching direction. In the case of (001)[$\bar{1}$00] combination, the dislocations can be formed and propagate at −45° to the scratching direction. Furthermore, the distance between (001) planes of copper even though smaller than that of (111) planes is relatively big (1.807 Å). Thus, it is possible to achieve a single atomic layer removal for the Cu(001) workpiece in the [$\bar{1}$00] scratching direction.

5 Conclusions

In this work, MD simulations of nanoscratching are performed to investigate the minimum depth of material removal of single crystalline copper workpiece in nanoscale mechanical machining. It is demonstrated that the minimum depth of material removal of single crystalline copper workpiece can achieve a single atomic layer under certain machining conditions in nanoscale machining process. The minimum depth of material removal is closely related to the crystal orientation and scratching direction of copper workpiece. For the (001)[$\bar{1}$00] and (111)[$\bar{1}$10] combinations, the minimum depth of material removal can achieve a single atomic layer of workpiece

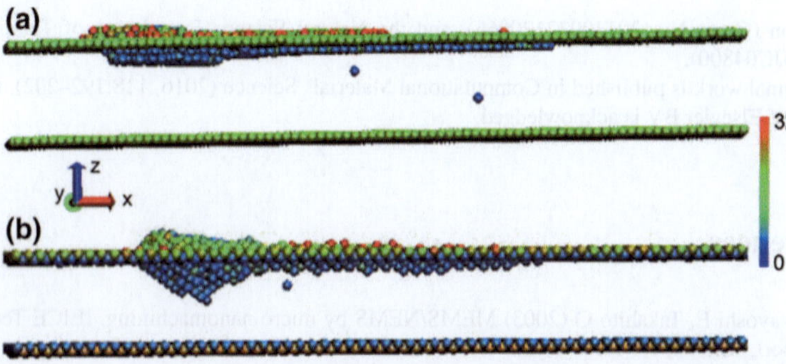

Fig. 15 Typical defect structures of the Cu(111) workpiece **a** and the Cu(110) workpiece **b** at the scratching distance of 8 nm. The radius of tool is 4 nm. Atoms are colored according to their CSP values. For the Cu(111) workpiece (**a**), the scratching depth is 0.12 nm with a weak adhesion between tool and workpiece ($D = 0.1D^*$). For the Cu(110) workpiece (**b**), the scratching depth is 0.02 nm with a standard adhesion between tool and workpiece

under perfect machining conditions with a perfectly ordered crystalline surface generated in the machined groove along the scratch path. Moreover, it is much easier for a (111)[$\bar{1}$10] combination to realize a single atomic layer removal than a (001)[$\bar{1}$00] combination. However, it is impossible to realize a single atomic layer removal for a (110)[001] combination. It is also found that the scratching velocity and tool size have little effect on the realization of a single atomic layer material removal in the scratching process, while reduction in adhesion between the tool and workpiece plays an important role.

Even when the depth of material removal for the Cu(001) and Cu(111) surfaces is a single atomic layer of workpiece, the workpiece material is not removed in a layer-by-layer fashion. In the scratching process, plastic deformation takes place in the workpiece as dislocations (or dislocation embryos) are generated within the workpiece. When a single atomic layer removal of the Cu(001) and Cu(111) surfaces appears to occur, some atoms in the topmost layer of workpiece are inevitably implanted into the second topmost atomic layer and some atoms in the second topmost layer of workpiece are displaced and incorporated into the initially topmost layer along the scratch path. The seemingly single atomic layer material removal process usually involves two topmost atomic layers of workpiece. Our results reject the hypothesis that single crystalline metal materials can be continuously and stably removed one layer of atoms after another in nanoscale mechanical machining.

The knowledge gained not only provides insights into the control and optimization of nanoscale machining process of single crystalline copper, but also would enrich our understanding of material removal mechanism of a variety of single crystalline metal materials in nanoscale mechanical machining process.

Acknowledgements This work is supported by the National Natural Science Foundation of China (Grant No. 51405337), the Specialized Research Fund for the Doctoral Program of Higher

Education (Grant No. 20130032120065), and the Natural Science Foundation of Tianjin (No. 15JCQNJC04800).
The original work is published in Computational Materials Science (2016, 118:192–202), and the support of Elsevier BV is acknowledged.

References

1. Masayoshi E, Takahito O (2003) MEMS/NEMS by micro nanomachining. IEICE Technical Report, vol 103, 13–18
2. Schumacher HW, Keyser UF, Zeitler U, Haug RJ, Ebert K (2000) Controlled mechanical AFM machining of two-dimensional electron systems: fabrication of a single-electron transistor. Phys E 6:860–863
3. Sohn LL, Willett RL (1995) Fabrication of nanostructures using atomic-force-microscope-based lithography. Appl Phys Lett 67:1552–1554
4. Dornfeld D, Min S, Takeuchi Y (2006) Recent advances in mechanical micromachining. CIRP Ann 55:745–768
5. Brinksmeier E et al (2010) Ultra-precision grinding. CIRP Ann 59:652–671
6. Tseng AA (2011) Removing material using atomic force microscopy with single- and multiple-tip sources. Small 7:3409–3427
7. Fang FZ, Wu H, Liu YC (2005) Modeling and experimental investigation on nanometric cutting of monocrystalline silicon. Int J Mach Tools Manuf 45:1681–1686
8. Fang FZ, Wu H, Zhou W, Hu XT (2007) A study on mechanism of nano-cutting single crystal silicon. J Mater Process Technol 184:407–410
9. Pei QX, Lu C, Lee HP (2007) Large scale molecular dynamics study of nanometric machining of copper. Comput Mater Sci 41:177–185
10. Komanduri R, Varghese S, Chandrasekaran N (2010) On the mechanism of material removal at the nanoscale by cutting. Wear 269:224–228
11. Ikawa N, Shimada S, Tanaka H (1992) Minimum thickness of cut in micromachining. Nanotechnology 3:6–9
12. Shimada S, Ikawa N, Tanaka H, Ohmori G, Uchikoshi J (1993) Feasibility study on ultimate accuracy in microcutting using molecular dynamics simulation. CIRP Ann 42:117–120
13. Yuan ZJ, Zhou M, Dong S (1996) Effect of diamond tool sharpness on minimum cutting thickness and cutting surface integrity in ultraprecision machining. J Mater Process Technol 62:327–330
14. Liu X, Devor RE, Kapoor SG (2006) An analytical model for the prediction of minimum chip thickness in micromachining. J Manuf Sci Eng 128:474–481
15. Son SM, Lim HS, Ahn JH (2005) Effects of the friction on the minimum cutting thickness in micro cutting. Int J Mach Tools Manuf 45:529–535
16. Lai XM, Li HT, Lin CF, Lin ZQ, Ni J (2007) Modeling and analysis of micro scale milling considering size effect, micro cutter edge radius and minimum chip thickness. Int J Mach Tools Manuf 48:1–14
17. Malekian M, Mostofa MG, Park SS, Jun MB (2012) Modeling of minimum uncut chip thickness in micro machining of aluminum. J Mater Process Technol 212:553–559
18. Li Z, Huang Y, Zhang J, Yan Y, Sun Y (2013) Atomistic insight into the minimum wear depth of Cu(111) surface. Nano. Res. Lett. 8:514
19. Luan BQ, Robbins MO (2005) The breakdown of continuum models for mechanical contacts. Nature 435:929–932
20. Luan B, Robbins MO (2006) Contact of single asperities with varying adhesion: comparing continuum mechanics to atomistic simulations. Phys Rev E 74:026111
21. Urbakh M, Klafter J, Gourdon D, Israelachvili J (2004) The nonlinear nature of friction. Nature 430:525–528

22. Szlufarskal I, Chandross M, Carpick RW (2008) Recent advances in single-asperity nanotribology. J Phys D Appl Phys 41:123001
23. Si LN, Guo D, Luo JB, Lu XC (2010) Monoatomic layer removal mechanism in chemical mechanical polishing process: a molecular dynamics study. J Appl Phys 107:064310
24. Custance O, Perez R, Morita S (2009) Atomic force microscopy as a tool for atom manipulation. Nat Nanotech 4:803–810
25. Morita S (2011) Atom world based on nano-forces: 25 years of atomic force microscopy. J Electron Spectrosc 60:S199–S211
26. Kawai K et al (2014) Atom manipulation on an insulating surface at room temperature. Nat Common 5:4403
27. Gotsmann B, Lantz M (2008) Atomistic wear in a single asperity sliding contact. Phys Rev Lett 101:125501
28. Bhaskaran H et al (2010) Ultralow nanoscale wear through atom-by-atom attrition in silicon-containing diamond-like carbon. Nat Nanotech 5:181–185
29. Jacobs T, Carpick RW (2013) Nanoscale wear as a stress-assisted chemical reaction. Nat Nanotech 8:108–112
30. Moriwaki T, Okuda K (1989) Machinability of copper in ultra-precision micro diamond cutting. CIRP Ann 38:115–118
31. Lucca DA, Seo YW, Rhorer RL (1994) Energy dissipation and tool-workpiece contact in ultra-precision machining. Tribol Trans 37:651–655
32. Komanduri R, Raff LM (2001) A review on the molecular dynamics simulation of machining at the atomic scale. Proc I Mech E Part B 215:1639–1672
33. Yan YD, Sun T, Dong S, Luo XC, Liang YC (2006) Molecular dynamics simulation of processing using AFM pin tool. Appl Surf Sci 252:7523–7531
34. Fang T, Weng C (2000) Three-dimensional molecular dynamics analysis of processing using a pin tool on the atomic scale. Nanotechnology 8:148–153
35. Zhu PZ, Hu YZ, Ma TB, Wang H (2010) Study of AFM-based nanometric cutting process using molecular dynamics. Appl Surf Sci 256:7160–7165
36. Shi J, Shi Y, Liu CR (2010) Evaluation of three dimensional single point turning at atomistic level by molecular dynamics simulation. Int J Adv Manuf Technol 8:161–171
37. Tong Z, Liang Y, Jiang X, Luo X (2014) An atomistic investigation on the mechanism of machining nanostructures when using single tip and multi-tip diamond tools. Appl Surf Sci 290:458–465
38. Li J, Fang QH, Liu YW, Zhang LC (2014) A molecular dynamics investigation into the mechanisms of subsurface damage and material removal of monocrystalline copper subjected to nanoscale high speed grinding. Appl Surf Sci 303:331–343
39. Hu CK et al (1995) Copper interconnection: integration and reliability. Thin Solid Films 262:84–92
40. Lyshevski SE (2002) MEMS and NEMS: systems, devices, and structures. CRC Press, Boca Raton, FL
41. Plimpton S (1995) Fast parallel algorithms for short-range molecular dynamics. J Comput Phys 117:1–19
42. Ziegenhain G, Urbassek HM, Hartmaier A (2010) Influence of crystal anisotropy on elastic deformation and onset of plasticity in nanoindentation: a simulational study. J Appl Phys 107:061807
43. Komanduri R, Chandrasekaran N, Raff LMMD (2000) Simulation of nanometric cutting of single crystal aluminum-effect of crystal orientation and direction of cutting. Wear 242:60–88
44. Rapaport DC (1995) The art of molecular dynamics simulation. Cambridge University Press, Cambridge
45. Zhou XW, Johnson RA, Wadley HNG (2004) Misfit-energy-increasing dislocations in vapor-deposited CoFe/NiFe multilayers. Phys Rev B 69:144113
46. Daw MS, Baskes MI (1984) Embedded atom method: derivation and application to impurities, surfaces and other defects in metals. Phys Rev B 29:6443

47. Zhou XW et al (2001) Atomic scale structure of sputtered metal multilayers. Acta Mater 49:4005
48. Khomenko AV, Prodanov NV, Persson BNJ (2013) Atomistic modelling of friction of Cu and Au nanoparticles adsorbed on graphene. Condens Matter Phys 16:33401
49. Zhu PZ, Hu YZ, Ma TB, Wang H (2011) Molecular dynamics study on friction due to ploughing and adhesion in nanometric scratching process. Tribol Lett 41:41–46
50. Maekawa K, Itoh A (1995) Friction and tool wear in nano-scale machining—a molecular dynamics approach. Wear 188:115–122
51. Shimizu J, Eda H, Zhou L, Okabe H (2008) Molecular dynamics simulation of adhesion effect on material removal and tool wear in diamond grinding of silicon wafer. Tribol Online 3:248–253
52. Zhu PZ, Fang FZ (2012) Molecular dynamics simulations of nanoindentation of monocrystalline Germanium. Appl Phys A 108:415–421
53. Rentsch R, Inasaki I (1994) Molecular dynamics simulation for abrasive processes. CIRP Ann 43:327–330
54. Li Q, Dong Y, Perez D, Martini A, Carpick RW (2011) Speed dependence of atomic stick-slip friction in optimally matched experiments and molecular dynamics simulations. Phys Rev Lett 106:126101
55. Egberts P et al (2013) Environmental dependence of atomic-scale friction at graphite surface steps. Phys Rev B 88:035409
56. Çakīr O, Yardimeden A, Ozben T, Kilickap E (2007) Selection of cutting fluids in machining processes. J Achieve Mater Manuf Eng 25:99–102
57. Zhou LB, Hosseini BS, Tsuruga T, Shimizu J, Eda H (2007) Fabrication and evaluation for extremely thin Si wafer. Int J Abras Technol 1:94–105
58. Hu XL, Sundararajan S, Martini A (2014) The effects of adhesive strength and load on material transfer in nanoscale wear. Comput Mater Sci 95:464–469
59. Barthel AJ, Al-Azizi A, Surdyka ND, Kim SH (2014) Effects of gas or vapor adsorption on adhesion, friction, and wear of solid interfaces. Langmuir 30:2977–2992
60. Ryan KE et al (2014) Simulated adhesion between realistic hydrocarbon materials: effects of composition, roughness, and contact point. Langmuir 30:2028–2037
61. Kelchner CL, Plimpton SJ, Hamilton JC (1998) Dislocation nucleation and defect structure during surface indentation. Phys Rev B 58:11085–11088
62. Zhu PZ, Hu YZ, Wang H, Ma TB (2011) Study of effect of indenter shape in nanometric scratching process using molecular dynamics. Mater Sci Eng, A 528:4522–4527
63. Dieter GE (1986) Mechanical metallurgy. McGraw-Hill, New York

Thermally Assisted Microcutting of Calcium Fluoride Single Crystals

Yan Jin Lee, Akshay Chaudhari, Jiong Zhang and Hao Wang

Abstract Optical materials, such as calcium fluoride single crystals, are widely used across various industries for their light transmission capabilities. These materials possess excellent mechanical, physical, and chemical resistant properties but also tend to be very brittle, which poses a challenge to the microcutting of complex freeform shapes with optical surface quality. Ultraprecision single-point diamond turning is commonly used in ductile-regime machining of hard and brittle materials, and polishing is employed as a secondary process to achieve optical quality surfaces, which can be time consuming. To improve the machining efficiency of high-strength materials, hot machining techniques have been developed to improve workpiece plasticity and surface integrity after machining. Surface quality and subsurface damage evaluation of the machined material along with finite element analysis allow a deeper understanding of the effectiveness in heat-assisted machining. In this chapter, an introduction to ultraprecision single-point diamond turning and the fundamentals of ductile-regime machining of hard and brittle materials will be discussed, followed by its application in fabrication of calcium fluoride single crystal lenses. Subsequently, the anisotropic characteristics of calcium fluoride single crystal will be investigated through experimentally validated crack formation models and surface generation morphology to gain detailed appreciation of the challenges faced during production of brittle single-crystal materials. To conclude the chapter, the effects of elevated temperatures on the material properties and machinability will be evaluated using experimental and numerical solutions.

1 Optics Fabrication Methods

Different types of materials are employed for various applications depending on the desired material properties such as electrical conductivity, wear resistance, heat con-

Y. J. Lee · A. Chaudhari · J. Zhang · H. Wang (✉)
Department of Mechanical Engineering, Faculty of Engineering,
National University of Singapore, 9 Engineering Drive 1, Singapore 117575, Singapore
e-mail: mpewhao@nus.edu.sg

J. Zhang et al. (eds.), *Simulation and Experiments of Material-Oriented Ultra-Precision Machining*, Springer Tracts in Mechanical Engineering,
https://doi.org/10.1007/978-981-13-3335-4_4

ductivity, thermal expansion, light transmissivity, etc. In modern-day industries, there is an increasing demand for precision and accuracy for product development. Excellent surface finish is required for the lenses made of magnesium fluoride (MgF_2) and high-purity fused silica (SiO_2) in the optics industry for efficient light transmission. Lens materials used for lithography often have stringent requirements of refractive index homogeneity and low residual stress-induced birefringence. Calcium fluoride (CaF_2) single crystal is the most economical choice for lithography lens applications with high transmission or low absorption properties less than 0.002 cm^{-1} [1, 2]. CaF_2 is also used for infrared and visible wavelength applications such as astronomical instrumentation due to its high transmission from 120 nm to 20 μm [3, 4]. However, the crystal is considered to have a poor machinability due to its brittle nature that can result in structural defects during processing.

The manufacturing of high-quality CaF_2 encompasses a few stages of processes such as raw material purification, crystal growth through the Czochralski or Bridgman–Stockbarger methods, post-growth annealing, and surface improvement [5–7]. Each stage plays a crucial role in achieving the desired optical properties of the end product. The latter process seeks to reduce surface roughness and remove subsurface damage (SSD) through conventional low-material-removal-rate methods of grinding, lapping, and polishing [8, 9]. However, these machining techniques pose a challenge to fabricate complex-shaped lenses such as an aspherical lens, which has led to the investigation of employing ultraprecision single-point diamond turning (SPDT) as an alternative to achieve complex geometries [10]. Over the last few decades, SPDT has been explored and applied toward the precision machining of brittle materials such as glass, germanium, and silicon [11–13]. These brittle materials commonly fail by fracture with little plastic deformation but can still be deformed plastically on the condition that the stress intensity factor at the machining interface is below the material fracture toughness. This is often determined by the machining parameter "critical depth of cut", generally associated in the nanometric range of 10–100 nm. Therefore, given no other specially developed technique being employed, the central task in performing ductile-regime cutting of brittle materials like CaF_2 has been concluded as the determination of the critical depth of cut for a specific work material.

2 Machinability of Calcium Fluoride

2.1 Slip Model

Like most single crystals, the mechanical properties of CaF_2 ionic crystal are defined by the formation of cracks through dislocation interactions and anisotropic crystallography behavior [14]. This phenomenon can be explained by the well-known slip line theory that is easily observed in anisotropic materials. Plastic deformation along the slip direction and the cleavage fracture on the cleavage plane are the two fundamental modes of material deformation [11, 15–17]. Mechanical properties such as hardness and fracture toughness can therefore be determined through the study of slip systems to understand the ductile-mode machining of CaF_2. The CaF_2 single crystal is reported to cleave on the {111} and partially {110} planes, with {100} <110> primary slip systems and secondary slip systems at {110} <1-10> [14, 17, 18]. Hardness tests by nanoindentation, Vickers or Knoop microhardness tests are commonly used to assess these mechanical properties. The plastic deformability that is a ratio between Young's modulus and hardness value (*E/H*) can be studied using the Vickers nanoindentation experimental method over varying loads ranging from 20 to 100 mN. The *E/H* values of the (100), (110), and (111) planes were calculated to be 70.5, 57.9, and 69.0, respectively, attributing the differences to the indentation orientation with respect to the slip planes [19]. Different crystallographic plane orientations are expected to produce different measures of deformability. According to the slip model, higher plastic deformability should be observed when the applied force direction is parallel to the slip plane.

In micro-Vickers hardness testing, cracks can be observed at the peripherals of the indentations. Testing on the (110) plane has the highest propensity for crack initiation at relatively low loads, which can be explained using the slip model shown in Fig. 2. In this case, the cleavage plane is parallel to the indentation load, and the resultant forces induced by the pyramidal-shaped indenter causes cleavage fracture to occur prior to shearing along the slip planes. Higher loads of 1.0 and 5.0 N were required for crack initiation on the (111) and (100) plane orientations, respectively [20]. The higher plasticity observed when performing indentations on the (100) plane is due to the loading direction being parallel to the slip plane.

Although single crystals commonly exhibit superior mechanical and optical properties, the existence of slip systems in a single crystal results in the anisotropic effect that is a difficult issue to resolve in machining, i.e., directional-dependent material properties. A common observation for anisotropy during machining is the different symmetrical properties when machining on different plane orientations. Multiple works on the (100), (110), and (111) plane orientations have exhibited fourfold, twofold, and threefold symmetries on the machined surface [17, 21–23]. Figure 1 displays a typical outcome of the threefold symmetry on the (111) plane orientation after face machining of CaF_2. Each form of symmetry can be observed through a systematic study of the machined surface quality through face machining on each plane orientation and the critical undeformed chip thickness through progressive grooving along different directions of the crystal. These studies on brittle materials are often coupled with force measurements and crack morphology analysis. The detailed experimental design is out of the scope of this chapter and will not be discussed here.

Similarly, indentation tests with the indenter positioned along different crystallographic directions offer another type of anisotropic analysis. The fracture toughness, K_c, along different crystallographic directions of a single crystal CaF_2 sample can be determined using the nanoindentation and microhardness indentation techniques [22]. In these tests, the pyramidal-shaped indenter is oriented along different directions and the crack length, c, is measured for calculation as shown in Fig. 3. The K_c value can be calculated by Eq. 1 [21, 23]:

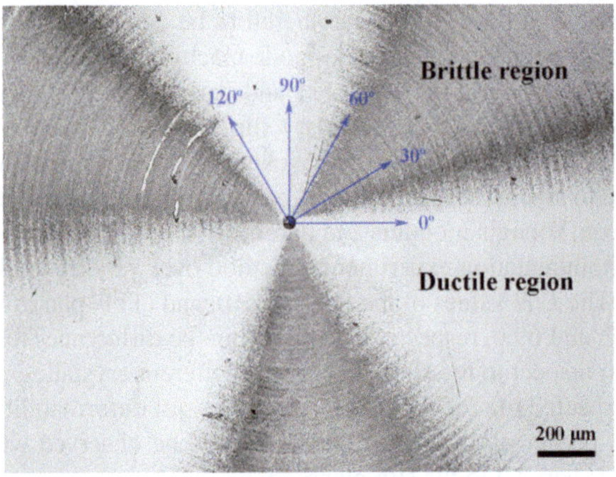

Fig. 1 Surface morphology of a (111) CaF_2 sample exhibiting the threefold symmetry [22] (reprinted with permission from Springer © 2017)

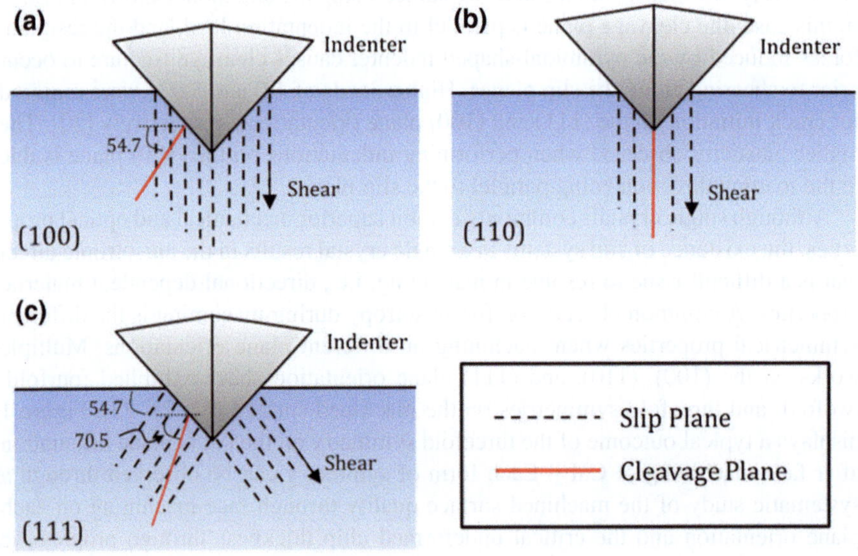

Fig. 2 Schematic overview of crack initiation mechanism: **a** (100); **b** (110); **c** (111) [24]

$$K_c = (0.016 \pm 0.002)\left(\frac{E}{H}\right)^{\frac{1}{2}} \frac{P}{c^{\frac{3}{2}}} \quad (1)$$

where E and H are Young's modulus and hardness measured from nanoindentation experiments, P is the applied load in the microhardness test.

The K_c values were reported to range between 0.325 and 0.675 MPa $m^{1/2}$ at room temperature and were in direct correlation with the relative changes in critical

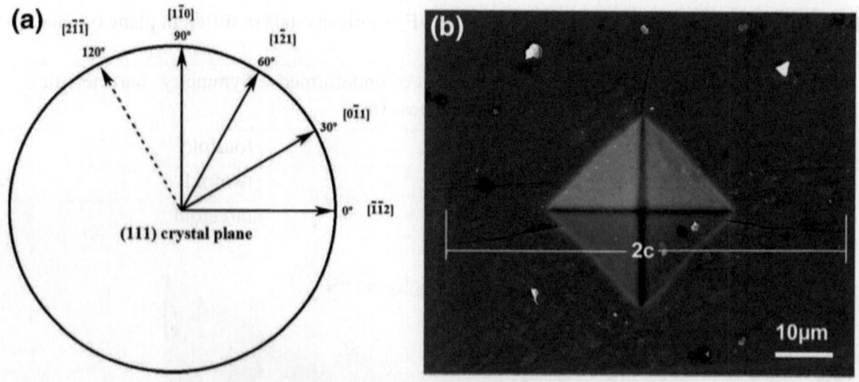

Fig. 3 **a** schematic of crystallographic directions; **b** SEM image of microhardness indentation and crack formation [22] (reprinted by permission from Springer © 2016)

undeformed chip thickness [22]. It must be noted that different materials exhibit different types of cracking sequence and behaviors [25]. Therefore, this should not be used as a universal method for fracture toughness measurements of brittle materials.

2.2 *Critical Undeformed Chip Thickness*

Optical grade surfaces defined by the fracture-free machined surfaces or mirror-like surfaces have been made possible to machine with technological advances in improving the stiffness and precision controls of conventional machine centers. With the development of ultraprecision machining (UPM) technology, the ductile-regime machining has been realized for hard and brittle materials by accurate control of the cutting process parameters such as the undeformed chip thickness, feed rate, and cutting speed. Of all the tunable process parameters, the undeformed chip thickness is the leading factor that allows ductile-mode machining. Transition to brittle failure occurs when the energy required for plastic deformation exceeds that required for crack initiation [26]. The critical undeformed chip thickness values for machining on different plane orientations of CaF_2 at room temperature are reflected in Table 1. A range is provided due to the resolved stresses acting along particular crystallographic directions that promote cleavage fractures instead of slipping. In progressive grooving of CaF_2 along different directions on a defined plane, different types of crack morphologies can be observed. Till date, there has not been any report on the influence of thermal effects on the crack morphologies.

There are a few methods to experimentally determine the undeformed chip thickness values of brittle materials in dynamic machining processes such as plunge cuts. Progressive grooving along the different crystallographic orientations using a diamond cutting tool on the UPM has been the most commonly practiced method

Table 1 Critical undeformed chip thickness of CaF_2 single crystals of different plane orientations [17]

Plane	Lower limit of undeformed chip thickness (nm)	Upper limit of undeformed chip thickness (nm)	Symmetry characteristic
(100)	140	280	fourfold
(110)	80	300	twofold
(111)	70	380	threefold

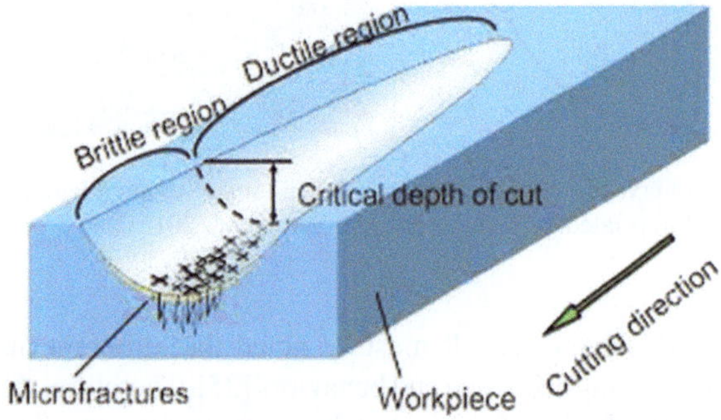

Fig. 4 Schematic of a progressive microgroove [27] (reprinted with permission from Elsevier © 2009)

for quick assessment. The concept of progressive scratching enables a systematic approach to determine the undeformed chip thickness and visually reveals the types of surfaces generated and crack morphologies corresponding to the slip systems of the single crystal. In Fig. 4, two features are commonly observed in a progressive scratch—the ductile region and the brittle region. The ductile–brittle transition depends on the gradient of the scratch. Based on basic geometry, the high gradient will show an immediate change from ductile to brittle and a gradual tilt will show the progressive change in surface generation. Therefore, proper machine tool precision control and selection of grooving parameters should be considered to study the ductile–brittle transition zone.

When considering the influence of the slip theory in ductile-mode machining, cutting speeds are important machining parameters to control to accommodate the speed of dislocation movement in the crystal at room temperature and elevated temperatures. However, cutting speeds are often the limitations of orthogonal grooving experiments due to the mechanical feasibility and do not provide the most accurate representation of the material removal process in an actual cutting process. Excellent machined surface finish on the (111) CaF_2 single crystals with an arithmetic mean height of 1.832 nm can be achieved at maximum cutting speeds of over 35 m/min [21]. However, these speeds are much higher than the allowable one-directional movement speed of 400 mm/min for an average UPM. Alternate UPM setups can be specially constructed such as high-speed shapers that use reciprocating speeds of up to 5.4 Hz and are capable of achieving cutting speeds of 130 m/min [28]. In other

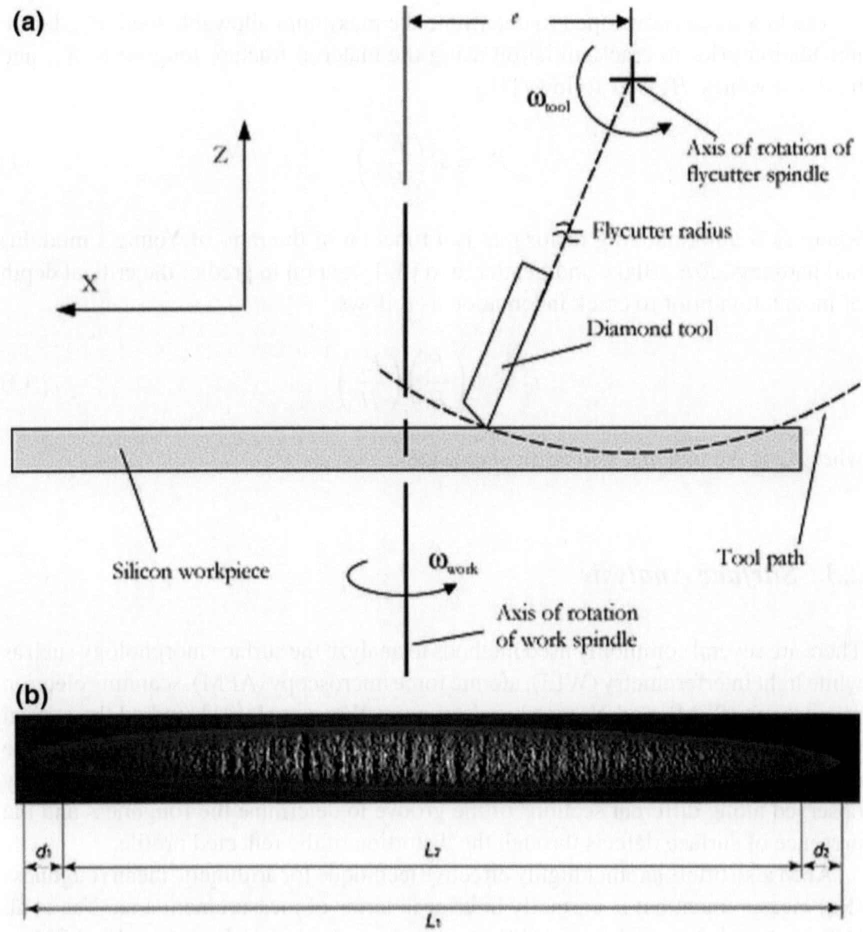

Fig. 5 Fly cutting approach [29] reprinted with permission from Elsevier © 2017. **a** Cross-sectional schematic of fly cutting scratch on silicon; **b** geometry of scratch made by fly cutting

cases, the fly cutting approach may also be used at realistic cutting speeds to produce similar results to determine the undeformed chip thickness by considering both ends of the groove as shown in Fig. 5.

While indentation tests are commonly used to determine mechanical properties of a material using the slip theory for single crystals, the depth of the indentation is another important result to exploit in determining the critical undeformed chip thickness. Shear stress models resolved along the slip or cleavage planes are commonly used on single crystals when considering the material properties and cutting conditions to describe ductile-regime machining [5, 8, 10].

The first model developed to determine the maximum allowable load, P_c, during indentation prior to crack initiation using the material fracture toughness, K_c, and hardness values, H, is as follows [16]:

$$P_c = \Omega\left(\frac{K_c^4}{H^3}\right) \tag{2}$$

where Ω is a normalizing factor that is a function of the ratio of Young's modulus and hardness, *E/H*. Blake and Scattergood [30] went on to predict the critical depth of indentation prior to crack indentation as follows:

$$t_c = \psi\left(\frac{E}{H}\right)\left(\frac{K_c}{H}\right)^2 \tag{3}$$

where ψ is the indenter geometrical constant.

2.3 Surface Analysis

There are several commonly used methods to analyze the surface morphology such as white light interferometry (WLI), atomic force microscopy (AFM), scanning electron microscopy (SEM), and Nomarski microscopy. Wang et al. [21] studied the critical undeformed chip thickness along different crystallographic orientations using the contactless WLI measurement as shown in Fig. 6. Surface profiles can be effectively observed along different sections of the groove to determine the roughness and the presence of surface defects through the distortion of the reflected profile.

AFM also offers another highly effective technique for arithmetic mean roughness (Ra) measurement but is normally inferior in terms of measurement area. Yan et al. [10] analyzed the crack-free (111) machined surface of CaF_2 using the AFM to observe the feed marks and final surface roughness in Fig. 7. They also used the Nomarski microscope to study the critical undeformed chip thickness by reducing the feed rate during end face turning of the sample. Brittle fracture regions can be easily observed at the higher feed rates attributing to the higher undeformed chip thickness values in Fig. 8.

Crack-free surface generation is an important factor in machining optical lenses, but the optimal undeformed chip thickness parameter selected for machining has to be translated to reduced subsurface damage (SSD). The chemical etching method is effective in showing the subsurface dislocation distributions. In fact, a surface prepared by diamond turning reveals more evenly distributed dislocations as compared to a surface produced by polishing. Figure 9 shows the WLI images of etched surfaces that were prepared by different conventional machining methods.

Transmission electron microscopy (TEM) is an excellent equipment for a visual understanding and analysis of subsurface damage induced after machining processes. Lattice rotations, nanocrystal formations, and dislocation activities as a result of the

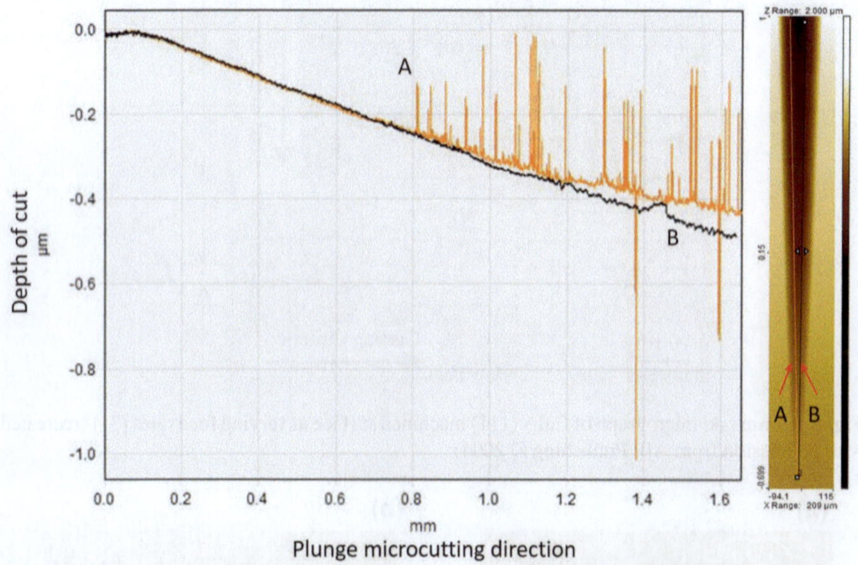

Fig. 6 WLI image of the groove profile made on CaF_2 [21] (recreated with permission from Elsevier © 2016)

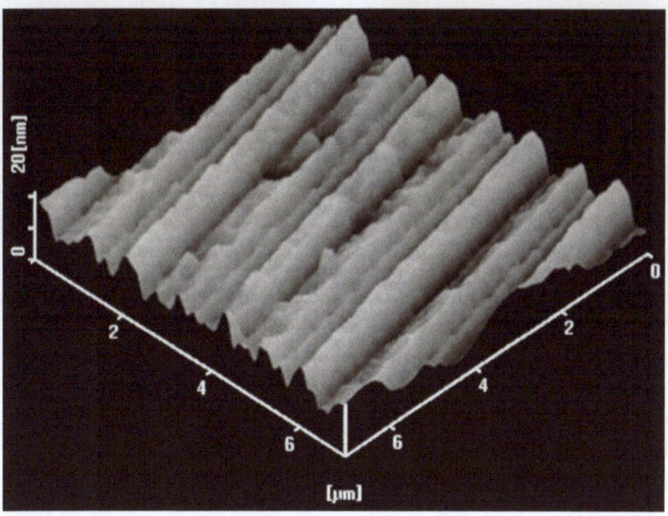

Fig. 7 AFM image of feed lines on the CaF_2 surface with Ra 3.3 nm after machining [31] (recreated with permission from AIP Publishing © 2004)

stresses induced by the machining process can be observed using the high-resolution TEM (HR-TEM) [32]. For effective analysis of the strained lattice structure, TEM samples have to be prepared using a focused ion beam (FIB) system at a crack-free

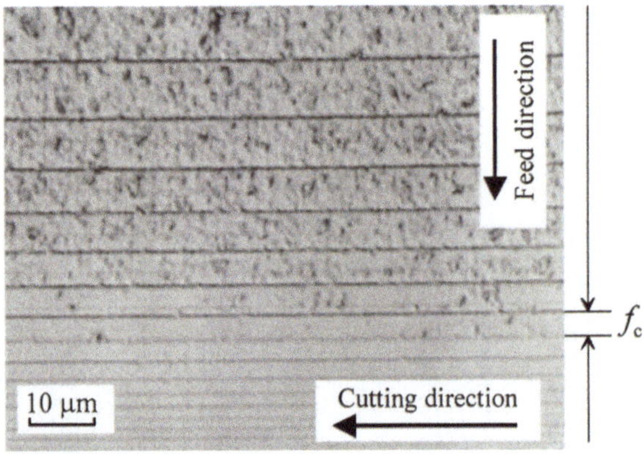

Fig. 8 Normarski micrograph of CaF_2 (111) machined surface at varying feed rates [31] (recreated with permission from AIP Publishing © 2004)

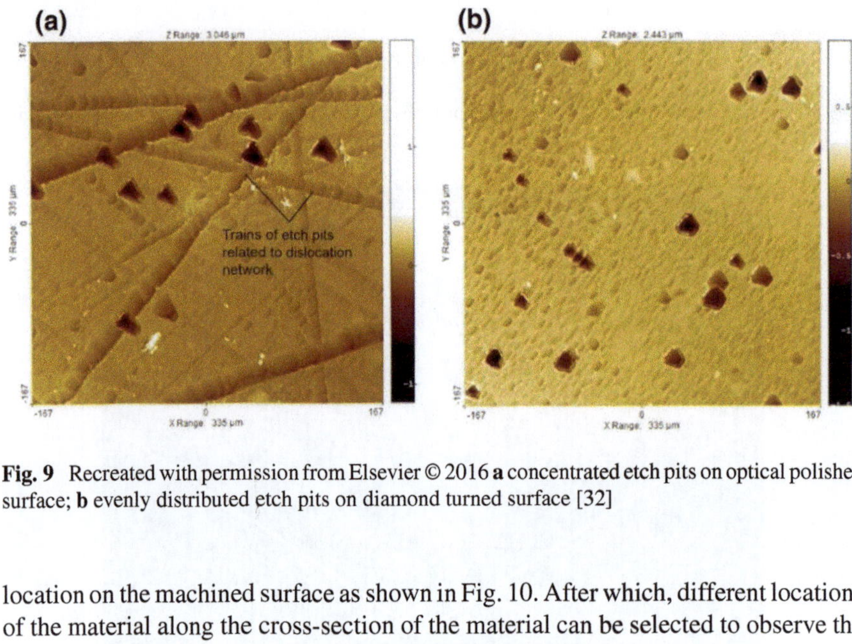

Fig. 9 Recreated with permission from Elsevier © 2016 **a** concentrated etch pits on optical polished surface; **b** evenly distributed etch pits on diamond turned surface [32]

location on the machined surface as shown in Fig. 10. After which, different locations of the material along the cross-section of the material can be selected to observe the lattice strain and defects using the HR-TEM shown in Fig. 11.

2.4 *Theoretical Modeling*

Various types of theoretical modeling simulations can be used to study the ductile–brittle transition mechanism depending on the scale of simulation. Most macro-sized modeling can be done using the crystal plasticity finite element method

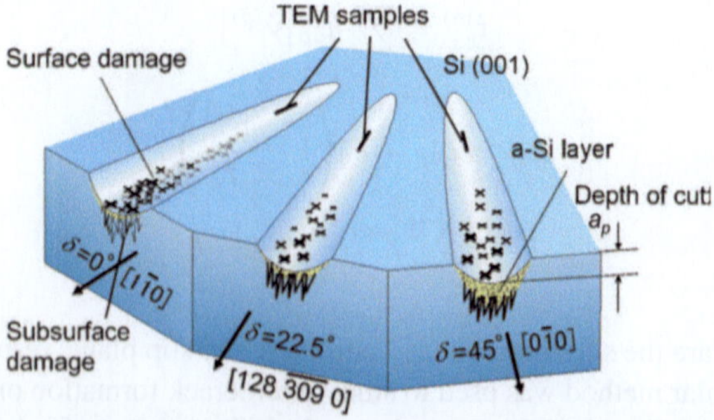

Fig. 10 Schematic of progressive scratches along different directions and TEM sample extraction locations [33] (recreated with permission from Elsevier © 2012)

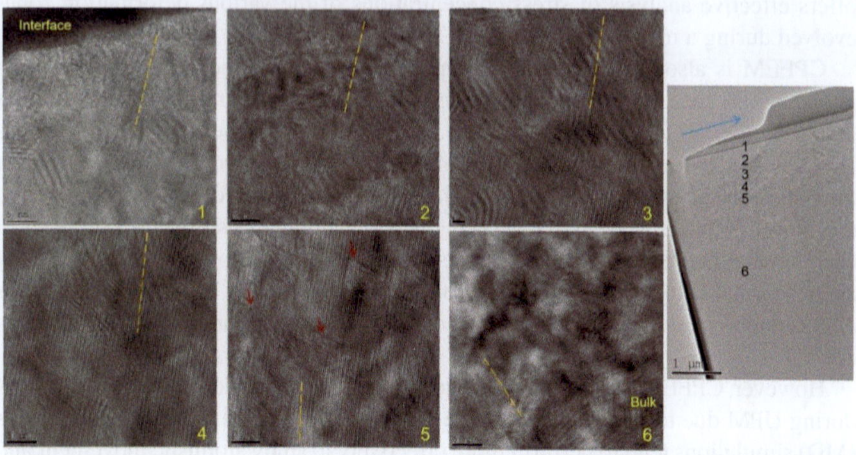

Fig. 11 HR-TEM images of a machined CaF_2 from the surface (position 1) to the bulk material (position 2). The blue arrow indicates the cutting direction of the groove [32] (recreated with permission from SAGE Publications © 2016)

(CPFEM) and is highly effective when dealing with ductile materials with higher strain rates. Under certain conditions, CPFEM may also be used to study microscale simulations of brittle materials using commercial finite element method (FEM) software ABAQUS/Standard equipped with the UMAT subroutine. Rigid analytical surfaces of the cutting tool are essential to be used in this type of study to reduce computational costs and neglect the influence of tool wear. Crystal plasticity Eqs. 4–7 characterized the crystallographic slip systems governed by Schmid's law. The slip rate $\dot{\gamma}^{(\alpha)}$ of slip system (α) can first be calculated using the critical resolved shear stress ($\tau_c^{(\alpha)}$), rate sensitive factor (m), and strain hardening factor ($h_{\alpha\beta}$) to determine the plastic shear rate ($\dot{F}^p$) as follows:

$$\dot{\gamma}^{(\alpha)} = \dot{\gamma}_0^{(\alpha)} \left| \tau^{(\alpha)} / \tau_c^{(\alpha)} \right|^{1/m} \operatorname{sgn}\left(\tau^{(\alpha)}\right) \tag{4}$$

$$\dot{\tau}_c^{(\alpha)} = \sum_{\beta} h_{\alpha\beta} \left| \dot{\gamma}^{(\beta)} \right| \tag{5}$$

$$h_{\alpha\beta} = q_{\alpha\beta} \cdot h_0 \left(1 - \frac{\tau_c^{(\beta)}}{\tau_s} \right)^a \tag{6}$$

$$\dot{F}^p = \sum_{\alpha} \dot{\gamma}^{(\alpha)} s^{(\alpha)} \otimes n^{(\alpha)} F^p \tag{7}$$

where s and n are the slip direction and normal to the slip plane, respectively.

This particular method was used to study microcrack formation on the machined surface by calculating the lattice rotation and tensile stresses of selected nodes. By applying the slip model as explained in Sect. 2.1, tensile stresses exceeding a critical value on cleavage plane prior to slip initiation would result in brittle fracture. CPFEM offers effective analysis of stress concentrations of the various deformation zones evolved during a machining process.

CPFEM is also capable of simulating the asymmetric ductile–brittle transition phenomenon seen in orthogonal cutting of CaF_2 by observing the tensile stresses acting on the tertiary deformation zone. This is done by observing the differences in stress concentrations on both sides of the thin slab that would provide a hint of the nature of the surface generated [21]. This method is preferable as compared to the conventional cubic deformable body method. Cumulative shear strain analysis of each node explains the involvement of slip systems and dislocation activity during an orthogonal cutting process. In general, a higher number of activated slip systems allows ductile material flow that can be observed at the primary plastic deformation zone.

However, CPFEM has its setbacks when modeling the plasticity of brittle materials during UPM due to the microscale time steps in computation. Molecular dynamics (MD) simulations tend to perform more effectively to study atomistic movement and reactions when using various potentials for calculations. Xiao et al. [34] describe in detail the various types of potentials applied and their limitations toward the study of the ductile–brittle phenomenon. It was explained that the Tersoff potential is not suitable in the modeling of brittle materials due to it being a short-ranged bond order potential that often predicts ductile failure rather than to accurately present brittle fracture. On the other hand, the Stillinger–Weber potential and modifications are capable of predicting the brittle failure accurately.

MD simulations offer analysis of the point of crack initiation, direction of crack propagation, and stress distribution around the machining interface. When modeling the orthogonal cutting of 6H SiC, Xiao et al. [34] observed the crack initiation and directionality at sufficiently high undeformed chip thickness settings and discovered that crack formation ahead of the cutting tool is the main cause of crater formation on machined surfaces. This spectacle can be seen in Fig. 12 where the nature of crack initiation and the direction of the crack propagation are dependent on the crystallographic orientations of the machined workpiece.

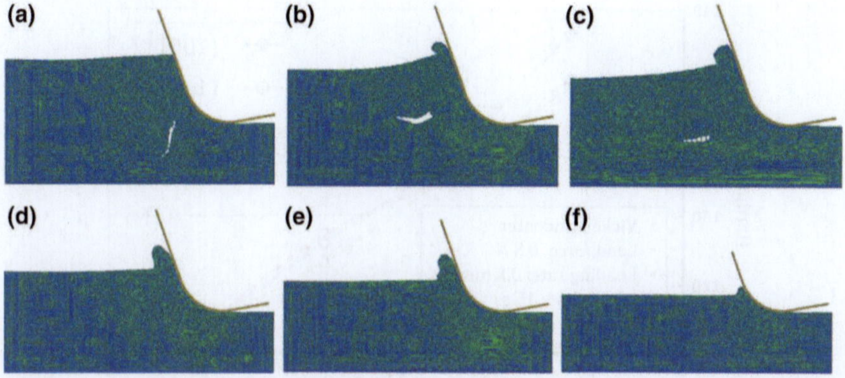

Fig. 12 MD simulation of orthogonal cutting of 6H SiC at varying undeformed chip thickness parameters [34] (recreated with permission from Elsevier © 2015)

3 Analysis of Thermal Assistance

3.1 Thermal Effects on Material Properties

In Sect. 2.1, slip model, it was explained that indentation tests were capable of verifying the mechanical properties and relating the influence of the slip and cleavage planes to crack initiation. Generally, hardness and modulus values tend to decrease with increase in temperature for both ductile and brittle materials [33, 34]. Similarly, experimental work of indentations at the elevated temperatures provides insights into the influence of the additional heat on the crack initiation mechanisms. Vickers microhardness tests conducted at elevated temperatures show decreased material hardness and an increase in indentation depths prior to crack initiation.

Deadmore and Sliney [35] applied a 0.5 N load at a constant indentation rate and dwell time to CaF_2 and BaF_2 single crystals over a range of temperatures from 25 to 670 °C. They observed a 40% decrease in hardness at 100 °C and the absence of indentation crack formation at temperatures above 85 °C. Another study on thermally assisted microhardness indentation studies on CaF_2 single crystals of different plane orientations (100), (110), and (111) with a load of 0.5 N showed a conversion of hardness values at 100 °C in Fig. 13. Thermally assisted indentation shows the potential in reducing the anisotropic effects that can be highly beneficial in an evenly distributed residual stresses and subsurface damage on the machined surface.

The material hardness dropped remarkably by approximately 50% at 100 °C in this test. Differences in results by different authors could be attributed to the different indentation orientations that have an effect on the resolved stresses relative to the slip and cleavage planes as discussed in Sect. 2.1. As expected, elevated temperatures promote slip activation prior to cleavage fracture and increase the indentation depth required for crack formation. Figure 14 shows the improvement in crack initiation

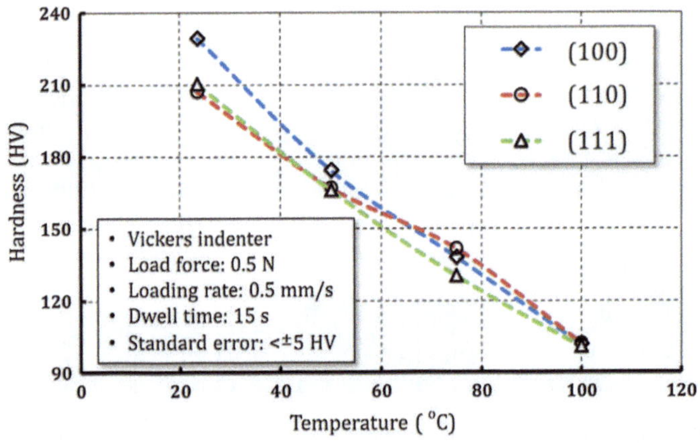

Fig. 13 Decrease in hardness of CaF_2 single crystals with increase in temperature [24]

resistance on the (110) plane-oriented CaF_2 sample that has the highest tendency for brittle failure by cleaving shown by Azami et al. [20]. Additionally, based on Eq. 1, it can be understood that a trend of increasing fracture toughness is a resultant cause of the rise in material temperature and therefore enables the material to undergo enhanced plastic deformation at higher yield stresses.

3.2 *Heat-Assisted Machining Simulations*

More than one slip system can be activated at higher temperatures when conducting compression tests along different crystallographic directions of CaF_2 single crystals [36]. Conventional theories on the motion of dislocations are necessary to understand the influence of slip systems and plastic deformation. The motion of a dislocation occurs through the shifts of the adjacent atoms but encounters obstacles like the short-ranged Peirels–Nabarro energy barrier. However, the energy required for the displacement of atoms can be easily overcome with a thermally assisted applied force and promotes plastic deformation [37]. Therefore, the energy for slip to occur would be reduced to precede the occurrence of brittle fracture. In compression tests at room temperature, the {100} plane of CaF_2 is the only slip plane [36]. Subsequently when heated to temperatures between 90 and 400 °C, the {111} slip plane was activated and a third {110} slip plane was activated at temperatures above 400 °C [36].

Theoretical simulations for orthogonal microcutting can be developed for establishment of the theoretical foundation for the material removal process at room and elevated temperatures associated with the experimentally proven number of slip planes that were thermally activated. Wang et al. [32] developed a CPFEM model to theoretically describe the effect of the activated slip systems during orthogonal cutting. In their work, three different models were created to simulate the cutting of

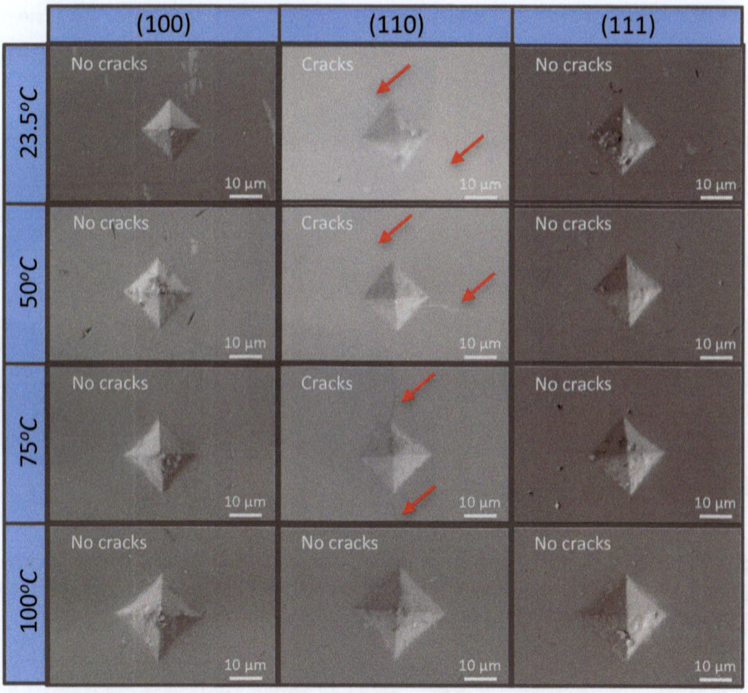

Fig. 14 SEM images of microhardness indentations on the (100), (110), and (111) plane orientations of CaF_2 single crystals over a range of temperatures [24]

CaF_2 single crystals at the room temperature, 200 °C, and 400 °C. The cumulative shear strain plots of each model simulated for the different number of activated systems in Fig. 15 show the increase in plasticity due to the dislocation activities ahead of the cutting tool. Defects in the material act as nuclei for cleavage fracture, and a larger stress field at the cutting edge could allow a larger density of defects to initiate cleavage fracture [15]. Under room temperature conditions, the size effect is commonly controlled by the undeformed chip thickness where a small undeformed chip thickness would result in a smaller critical stress field. At elevated temperatures, the critical stress field size is simulated to decrease, thus reducing the interaction with subsurface defects.

CPFEM is also capable of verifying the trend observed in the experimental results from Sect. 3.1, where a reduction in anisotropic characteristics is realized for elevated temperatures. Two sides of the model can be shown to review the influence of the material anisotropy that would result in the asymmetric deformation. The asymmetric ductile–brittle transition (aDBT) is an important aspect of the ductile–brittle transition study in view of the lateral forces induced during different types of crack formations. An asymmetric orthogonal groove results in unbalanced lateral force fluctuations with a feature of lamellar cracks on one side of the groove as shown in Fig. 16 [21]. Balanced forces can be observed with more symmetrical types of

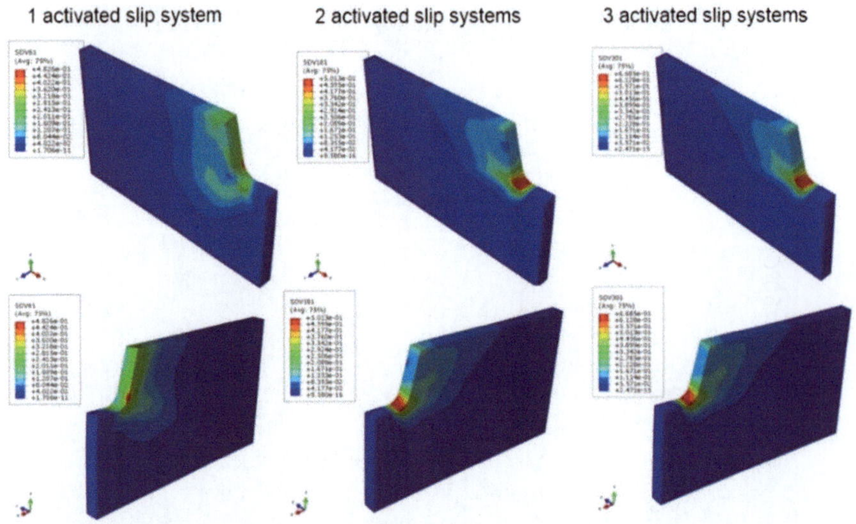

Fig. 15 CPFEM simulations of cumulative shear strain on all slip systems during ductile-regime cutting for one, two, and three activated systems at 25, 200, and 400 °C, respectively [32] (recreated with permission from SAGE Publications © 2016)

grooving that feature different types of crack morphology due to the relative cutting direction with respect to the cleavage plane.

The simulated result for one activated slip system in Fig. 15 shows the differences in the cumulative shear strains on each side of the model attributed to the asymmetric feature. At higher temperatures, the asymmetric feature loses prominence when more slip systems are activated, and the anisotropic effect is severely reduced. Simulated results of the 200 and 400 °C case with two and three activated slip systems show similarities in the cumulative shear strain plots on both sides of the model, signifying the transition to an isotropic material at higher temperatures.

As expected, the simulated lateral forces in Fig. 17a are further reduced and rendered negligible at elevated temperatures where two and three sets of slip system are active. Simulated cutting and thrust forces also show a reduction of approximately 50% in Fig. 17b when heated from room temperature with one activated slip system to 200 and 400 °C with more slip systems thermally activated. The thermally activated slip systems allow the movement of dislocations and thus increasing the plasticity of the material being removed.

Therefore, it can be understood from theoretical simulations that thermally assisted machining has multiple benefits to enhance the machinability of the brittle material. The benefits observed from experimental and theoretical results are as summarized below:

- Linear reduction in hardness,
- Activation of secondary and tertiary slip systems,
- Reducing anisotropic properties of single crystals.

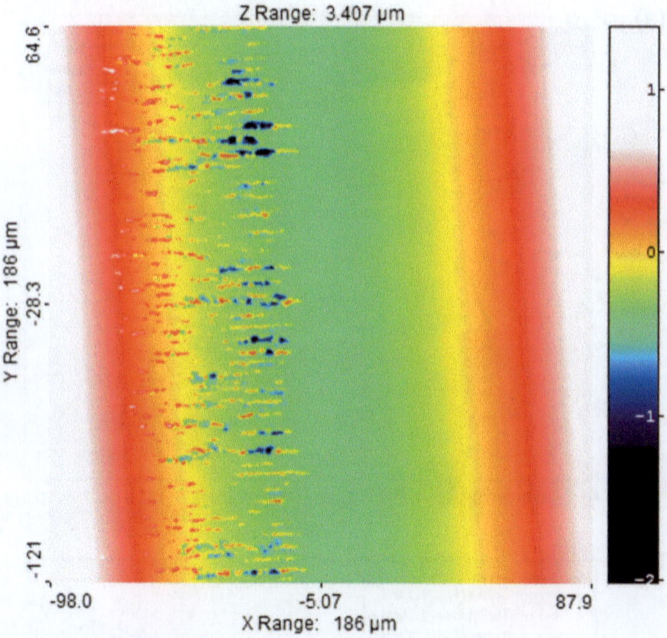

Fig. 16 WLI image of asymmetric ductile–brittle transition along a groove [21] (recreated with permission from SAGE Publications © 2016)

4 Thermally Assisted Machining Technology

4.1 Current State of Technology

Many strategies have been investigated to improve the machinability of difficult-to-machine materials. Thermally assisted machining is one of the advancing methods used in the aerospace and automotive industry for difficult-to-machine superalloy materials. Although there are many types of heating methods, the induction heating, laser heating, and plasma arc heating are the most promising methods to be adopted in conventional machining. Each method generally aims to reduce the cutting forces during machining and improve the material removal rate efficiencies while prolonging cutting tool life. The detailed comparison between each method is described in Table 2. The effectiveness of thermally assisted machining depends on the conventional cutting parameters such as cutting speed and heating parameters such as the distance between the localized heating zone and the tool cutting edge and the amount of supplied heat energy. The common benefits of the thermally assisted machining techniques are summarized below [38–44]:

- Improved surface finish,
- Improved cutting tool life,

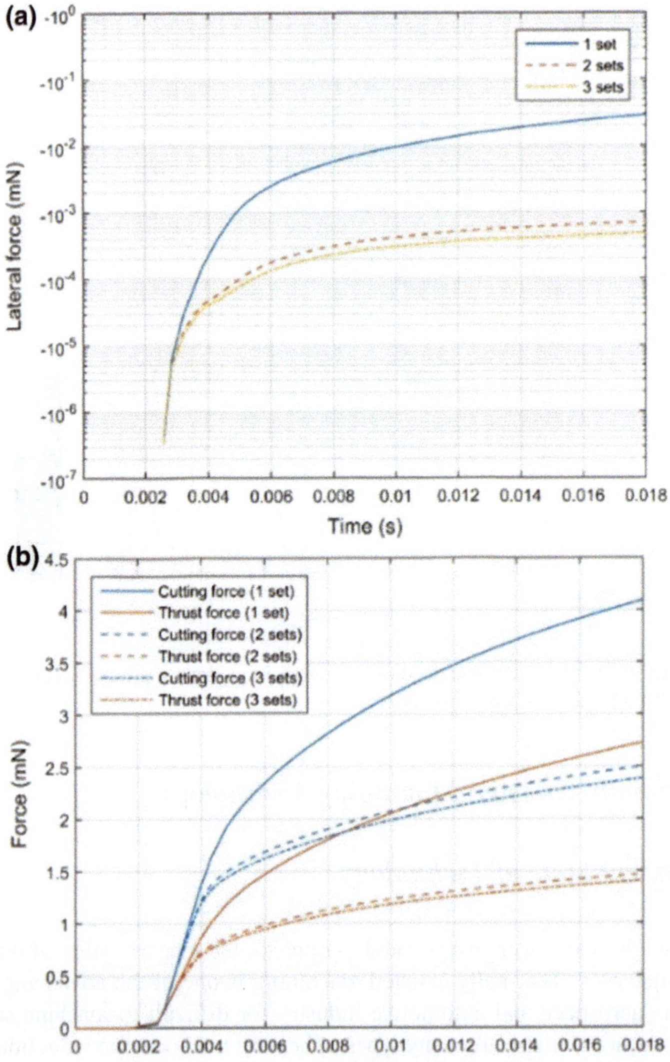

Fig. 17 Simulation forces during orthogonal cutting of one, two, and three activated slip system sets: **a** lateral force; **b** cutting and thrust forces [32] (recreated with permission from SAGE Publications © 2016)

- Increased material removal rates,
- Reduced machining forces,
- Reduced residual stresses,
- Reduced chattering and machining vibrations.

Table 2 Thermally assisted machining methods [45–47]

	Induction heating	Laser heating	Plasma heating
Heat source	Electric current	Laser	Plasma arc
Source mobility	Rigid	Flexible	Flexible
Temperature range	Low	High	High
Advantages	• Easy to regulate • Stable temperatures	• Stable and flexible heat distribution	• No metallurgical damage to workpiece • Economical
Disadvantages	• Heat induction in a line (delocalized)	• Costly • Material-dependent absorption rates	• Inaccurate control of localized heating temperature

4.2 *Laser-Assisted Machining*

Over the last few decades, much work has been done on thermally assisted machining that has been dedicated to conventional macroscale machining processes such as turning, grinding, and milling. But with the advancement in optical technology, thermal assistance has thus shifted its focus toward laser-assisted machining (LAM). LAM was first introduced by Bass et al. [48] in 1978, inspired by the adoption of gas torch and induction heating for heat-assisted machining. A common heat-assisted setup as illustrated in Fig. 18 involves installing a conventional setup with an external heat source to produce a localized heat spot to soften the mechanically hard workpiece ahead of the cutting tool [39, 42]. Noncontact thermal sensing equipment, such as an infrared camera, is often used to measure the degree of heating of the workpiece. When the high-power laser heats the workpiece, the material yield strength and hardness values decrease below the fracture strength, thus allowing the material to be removed in the ductile-regime. LAM commonly employs the use of the CO_2 laser and Nd:YAG laser, but the latter is a better candidate due to its higher absorption rates of laser energy [49].

From conventional machining theories, a higher feed rate increases the volume of material removed per revolution, which is proportional to the increase in cutting forces. However, the opposite is observed when machining with LAM. Few differences in the cutting forces at low feeds were reported when machining Ti-6Cr-5Mo-5V-4Al titanium alloy with and without LAM, but a 12% reduction in cutting forces with LAM was measured upon increasing the feed rate [44]. Therefore, it is obvious that LAM allows more productive machining efficiencies in terms of material removal rates and tool life when handling difficult-to-machine materials.

Much work on the adoption of LAM has been focused on improving the machinability of difficult-to-machine materials in the macroscale. Micro-laser-assisted machining (μ-LAM) techniques are the latest advancement of heat addition methods by laser irradiation. Unlike conventional methodologies of heating a spot of the workpiece a distance ahead of the cutting tool, μ-LAM process beams an IR diode

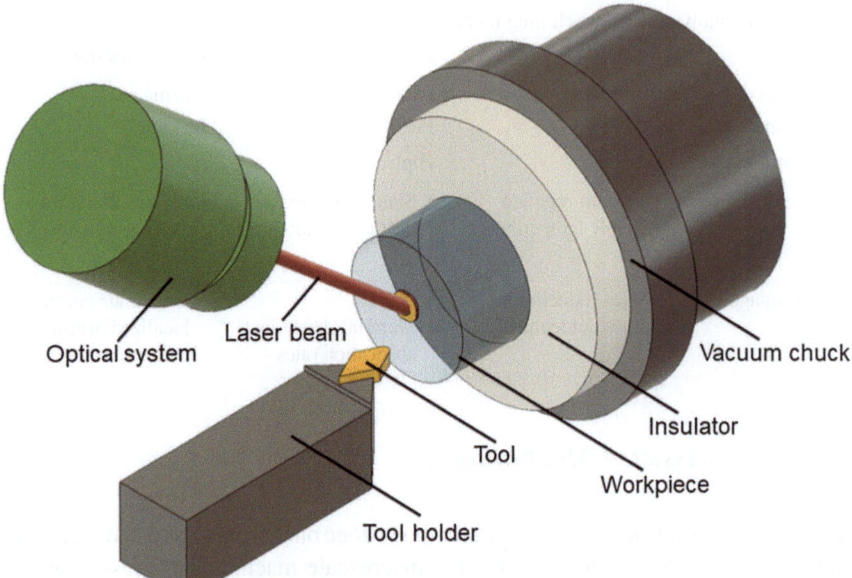

Fig. 18 Laser-assisted machining turning schematic

laser ($\lambda = 1070$ nm) through the diamond cutting tool to the tool–workpiece interface as shown in Fig. 19 [50]. The thermal softening efficiency of this method during an actual machining process can be observed through basic tribology tests. Shayan et al. [51] equipped a universal micro-tribometer (UMT) with the μ-LAM technique and conducted scratch tests on 4H SiC with a diamond stylus under varying speeds but a constant applied load. They reported a minor increase in groove depth by 5 nm when testing at higher cutting speeds. However, at extremely low cutting speeds, the average groove depth increased significantly from 54 to 90 nm. The thermal softening effect can therefore be concluded to be less prominent at higher cutting speeds when using the μ-LAM technique.

Optimization of the heating parameters is essential to achieve the desired surface quality. Surprisingly, there exists a critical value of laser power for ductile-regime machining, and an excess laser power applied could result in poorer surface finishing. Mohammadi et al. [52] employed the μ-LAM on an SPDT process of silicon single crystal and found an increase in Ra value by a factor of 1.6 when the laser power was increased from 20 to 30 W. It was also added that the advantages of this technique were the capability of conducting a thermally assisted machining process accompanied with cutting fluids. This is one of the major advantages of the fast contactless heating method as shown in Fig. 19. With the μ-LAM technique, they were able to conduct ductile-mode machining at a higher undeformed chip thickness value to achieve a mirror-like surface finishing on the silicon single crystal as shown in Fig. 20.

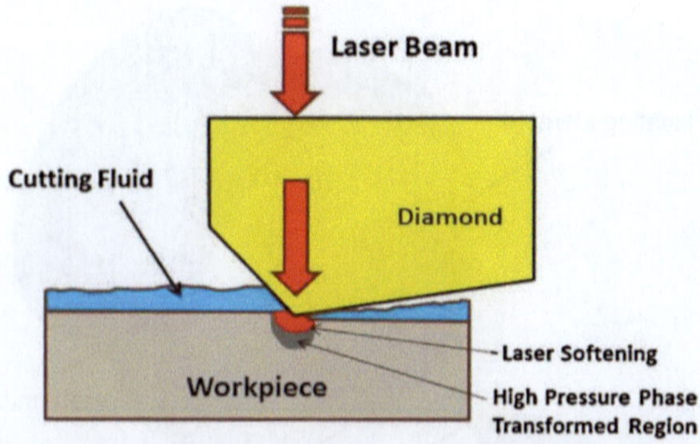

Fig. 19 Schematic of the μ-LAM process with cutting fluids [52] (recreated with permission from Elsevier © 2015)

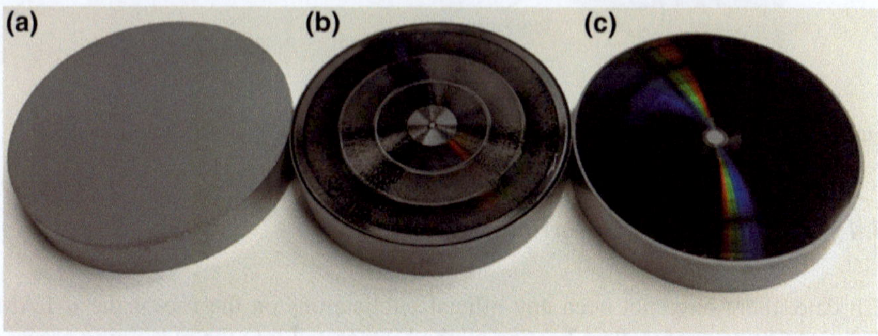

Fig. 20 **a** As received; **b** conventional micromachining; **c** with μ-LAM technique [52] (recreated with permission from Elsevier © 2015)

4.3 Alternate Heating Methods

While most thermally assisted machining methods use a localized heat source shown by the schematic in Fig. 18, it is interesting to note that most static experimental tests such as the heated nanoindentation test can be conducted with a globalized heat source. The global heating method may be defined by applying heat to the sample and to achieve an evenly heated sample in Fig. 21.

In some cases, heat is also applied to the machine tool to reduce abrupt fluctuations in heat transfer. Korte et al. [53] developed a customized hot nanoindentation setup that operates in vacuum to study the thermal stability and instrument noises caused by the impurities of inert gases during high-temperature testing. Resistive heaters were applied to both the indenter and the sample. A drastic temperature difference between a room temperature indenter and heated sample may cause unwanted undesired expansion and contraction at the interface, causing dimensional changes that disrupt the experiment [53].

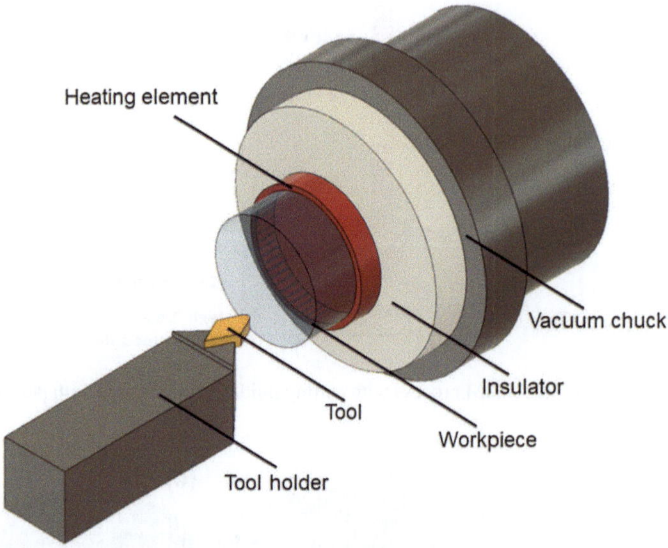

Fig. 21 Orthogonal cutting schematic with **a** localized heating and **b** global heating methods

4.4 Challenges in Hot Machining

Till date, there have not been any official publications on the use of the μ-LAM technique or any contactless heating method integrated into the machining process of CaF_2. Different types of heat sources are still being explored to improve the machinability of CaF_2. In general, several complications have arisen in the use of thermally assisted machining. Experimentation and industrial use of hot machining in ambient environments potentially promote the formation of an oxide layer on the pre-machined or machined surface, which could substantially affect the material removal mechanisms in the submicron range. Oxidation proof coatings on the sample surface or experimentation in an inert environment help with the oxidation problem. However, different materials oxidize at different temperatures and environments. For example, it is recommended to coat copper with a thin layer of platinum to prevent oxidation during nanoindentation tests in the lower range of elevated temperatures (>200 °C) [54]. Oxygen is known to penetrate CaF_2 at elevated temperatures and begin precipitation when cooled below 725 °C [55].

High-accuracy temperature control equipment is extremely important in this aspect when taking material thermal expansion into consideration. This is particularly crucial when machining with submicron undeformed chip thickness values. As seen in Fig. 22, a slight deviation in temperature would cause a material expansion or contraction at a magnitude similar to the desired undeformed chip thickness. A basic temperature control setup would commonly include temperature controllers, a heat source, and temperature measurement devices.

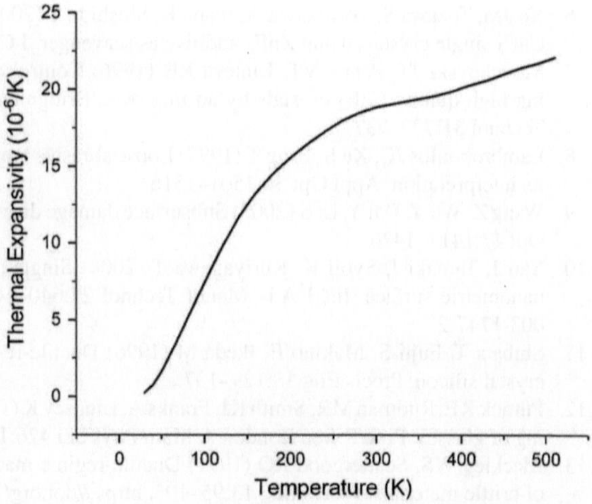

Fig. 22 Thermal expansion of CaF_2 [56, 57]

In summary, thermally assisted machining has been proven to improve the machining process of difficult-to-machine and brittle materials in various ways. A detailed description of a systematic experimental process had been discussed with the ductile–brittle transition explained using the well-established slip theory. However, the application of such technology on CaF_2 single crystals has only been limited to the preliminary studies using indentation tests at elevated temperatures. The challenges in achieving thermally assisted ductile-mode machining of CaF_2 have been briefly explained to serve as a foundation for considerations in the future works of this field.

Acknowledgements The authors are grateful for the financial support from National University of Singapore Start-up Grant (Grant No.: R-265-000-564-133) and Singapore Ministry of Education Academic Research Fund Tier 1 (Grant No.: R-265-000-593-114).

References

1. Dao GT, Borodovsky YA (2001) 157-nm lithography for 100-nm generation and beyond: progress and status. In: Proceeding of SPIE 4186, 20th annual BACUS symposium on photomask technology
2. Rothschild M, Bloomstein TM, Fedynyshyn TH, Kunz RR, Liberman V, Switkes M, Efremow NN, Palmacci ST, Sedlacek JHC, Hardy DE, Grenville A (2003) Optical lithography. Lincoln Lab J 14:221–236
3. Hahn D (2014) Calcium fluoride and barium fluoride crystals in optics: multispectral optical materials for a wide spectrum of applications. Opt Photonik 9:45–48
4. Nachimuthu S, Aoshima M, Shimizu S, Sumiya K, Ishibashi H (2007) Improvement in optical properties of CaF_2 single crystals used for nanolithography
5. Elswie HI, Lazarevi ZŽ, Radojevi V, Gili M, Rabasovi M, Ševi D, Rom NŽ (2016) The Bridgman method growth and spectroscopic characterization of calcium fluoride single crystals. Sci Sinter 48:333–341. https://doi.org/10.2298/SOS1603333E

6. Ko JM, Tozawa S, Yoshikawa A, Inaba K, Shishido T (2001) Czochralski growth of UV-grade CaF_2 single crystals using ZnF_2 additive as scavenger. J Cryst Growth 222:243–248
7. Mouchovski JT, Penev VT, Kuneva RB (1996) Control of the growth optimum in producing high-quality CaF_2 crystals by an improved Bridgman-Stockbarger technique. Cryst Res Technol 31:727–737
8. Lambropoulos JC, Xu S, Fang T (1997) Loose abrasive lapping hardness of optical glasses and its interpretation. Appl Opt 36:1501–1516
9. Wang Z, Wu Y, Dai Y, Li S (2008) Subsurface damage distribution in the lapping process. Appl Opt 47:1417–1426
10. Yan J, Tamaki J, Syoji K, Kuriyagawa T (2004) Single-point diamond turning of CaF_2 for nanometric surface. Int J Adv Manuf Technol 24:640–646. https://doi.org/10.1007/s00170-003-1747-2
11. Shibata T, Fujii S, Makino E, Ikeda M (1996) Ductile-regime turning mechanism of single-crystal silicon. Precis Eng 18:129–137
12. Puttick KE, Rudman MR, Smith KJ, Franks A, Lindsey K (1989) Single-point diamond machining of glasses. Proc R Soc London A Math Phys Sci 426:19–30
13. Blackley WS, Scattergood RO (1991) Ductile-regime machining model for diamond turning of brittle materials. Precis Eng 13:95–103. https://doi.org/10.1016/0141-6359(91)90500-I
14. O'Neill JB, Redfern BAW, Brookes CA (1973) Anisotropy in the hardness and friction of calcium fluoride crystals. J Mater Sci 8:47–58. https://doi.org/10.1007/BF00755582
15. Nakasuji T, Kodera S, Hara S, Ikawa N (1990) Diamond turning of brittle materials for optical components. CIRP Ann Manuf Technol 39:89–92
16. Shimada S, Ikawa N, Toyoshiro I, Nobuhiro T, Hitoshi O, Toshio S (1995) Brittle-ductile transition phenomena in microindentation and micromachining. Ann CIRP 44:523–526
17. Mizumoto Y, Kakinuma Y (2018) Revisit of the anisotropic deformation behavior of single-crystal CaF_2 in orthogonal cutting. Precis Eng 53:9–16. https://doi.org/10.1016/j.precisioneng.2018.01.011
18. Lodes MA, Hartmaier A, Göken M, Durst K (2011) Influence of dislocation density on the pop-in behavior and indentation size effect in CaF2single crystals: experiments and molecular dynamics simulations. Acta Mater 59:4264–4273. https://doi.org/10.1016/j.actamat.2011.03.050
19. Mizumoto Y, Aoyama T, Kakinuma Y (2011) Basic study on ultraprecision machining of single-crystal calcium fluoride. Procedia Eng 19:264–269. https://doi.org/10.1016/j.proeng.2011.11.110
20. Azami S, Kudo H, Mizumoto Y, Tanabe T, Yan J, Kakinuma Y (2015) Experimental study of crystal anisotropy based on ultra-precision cylindrical turning of single-crystal calcium fluoride. Precis Eng 40:172–181. https://doi.org/10.1016/j.precisioneng.2014.11.007
21. Wang H, Riemer O, Rickens K, Brinksmeier E (2016) On the mechanism of asymmetric ductile-brittle transition in microcutting of (111) CaF_2 single crystals. Scr Mater 114:21–26. https://doi.org/10.1016/j.scriptamat.2015.11.030
22. Chen X, Xu J, Fang H, Tian R (2017) Influence of cutting parameters on the ductile-brittle transition of single-crystal calcium fluoride during ultra-precision cutting. Int J Adv Manuf Technol 89:219–225. https://doi.org/10.1007/s00170-016-9063-9
23. Marsh ER, John BP, Couey JA, Wang J, Grejda RD, Vallance RR (2005) Predicting surface figure in diamond turned calcium fluoride using in-process force measurement. J Vac Sci Technol B Microelectron Nanom Struct 23:84. https://doi.org/10.1116/1.1839897
24. Chaudhari A, Lee YJ, Wang H, Senthil Kumar A (2017) Thermal effect on brittle-ductile transition in CaF_2 single crystals. In: Proceedings of the 17th international conference of the European Society for precision engineering and nanotechnology, EUSPEN 2017
25. Cook RF, Pharr GM (1990) Direct observation and analysis of indentation cracking in glasses and ceramics. J Am Ceram Soc 73:787–817. https://doi.org/10.1111/j.1151-2916.1990.tb05119.x
26. Fang FZ, Chen LJ (2000) Ultra-precision cutting for ZKN7 glass. CIRP Ann Manuf Technol 49:17–20. https://doi.org/10.1016/S0007-8506(07)62887-X

27. Yan J, Asami T, Harada H, Kuriyagawa T (2009) Fundamental investigation of subsurface damage in single crystalline silicon caused by diamond machining. Precis Eng 33:378–386. https://doi.org/10.1016/j.precisioneng.2008.10.008
28. Kawasegi N, Niwata T, Morita N, Nishimura K, Sasaoka H (2014) Improving machining performance of single-crystal diamond tools irradiated by a focused ion beam. Precis Eng 38:174–182. https://doi.org/10.1016/j.precisioneng.2013.09.001
29. O'Connor BP, Marsh ER, Couey JA (2005) On the effect of crystallographic orientation on ductile material removal in silicon. Precis Eng 29:124–132. https://doi.org/10.1016/j.precisioneng.2004.05.004
30. Blake PN, Scattergood RO (1990) Ductile-regime machining of germanium and silicon. J Am Ceram Soc 73:949–957 . https://doi.org/10.1111/j.1151-2916.1990.tb05142.x
31. Yan J, Syoji K, Tamaki J (2004) Crystallographic effects in micro/nanomachining of single-crystal calcium fluoride. J Vac Sci Technol B Microelectron Nanom Struct 22:46. https://doi.org/10.1116/1.1633770
32. Wang H, Senthil Kumar A, Riemer O (2016) On the theoretical foundation for the microcutting of calcium fluoride single crystals at elevated temperatures. Proc Inst Mech Eng Part B J Eng Manuf 1–7. https://doi.org/10.1177/0954405416666907
33. Yan J, Asami T, Harada H, Kuriyagawa T (2012) Crystallographic effect on subsurface damage formation in silicon microcutting. CIRP Ann Manuf Technol 61:131–134. https://doi.org/10.1016/j.cirp.2012.03.070
34. Xiao G, To S, Zhang G (2015) Molecular dynamics modelling of brittle-ductile cutting mode transition: case study on silicon carbide. Int J Mach Tools Manuf 88:214–222. https://doi.org/10.1016/j.ijmachtools.2014.10.007
35. Deadmore DL, Sliney HE (1987) Hardness of CaF2 and BaF2 solid lubricants at 25 to 670 °C (NASA Technical Memorandum 88979)
36. Muñoz A, Domínguez-Rodríguez A, Castaing J (1994) Slip Systems and plastic anisotropy in CaF_2. J Mater Sci 29:6207–6211. https://doi.org/10.1007/BF00354561
37. Conrad H (1964) Thermally activated deformation of metals. J Met 16:582–588. https://doi.org/10.1007/BF03378292
38. López de Lacalle LN, Sánchez JA, Lamikiz A, Celaya A (2004) Plasma assisted milling of heat-resistant superalloys. J Manuf Sci Eng 126:274. https://doi.org/10.1115/1.1644548
39. Kumar M, Melkote S, Lahoti G (2011) Laser-assisted microgrinding of ceramics. CIRP Ann Manuf Technol 60:367–370. https://doi.org/10.1016/j.cirp.2011.03.121
40. Ding H, Shen N, Shin YC (2012) Thermal and mechanical modeling analysis of laser-assisted micro-milling of difficult-to-machine alloys. J Mater Process Technol 212:601–613. https://doi.org/10.1016/j.jmatprotec.2011.07.016
41. Ding H, Shin YC (2010) Laser-assisted machining of hardened steel parts with surface integrity analysis. Int J Mach Tools Manuf 50:106–114. https://doi.org/10.1016/j.ijmachtools.2009.09.001
42. Tian Y, Wu B, Anderson M, Shin YC (2008) Laser-assisted milling of silicon nitride ceramics and inconel 718. J Manuf Sci Eng 130:031013. https://doi.org/10.1115/1.2927447
43. Lei S, Shin YC, Incropera FP (2001) Experimental investigation of thermo-mechanical characteristics in laser-assisted machining of silicon nitride ceramics. J Manuf Sci Eng 123:639. https://doi.org/10.1115/1.1380382
44. Rahman Rashid RA, Sun S, Wang G, Dargusch MS (2012) An investigation of cutting forces and cutting temperatures during laser-assisted machining of the Ti-6Cr-5Mo-5 V-4Al beta titanium alloy. Int J Mach Tools Manuf 63:58–69. https://doi.org/10.1016/j.ijmachtools.2012.06.004
45. Madhavulu G, Ahmed B (1994) Hot machining process for improved metal removal rates in turning operations. J Mater Process Tech 44:199–206. https://doi.org/10.1016/0924-0136(94)90432-4
46. Jeon Y, Lee CM (2012) Current research trend on laser assisted machining. Int J Precis Eng Manuf 13:311–317. https://doi.org/10.1007/s12541-012-0040-4

47. Shams OA, Pramanik A, Chandratilleke TT (2017) Thermal-assisted machining of titanium alloys. In: Gupta K (ed) Advanced manufacturing technologies. Springer International Publishing, Berlin, pp 49–76
48. Bass M, Beck D, Copley SM (1979) Laser assisted machining. In: Proceeding of SPIE 0164, 4th European electro-optics conference, pp 233–240
49. Venkatesan K, Ramanujam R, Kuppan P (2014) Laser assisted machining of difficult to cut materials: Research opportunities and future directions - A comprehensive review. Procedia Eng 97:1626–1636. https://doi.org/10.1016/j.proeng.2014.12.313
50. Ravindra D, Ghantasala MK, Patten J (2012) Ductile mode material removal and high-pressure phase transformation in silicon during micro-laser assisted machining. Precis Eng 36:364–367
51. Shayan AR, Poyraz HB, Ravindra D, Patten JA (2009) Pressure and temperature effects in micro-laser assisted machining (μ-LAM) of silicon carbide. Trans North Am Manuf Res Inst SME 37:75–80
52. Mohammadi H, Ravindra D, Kode SK, Patten JA (2015) Experimental work on micro laser-assisted diamond turning of silicon (111). J Manuf Process 19:125–128. https://doi.org/10.1016/j.jmapro.2015.06.007
53. Korte S, Stearn RJ, Wheeler JM, Clegg WJ (2011) High temperature microcompression and nanoindentation in vacuum High temperature microcompression and nanoindentation in vacuum. J Mater Res 27:167–176. https://doi.org/10.1557/jmr.2011.268
54. Duan ZC, Hodge AM (2009) High-temperature nanoindentation: new developments and ongoing challenges. J Miner Met Mater Soc 62:32–36
55. Phillips WL, Hanlon JE (1963) Oxygen penetration into single crystals of calcium fluoride. J Am Ceram Soc 46:447–449
56. Batchelder DN, Simmons RO (1964) Lattice constants and thermal expansivities of silicon and of calcium fluoride between 6° and 322°K. J Chem Phys 41:2324–2329. https://doi.org/10.1063/1.1726266
57. Ballard SS, Brown SE, Browder JS (1978) Measurements of the thermal expansion of six optical materials, from room temperature to 250 °C. Appl Opt 17:1152. https://doi.org/10.1364/AO.17.001152
58. Beake BD, Smith JF (2002) High-temperature nanoindentation testing of fused silica and other materials. Philos Mag A Phys Condens Matter, Struct Defects Mech Prop 82:2179–2186. https://doi.org/10.1080/01418610208235727
59. Michel MD, Serbena FC, Lepienski CM (2006) Effect of temperature on hardness and indentation cracking of fused silica. J Non Cryst Solids 352:3550–3555. https://doi.org/10.1016/j.jnoncrysol.2006.02.113

Cutting Mechanism and Surface Formation of Ultra-Precision Raster Fly Cutting

Guoqing Zhang and Suet To

Abstract Ultra-precision raster fly cutting (UPRFC) is a discontinuous fly cutting, whereby the diamond cutting tool flies with spindle rotation and cuts the machined surface discontinuously. Accompanying the motion of spindle with a raster tool path covering the whole machined surface, the diamond tool can cut and form complex surface structures. The cutting mechanism of UPRFC makes it widely utilized in manufacturing non-rotational symmetric structures like pyramid array, free-form surface, F-theta lens, tetrahedron array, micro-lens array etc. The form accuracy of fabricated products could be down to submicron level, and surface roughness down to non-metric level. UPRFC is potentially used to fabricate products in multiple industry fields such as aerospace, automobile, laser, communication, optics. Study of the cutting mechanism of UPRFC and the influences thereof on surface integrity is a key topic since it helps to improve the machined surface quality. The intermittent cutting mechanism of UPRFC is quite different from other ultra-precision machining processes, e.g., single-point diamond turning, micro-milling; correspondingly, it may cause different effects on surface finish. This chapter will talk about the cutting mechanism and surface formation of UPRFC in five parts. Part 1 gives an overview to UPRFC technology and the application thereof on fabrication of optical products. Part 2 delivers a comprehensive knowledge on the material removal mechanism of UPRFC both in theoretical and experimental aspects. Part 3 talks about material sliding during chip formation, and it caused surface microwaves in UPRFC process; in this part, the appearance and influence factors of microwaves will be presented. Finally, tool wear features and their influences on the quality of machined surface were investigated in Part 4, and a short conclusion is summarized in Part 5.

G. Zhang (✉)
College of Mechatronics and Control Engineering, Shenzhen University, Nan-Hai Ave 3688, Shenzhen 518060, Guangdong, People's Republic of China
e-mail: zhanggq@szu.edu.cn

S. To
State Key Laboratory of Ultra-Precision Machining Technology, Department of Industrial and Systems Engineering, The Hong Kong Polytechnic University, Kowloon, Hong Kong, People's Republic of China

J. Zhang et al. (eds.), *Simulation and Experiments of Material-Oriented Ultra-Precision Machining*, Springer Tracts in Mechanical Engineering, https://doi.org/10.1007/978-981-13-3335-4_5

This chapter will give reader a comprehensive understanding of the cutting mechanism and surface generation in UPRFC.

1 Introduction

UPRFC, as a fly cutting process, is a discontinuous cutting, whereby diamond cutting tool flies with spindle rotation of machine tools and cuts the workpiece surface intermittently. In a rotary cut, a tool imprint is formed, and the pieces of tool imprint compose the desired surface structures [1]. Differing from the cutting kinematics of continuous cutting, e.g., single-point diamond turning, the diamond tool in UPRFC rotates around the spindle of machine tools, while the workpiece is fixed. UPRFC is also different from conventional or micro-end milling process, since it employs the single crystal diamond as the tool material, as is shown in Fig. 1. Totally, the difference of UPRFC includes: First, only one tool tooth was employed in UPRFC for cutting and spindle speed is usually fast; therefore, the cutting period in each rotation cycle is very short. Secondly, the chip morphology in UPRFC is short, thin, and leaf-shaped [2]. The special cutting mechanism of UPRFC has its own advantages: It is good at cutting non-rotational symmetric microstructures with high form accuracy and low roughness. It has become an essential method to fabricate precise products, especially optical components, and widely applied into the manufacturing of complex surface structures like pyramid structures, V-groove, free-form surfaces with submicron form accuracy and nanometric surface roughness [3, 4], as is shown in Fig. 2.

In UPRFC, the analysis of cutting mechanism and surface generation is quite significant, since it is contributed to the configuration of cutting conditions, the design of cutting plans, the determination of tool wear, and the improvement of surface quality, which are also important parameters for the ultra-precision machine tools' design [5, 6]. Unlike other precision cutting processes, e.g., single-point diamond turning and micro-milling, in UPRFC, the fly cutting of the cutting tools makes cutting chips thicker at their middle while thinner at their two sides. Owing to the thickness inconsistency of cutting chips along the cutting direction, material sliding behavior at different chip thickness during cutting and its influence on the surface roughness are different too. In UPRFC, the cutting direction changes all time, which

Fig. 1 UPRFC machine and cutting tools

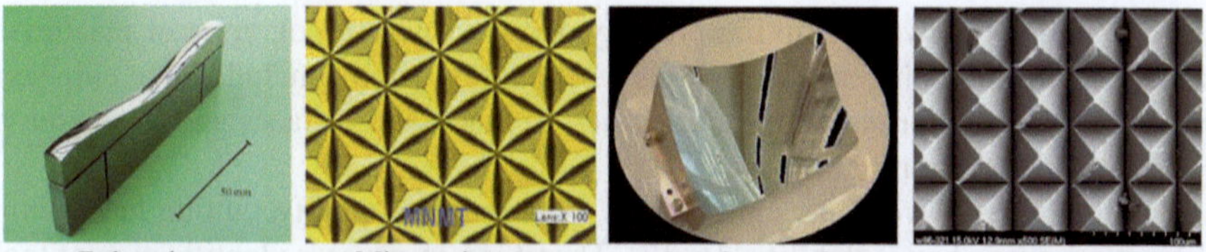

F-theta lens | Micro prism array | Freeform surface | Pyramid array

Fig. 2 Microstructures could be manufactured by UPRFC

always along the tensile direction of the tool path, the cutting kinematics of UPRFC makes the tool edge profile imprinted on both chip surface and machined surface, the cutting chips were then cut down in the following cutting and collected online for evaluation [7].

The intermittent cutting mechanism of UPRFC makes the surface generation quite different from other cutting processes, which also makes the surface quality-affecting factors different. The previous study found that material-sliding-caused vibration, tool-wear-caused surface quality changes have some influence on the machined surface quality. The following part will conduct review on these topics.

In UPRFC, the influence of vibration on the surface generation is significant, since it produce micro-waves or patterns on the machined surface whereby affects the optical function of fabricated products. In UPRFC, many vibration sources exist, mainly including tool-tip vibration, spindle vibration, background vibration, self-excited vibration, and material-caused vibration, which have unequal influences on the surface topography and roughness: Tool-tip vibration with ultra-high natural frequency influences surface topography and roughness at a nanometric level that leads to the formation of the surface profile fluctuation with certain period and a particular spatial frequency, as well as irregular zebra-stripe-like structures on the surface being machined [8–10]. Spindle vibration also owns a significant effect on the surface topography and roughness of UPRFC. Usually, axial vibration causes periodic concentric, concave, twofold, spiral structures on machined surface, coupled tilting vibration forms irregular patterns and ribbon stripe whereby plays a principle impact on surface topography of the machined surface [11, 12]. Material property (e.g., material anisotropy and material spring back) has some effects on the machined surface topography, and it usually causes a variation of the shear angle, cutting force, and rough surface, which leads to the formation of periodical structures on the surface being machined [13, 14]. Material sliding in the chip generation is a not negligible origin of vibration that influences the roughness of machined surface. Material sliding is a periodic behavior during chip generation, which is generated from the cutting energy accumulation and release, and has some influences on the surface topography and roughness of machined surface. However, little studies have been reported before on the material sliding behavior in chip generation and its effects on surface topography and surface roughness changes. Until now, Su et al. (2015) have conducted the most comprehensive study into the relation between material-sliding-caused sawtooth and surface waves in high-speed cutting [15]. In their investigation, the correlations between chip morphology and the micro-topography of machined surface at different chip serration levels were explored and concluded that chip serration leads to the formation of microwaves on machined surfaces and the increase in machined surface roughness.

In UPRFC, tool wear has some effects on the surface quality, cutting force, chip formation, even chatter stability [16, 17]. Therefore, in-process identification of tool wear and thereof influences on surface finish is important, which can identify the deteriorated machined surface under tool wear timely, avoid the unacceptable cutting, and remedy the deteriorated surface so as to improve the cutting efficiency. Until now, the study of tool wear evaluation and its effects on machined surface quality has been found a lot on traditional machining, micro-milling, and single-point diamond turning. Generally, the tool wear methods employed in previous studies include direct and indirect methods. Direct tool wear evaluation methods are a kind direct inspection method, which were usually performed by optical inspection [18, 19]. However, it is difficult to be realized in-process tool wear evaluation for optical inspection method since cutting tools are usually covered by chips or lubricant during cutting. In this case, many indirect evaluation methods for tool wear were employed, e.g., tool wear monitor during cutting by using indirect signals. The indirect signals could reflect the cutting tools' gradual wear [20–22]; however, it failed to precisely identify the tool wear characteristics. Besides, the evaluation of surface quality is also an essential topic in UPRFC [23, 24], especially for online surface quality evaluation [25, 26]. However, the in-process evaluation of machined surface considering tool wear effects in UPRFC has not been reported in the previous study.

As is talked above, UPRFC is an essential discontinuous cutting process, and the complex cutting kinematics of UPRFC makes its cutting mechanism and surface generation different as compared to single-point diamond turning. Most of prior studies about surface generation in UPRFC are limited to consider the effect of cutting strategy, cutting parameters, spindle vibration, kinematics error, and workpiece materials [27–29], while little attention has been paid on the influence of material sliding, size effect, tool wear on the surface generation. Material sliding, size effect, and tool wear have significant effects on the chip generation and surface formation; for example, material sliding will cause surface microwave formation, size effect leads to the formation of surface rough pattern, while tool wear will change tool geometry and further affect the surface roughness. In this chapter, the cutting mechanism of UPRFC and surface generation under considering influences of material sliding, size effect, and tool wear will be studied, and some useful conclusions were summarized. This research provides a deep undemanding of the cutting mechanism and surface formation in UPRFC and provides reference for the understanding of material sliding, size effect, and tool wear effect on the surface generation; this research is potentially used to improve the surface quality of UPRFC.

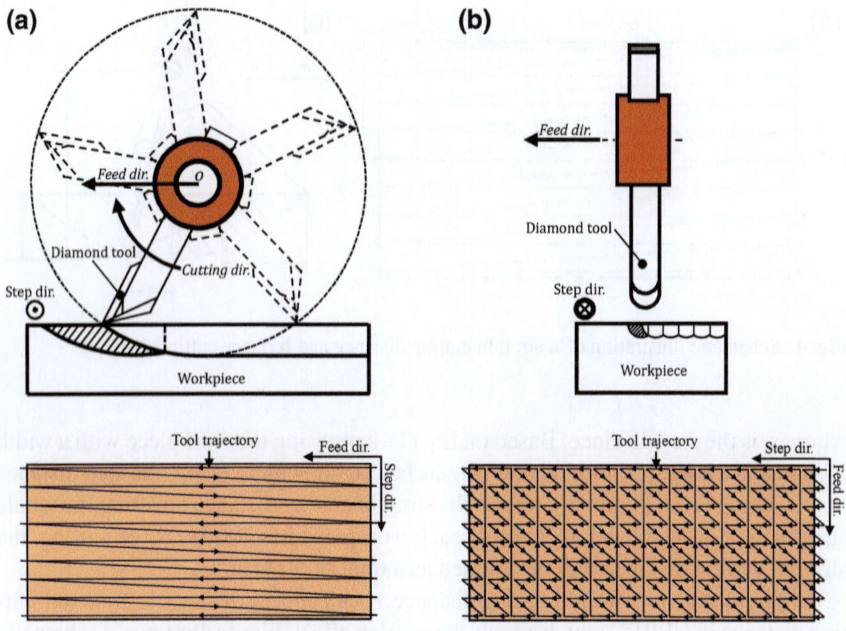

Fig. 3 Cutting strategies of UPRFC and the related tool trajectory

2 Material Removal Mechanism of UPRFC

2.1 Material Removal Mechanism and Cutting Distance Calculation

In UPRFC, two cutting strategies exist: vertical cutting and horizontal cutting. In horizontal cutting strategy, feed direction is perpendicular to step direction, and both directions in the horizontal cutting are opposite to that in the vertical cutting, as is shown in Fig. 3. Therefore, the cutting chips are enveloped by the tool geometry at intervals of the tool step distance and the feed rate.

Corresponding to the spindle motion and cutting tool rotation, there are two cutting distance-counting methods: straight cutting distance and rotary cutting distance. Straight cutting distance only is a nominal cutting distance, which considers the total length of spindle trajectory, while it does not consider the rotary trajectory of cutting tools during the cutting, as is shown in Fig. 4a. Therefore, if a workpiece has a rectangular surface area with its width w and length l, the counted straight cutting distance covering the whole surface area is calculated as:

$$\frac{lw}{s_t} \tag{1}$$

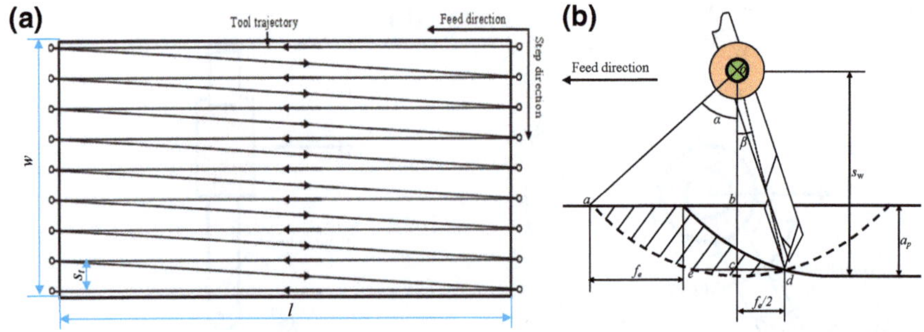

Fig. 4 Schematic illustration of **a** straight cutting distance and **b** rotary cutting distance

where s_t is the step distance. Based on Eq. (1), supposing the workpiece with a width 25 mm and a length 50 mm rectangular surface as shown in Fig. 4a, the step distance in the cutting is 0.025 mm; therefore, the straight cutting distance covering the whole surface area is calculated as 50 m. If each workpiece has 100 layers of cutting, the straight cutting distance will be counted as a total of 5000 m.

Different from the straight cutting distance, rotary cutting distance is the actual cutting distance in UPRFC which counts the overlap effect. Figure 4b shows a schematic illustration of the rotary cutting distance in UPRFC. The tool trajectory in rotary cutting is actually an arc *ad*, noted by $\widehat{t}$, the length of which can be calculated by:

$$\left|\widehat{t}\right| = (\alpha + \beta)s_w \tag{2}$$

where α is the angle of tool-entry surface, β is the angle of tool-exit surface, and s_w is the swing distance.

In triangle *oab* of Fig. 4b, angle α is calculated by:

$$\alpha = \arccos\left(\frac{s_w - a_p}{s_w}\right) \tag{3}$$

In triangular *ocd* of Fig. 4b, angle β is expressed as:

$$\beta = \arcsin\left(\frac{f_e/2}{s_w}\right) \tag{4}$$

where f_e is the feed rate in mm/rev, and it could be obtained by:

$$f_e = \frac{f_e'}{p} \tag{5}$$

where f_e' is the feed rate in mm/min, and p is the spindle speed in r/min.

For a workpiece with rectangle surface, to finish a complete surface cutting, the number of rotary cuttings needed is calculated by:

Table 1 Comparison of counted rotary cutting distance and straight cutting distance

Workpiece no.	No. 1	No. 2	No. 3	No. 4	No. 5
Straight cutting distance	1000	2000	3000	4000	5000
Rotary cutting distance	29,848	59,696	89,544	119,392	149,240

$$n = \frac{lw}{f_e s_t} \tag{6}$$

Combining Eqs. (2)–(6), the cutting distance to finish a rectangle surface cutting is obtained from:

$$e = \frac{lws_w}{f_e s_t}\left(\arccos\left(\frac{s_w - a_p}{s_w}\right) + \arcsin\left(\frac{f_e/2}{s_w}\right)\right) \tag{7}$$

From Eq. (7) and the given cutting parameters, cutting distance for completing a rectangular surface cutting is counted as 1492.4 m. Therefore, for a workpiece with 100-layer rectangular surface cutting, the total rotary cutting distance is calculated as 149,240 m.

Table 1 shows the comparison of cutting distances based on different calculation methods. It found that the rotary cutting distance is much longer than the straight cutting distance when they cut the rectangular surface. With the given cutting conditions and surface area, the counted rotary cutting distance is almost 29 times larger than the straight cutting distance.

2.2 Chip Formation in UPRFC

In the UPRFC, cutting chips are formed by three cutting steps: previous rotary cutting, previous step cutting, and current rotary cutting. Correspondingly, the cutting chip morphology equations are composed of equations of rough surface, previous rotary cutting formed surfaces, previous step cutting formed surface, and current rotary cutting formed surface. As is shown in Fig. 5a, b, the shaded area presents the chip area. For clear presentation, a coordinate system *o-xyz* is established with its origin at the intersection point of spindle axis and tool holder axis and its x-axis along the feed direction, z-axis upright.

Based on Fig. 5c, d, the 3D morphology of cutting chips was modeled and simulated by geometrical method. Since a cutting chip is compassed by four surfaces, the cutting chip model composed of four independent equations with each one presents one surface.

From Fig. 5c, cutting chips in both horizontal cutting and vertical cutting are composed of four surfaces: front surface (Fs), side surface (Ss), back surface (Bs), and rough surface (Rs). This chapter only talks about the chip modeling in horizontal

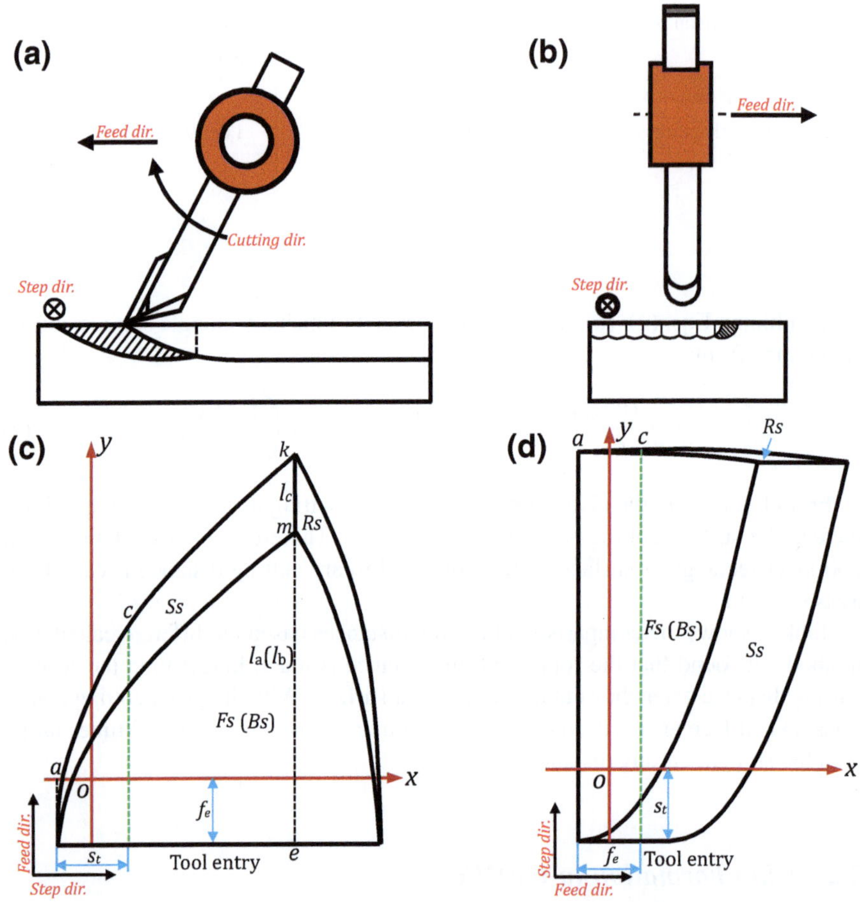

Fig. 5 Schematics of the cutting mechanism and chip formation of UPRFC in **a**, **c** a horizontal cutting strategy, and **b**, **d** a vertical cutting strategy

cutting, and the chip model in vertical cutting could be easily derived by referring chip model in horizontal cutting. Equations of the presented four surfaces are expressed as [30]:

$$\begin{cases} \text{Fs}: \left(\sqrt{(y+f_e)^2+z^2}+s_w-R\right)^2+x^2=R^2 \\ \text{Bs}: \left(\sqrt{y^2+z^2}+s_w-R\right)^2+x^2=R^2 \\ \text{Ss}: (z+s_w-R)^2+(x+s_t)^2=R^2 \\ \text{Rs}: z=a_p-s_w \end{cases} \tag{8}$$

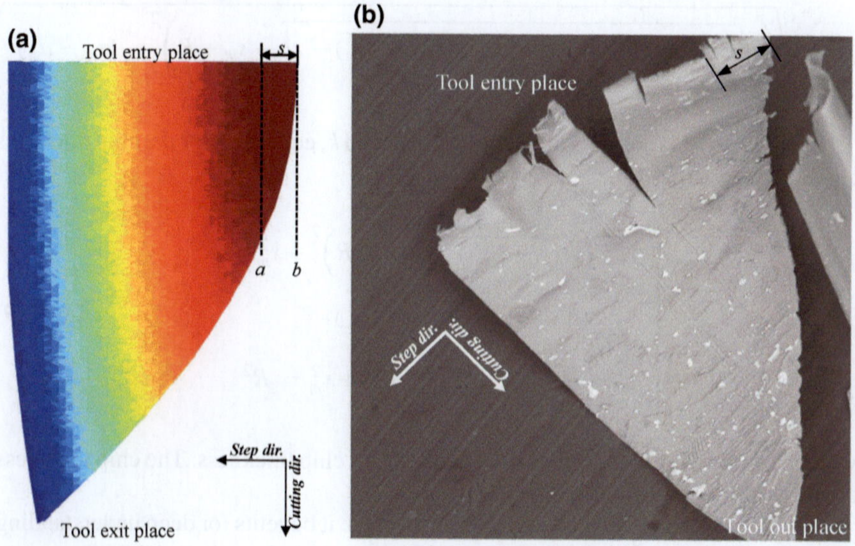

Fig. 6 Comparison of **a** simulated cutting chip figure. **b** SEM measured cutting chip figure

where f_e is the feed rate; R is the tool nose radius; s_t is the step distance; a_p is the depth of cut; and s_w is the swing distance.

Equations (8) are the geometry model of cutting chips. The 3D profile of cutting chip based on the given cutting parameters and tool geometry parameters was figured by MATLAB® in Fig. 6a, while the real captured chip figure is shown in Fig. 6b. From the comparison of two chip morphologies, it is found that they well concur each other in morphology. However, the chip was compressed by the tool rake face during cutting whereby cutting chip shrinks slightly along the cutting direction, whereby the real chip length along the cutting direction is a little shorter than the simulated one.

From Fig. 5c, the chip is thicker at its middle (point m) while thinner at its two sides along the feed direction. The geometric feature of a cutting chip makes it suitable for evaluation of tool flank wear. To clearly present the thickness changes of the cutting chips, a curve presenting the chip thickness is modeled within the plane mke in Fig. 2c, because cutting-edge segment here experiences the longest cutting distance.

From Eq. (8), the coordinates of point m and k are expressed as:

$$\left(x_m = \sqrt{a_p(2R - a_p)} - r,\right.$$

$$\left. y_m = \sqrt{\left(\sqrt{R^2 - a_p(2R - a_p) + 2r\sqrt{a_p(2R - a_p)} - r^2} - s_w + R\right)^2 - (s_w - a_p)^2 - f}\right)$$

$$\left(x_k = \sqrt{a_p(2R - a_p)} - r,\right.$$

$$y_k = \sqrt{\left(\left(\sqrt{R^2 - a_p(2R - a_p) + 2r\sqrt{a_p(2R - a_p)} - r^2} - s_w + R\right)^2 - (s_w - a_p)^2\right)}$$

Based on Eq. (1) and coordinates of point m and k, equations for the chip thickness curve are derived as:

$$\begin{cases} l_a : \left(\sqrt{(y+f)^2 + z^2} + s_w - R\right)^2 + x_m^2 = R^2 \\ l_b : z = a_p - s_w \quad y_m \le y \le y_k \\ l_c : \left(\sqrt{y^2 + z^2} + s_w - R\right)^2 + x_m^2 = R^2 \end{cases} \tag{9}$$

where l_a, l_b, and l_c are the three curve segments of chip thickness. The chip thickness at any point can be calculated from Eq. (9).

Derivation of the chip model is significant since it benefits for deep understanding of the cutting mechanism of UPRFC and chip generation. During the UPRFC process, all the cutting chip morphology changes are the true reflection of the tool status. However, only a part area of the chip surface can reflect the surface topography. By inspection of Fig. 6a, b, it is found that the chip area between line a and line b could be used to reflect surface quality under tool wear. Based on the geometric relation, the distance between line a and line b is the step distance.

3 Surface Microwaves in UPRFC

Material property has multiple effects on the surface generation and machined surface roughness, for example, the heat generated from UPRFC process changes the workpiece material property so that influence the machined surface quality [31], also, cutting process has effect on the changes of crystal orientation [32, 33]. Therefore, study of materials inducing surface roughness changes is significant.

The intermittent cutting of UPRFC makes the morphology of cutting chip thicker at its middle and thinner at its two sides. Owing to the ununiformed chip thickness along its length direction, the thickness of lamella structure generated in material sliding is also different, which has influence on the surface roughness. Based on the observed microwave formation on machined surfaces, this research conducted a research on the origin and suppressing of the formed microwaves, committed to provide a deep insight into the material sliding and cutting mechanism in UPRFC.

During the UPRFC process, the diamond cutting tool performs a linear feed motion and meantime a rotary cutting motion. The trajectory of diamond tool is clarified by combining the two tool motions, whereby the topography of machined surface is proposed, as is shown in Fig. 7a, b.

From prior cutting experiment results as shown in Fig. 8, the machined surface was found to be present a number of microwaves with very high space frequencies. The wavelength was in the scope between 2 and 4 μm, and the microwaves are accumulated together with almost the same distance in between; two neighboring groups' distance was about 42 μm [34].

The cutting parameters used in the experiment of microwave formation in Fig. 8 are: spindle speed of 4500 r/min and feed rate of 200 mm/min. Therefore, the tool feed length could be calculated based on the cutting parameters:

$$f_r = \frac{f_e}{s_p} = \frac{200\ \text{mm/min}}{4500\ \text{r/min}} = 0.044\ \text{mm/r} = 44\ \mu m/r$$

where f_r is the feed length within a rotary cutting, f_e is the feed rate in mm/min, and s_p is the spindle speed.

Because the distance of two neighboring microwave groups nearly equals to the tool feed length of a rotary cutting, the microwaves on the machined surface are believed to be caused by the rotary cutting mechanism of UPRFC.

Based on the given cutting parameters, the microwave frequency could be calculated. First, the linear velocity of the tool tip along the tangent direction of rotary cutting is calculated as:

$$v = \omega s_w = \frac{2\pi s_w s_p}{60} = \frac{2 \times 3.14 \times 0.02835\ \text{mm} \times 4500\ \text{rpm}}{60} = 13.35285\ \text{m/s}$$

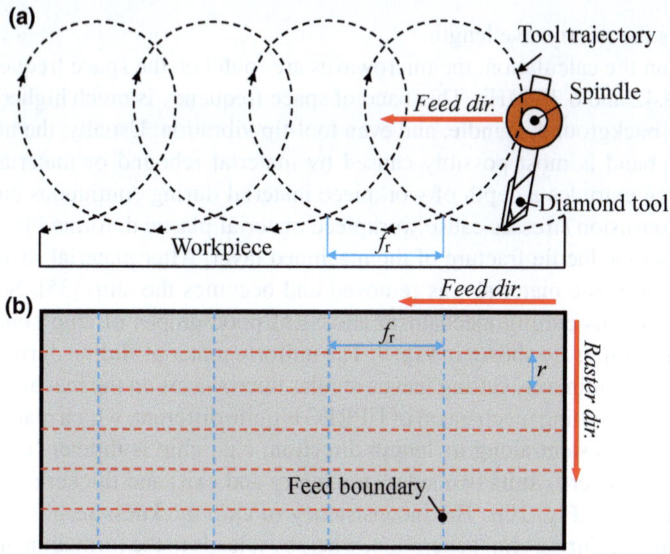

Fig. 7 Schematic of cutting kinematics of UPRFC (**a**) and the surface generation (**b**)

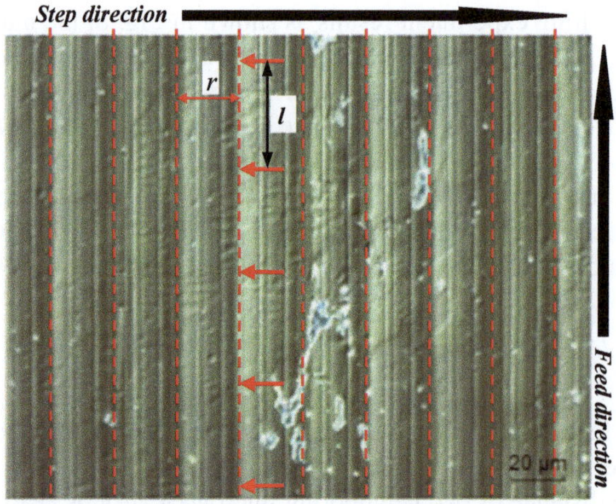

Fig. 8 Microwaves appear on the machined surface

where ω is the spindle angular velocity, and s_w is the swing distance.

The microwave space frequency is obtained as:

$$\begin{cases} f_{\min} = v/w = \frac{13.35285}{3.9\times10^{-6}} = 3.42\ \text{MHz} \\ f_{\max} = v/w = \frac{13.35285}{2.1\times10^{-6}} = 6.36\ \text{MHz} \end{cases}$$

where w is the microwave length.

Based on the calculation, the microwaves are found on the space frequency band between 3.42 and 6.36 MHz. This band of space frequency is much higher than that caused by background, spindle, and even tool-tip vibration. Usually, the high-space frequency band is most possibly caused by material rebound or material sliding. Cutting tool extrudes a depth of workpiece material during continuous cutting and whereby extrusion stresses cause workpiece material plastic deformed by shearing, and results in a ductile fracture of the machined layer. After material shearing, this layer of workpiece material was removed and becomes the chip [35]. Schematics of the continuous cutting mechanism and SEM photographs of chip generation in continuous cutting are shown in Fig. 9. The uniform material sliding during material shearing of a continuous cutting leaves regular microwaves on the machined surface [15]. However, cutting mechanism of UPRFC is quite different, which makes the chip thickness inconsistent along its length direction; e.g., chip is thinner (theoretically the thickness is zero) at its two sides (tool entry and exit) and thicker at its middle, as is presented in Fig. 10b. The inconsistency of chip thickness results in irregular material sliding during chip generation, whereby it leads to the formation of different pitch sawtooth chip.

Figure 11 shows the chip generation and morphology of UPRFC, in Fig. 11a, the chip beginning is denoted as surface-entry point (point b), while the chip ending is

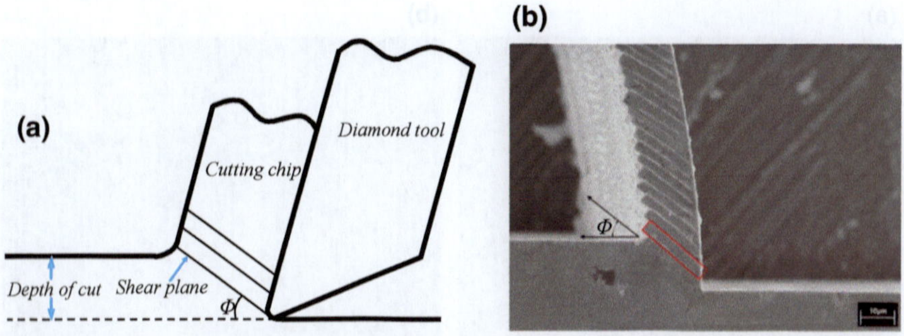

Fig. 9 Material sliding during straight cutting

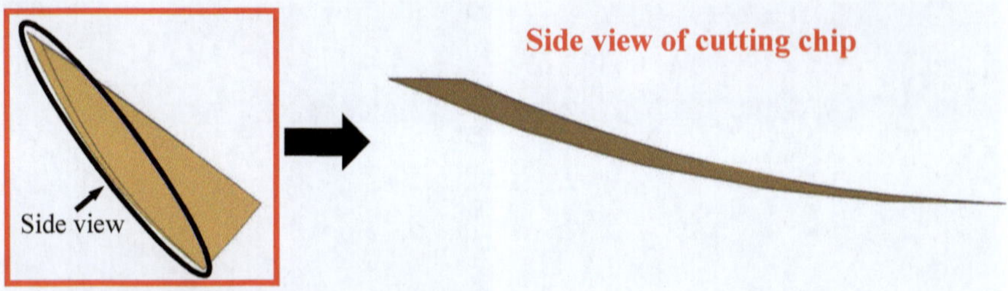

Fig. 10 Illustration of chip morphology of UPRFC

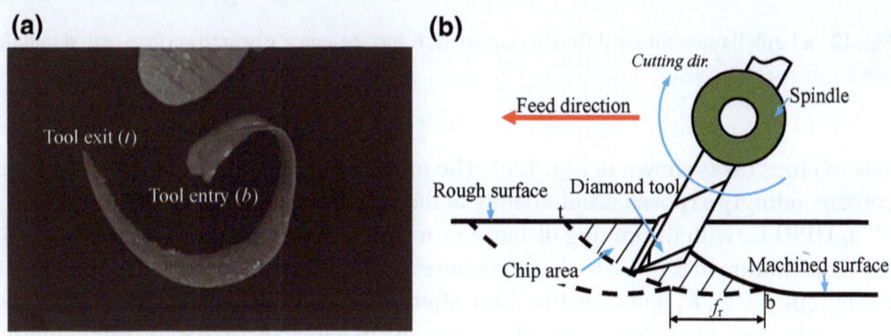

Fig. 11 Chip generation during UPRFC

defined as surface-exit point (point t). Theoretically, the distance between surface-entry point and surface-exit point equals to the feed length along the cutting direction, and it needs to note that the real chip length is shorter than the theoretical chip length (distance between point b and point t in Fig. 11b).

Figure 12 shows the lamella structures of a cutting chip (see Fig. 12a) at the different chip surface areas. Owing to the thin thickness at the surface-entry of chips and thick thickness at the surface-exit of chips, the pitch of lamella structure is smaller at surface-entry side of chips (see Fig. 12b) as compared to that at the surface-exit

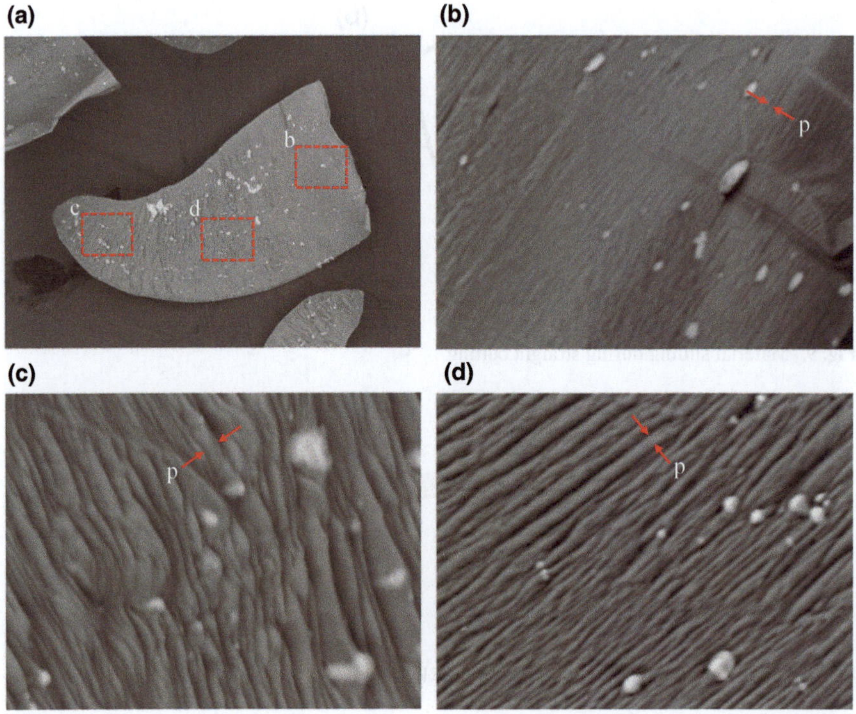

Fig. 12 **a** Lamella structure at different chip areas: **b** surface-entry, **c** near to surface-exit, **d** middle place

side of chips (as is shown in Fig. 12c). The microwaves on the machined surface are correspondingly exposed accumulating at the surface-exit of each feed imprint.

In UPRFC, with the cutting distance growing, tool wear features change as well. As is illustration in Fig. 13, tool wear feature at initial stage tends to be the fracture of cutting edge (e.g., F_1 and F_2 in Fig. 13a), after cutting a certain distance, a wear land is found to be formed on the cutting edge, the width of which increases correspondingly. Meanwhile, two fractures (F_1 and F_2) in Fig. 13a are flattened significantly and hard to be distinguished, as is shown in Fig. 13b.

The formation of wear land influences the microwaves' distribution. With the tool wear progress, the forms of cutting chip surface-entry change as well. In particular, as a smooth wear land is formed on the cutting edge, cutting chips truncate at both surface-entry and surface-exit sides; therefore, the flank wear land formation affects the chip generation in UPRFC. The relation between chip morphology and tool flank wear could be reflected by the microwave distribution on the machined surface. As is shown in Fig. 14, it is found that the tool flank wear changes the microwave distribution on the machined surface. With the tool flank wear progress, microwaves accumulate more at the surface-exit side of a feed length, and the wavelength becomes larger too. The phenomena are believed to be caused by the comparable thickness

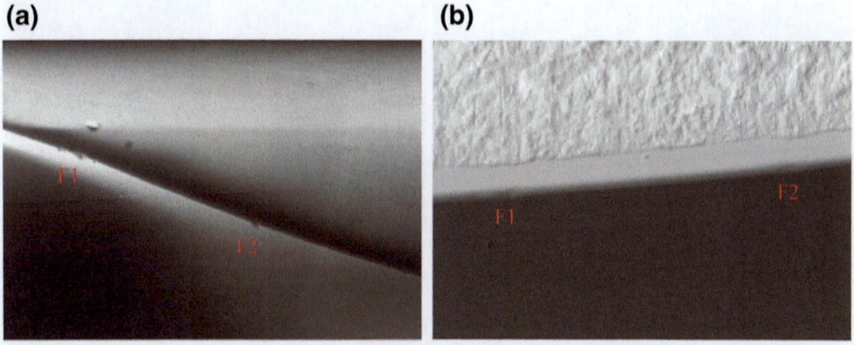

Fig. 13 **a** Initial tool wear and **b** flank wear land formation

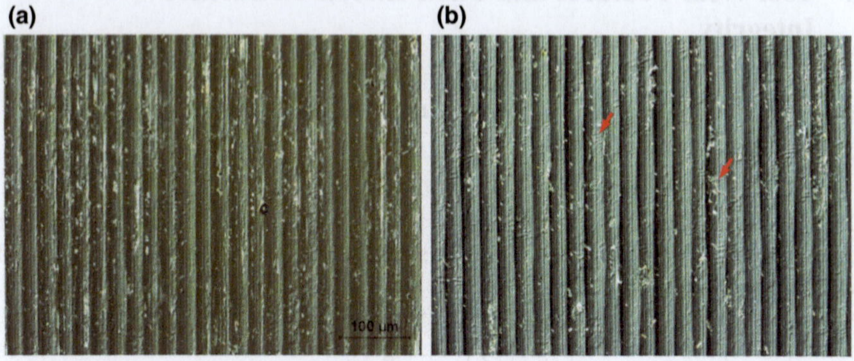

Fig. 14 Machined surface topography at different tool wear level

between cutting edge and chip. At the initial cutting, cutting tool with a sharp edge could cut down very thin chips; however, the progress of tool flank wear increases the cutting-edge radius, whereby it results in a plowing process instead of cutting, which causes the material accumulation at the surface-exit side of a feed imprint and forms microwaves on the machined surface. Accompanying the tool wear progress, the mean length of microwaves increases correspondingly; this is attributed to the material accumulation during material sliding. Moreover, with the tool wear progress, the microwave group length becomes shorter, indicating the microwaves are more concentrated, as is shown in Fig. 14.

From the discussion above, there are two methods to suppress the microwave appearance on the machined surface: first, reducing the chip thickness through changing cutting parameters; second, suppressing tool wear.

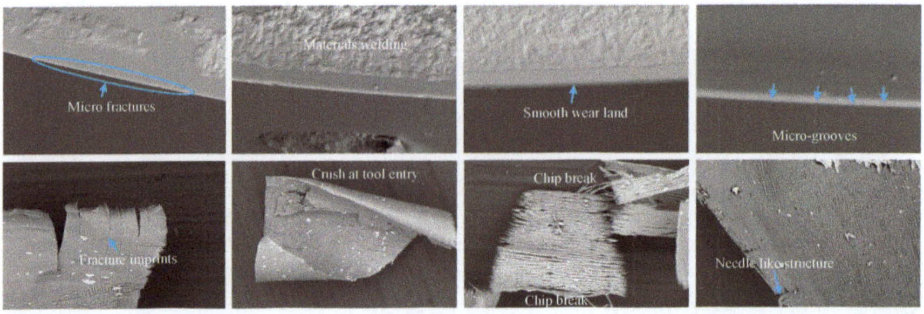

Fig. 15 Tool wear features and related chip morphologies

4 Tool Wear Features and Their Effects on Surface Integrity

4.1 Tool Wear Characteristics and Their Effect on Cutting Chips

The cutting mechanism of UPRFC makes the tool wear features different from other continuous cutting processes; the previous study about tool wear indicates that tool wear characteristics include tool fracture, material welding, wear land formation, micro-groove; tool wear features have different influences on the chip morphology, cutting force, and surface topography [36]. Figure 15 indicates the tool wear features at different cutting stages; at the initial cutting, some fractures appear on the cutting edge, and the distribution of the formed fractures is irregular and chaos. After cutting a certain distance, some workpiece material was welded on the tool rake face; meanwhile, small fractures had flattened and disappeared except some larger fractures. By further cutting, a smooth wear land is formed with its width about 1 μm, and the smooth wear land incorporates small fractures and flattens larger fractures. Moreover, several micro-grooves were found on the tool flank face with its meandering trajectory parallel to the cutting direction, which originated in several micro-fractures on the cutting edge.

In metal cutting, cutting chips have a closed relation with tool wear. The cutting experiment indicates that new cutting tool leads to the formation of feather-like structure at the tool-entry side of cutting chips. Tool fractures will be imprinted on the chip surface and will form some 'ridges'-like structure with its length along the cutting direction. Workpiece material welding on to the cutting tool rake face will increase the friction between cutting chip and tool rake face whereby it makes cutting chips crushed at their tool-entry side. Flank wear land on the cutting edge causes shutter-like structure formation at cutting chips' tool-entry area under the effect of minimum cutting depth. Micro-grooves formed on the cutting tool clearance face lead to the formation of some pin-like structures at the tool-entry area of cutting chips.

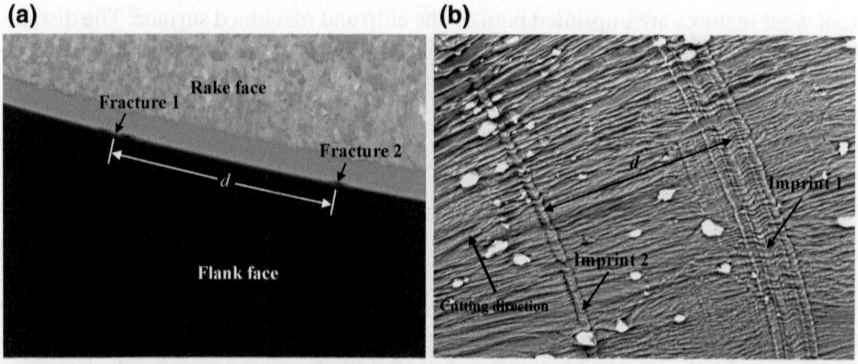

Fig. 16 Tool micro-crack and the related imprints on the chip surface

Fig. 17 Smooth wear land and the corresponding truncated chip

In UPRFC, two chip morphologies caused by tool wear features should be noted: tool fracture and smooth wear land on the cutting edge. Tool fracture on the cutting edge causes the ridge formation on chip surface, as is presented in Fig. 16, while smooth wear land on the cutting edge makes cutting chips truncated at their both sides, as is presented in Fig. 17.

4.2 Tool Wear Evaluation Using Cutting Chips

The appearance of tool fractures causes the imprints of ridge-like structures on the chip surface as a group of signatures. Tool-entry area of cutting chips is thin and smooth due to the small material stacking, the ridge-like signatures in this area are therefore clear and easy to be precisely measured.

For the evaluation of tool fracture using cutting chips, only a part of the cutting chip is usable (chip part area between line *a* and *b* in Fig. 6), because in this section tool wear features are imprinted both on the chip and machined surface. The distance between line a and line b equals to the cutting-edge length within a step distance, which could be obtained by:

$$\widehat{a} = 2R\arcsin(r/2R) \tag{10}$$

In order to identify the feature of tool fractures by using cutting chip morphology, the signatures' distance in between and its cross-sectional profile parameters are measured by a SEM with 3D viewer. The measured distances and parameterized cross-sectional profile of these signatures are then used to rebuild the cutting-edge 2D profile and further used to simulate the machined surface topography. The tool fracture evaluation process is listed as follows: firstly, parameterizing the measured cross-sectional profile of a signature into a geometric element. The parameterization is performed according to the similarity between the signature cross-sectional profile and the geometric profile of the basic element, and the geometric element used here includes isosceles triangle, isosceles trapezoid, and semicircle. Equations of these geometric elements are listed in Table 2 [37].

Secondly, the measured distances of signatures in between and the parameterized geometric elements are combined to rebuild the 2D profile of cutting edge. For easy description, two coordinate systems o-uw and o_i-u_iw_i are established to present the equations of geometric element. Usually, the diamond tool edge is an arc, and its equation is:

$$u^2 + w^2 = R^2 \tag{11}$$

The angle between two neighboring signatures can be calculated by the signature distance and tool nose radius through:

$$\theta_i = \mathrm{d}_i/R \tag{12}$$

The angle α in Fig. 18 could be obtained by:

$$\alpha = \arcsin(r/2R) \tag{13}$$

Therefore, each geometric element is then constituted to rebuild the 2D cutting edge by coordinate transformation:

$$\begin{bmatrix} u_i \\ w_i \\ 1 \end{bmatrix} = \begin{bmatrix} \cos\beta & -\sin\beta & R\sin\beta \\ \sin\beta & \cos\beta & R\cos\beta \\ 0 & 0 & 1 \end{bmatrix} \begin{bmatrix} u \\ w \\ 1 \end{bmatrix} \tag{14}$$

where $\beta = \alpha - \sum_1^i \theta_i$ is an angle.

Thirdly, after rebuilding the 2D profile cutting edge with tool fracture consideration, the tool wear patterns and their locations could be identified. Because the tool

Table 2 Equations of these geometric elements

Semicircle	Isosceles triangle	Isosceles trapezoid
$u_i^2 + w_i^2 = h_i^2 \quad (u_i > 0)$	$\begin{cases} \dfrac{2u_i}{t_i} + \dfrac{w_i}{h_i} = 1 & (u_i, w_i > 0) \\ -\dfrac{2u_i}{t_i} + \dfrac{w_i}{h_i} = 1 & (u_i < 0, w_i > 0) \end{cases}$	$\begin{cases} w_i = \dfrac{2h_i}{p_i - q_i}\left(u_i + \dfrac{p_i}{2}\right) & 0 \le w_i \le h_i \\ w_i = h_i & -\dfrac{q_i}{2} \le u_i \le \dfrac{q_i}{2} \\ w_i = -\dfrac{2h_i}{p_i - q_i}\left(u_i - \dfrac{p_i}{2}\right) & 0 \le w_i \le h_i \end{cases}$

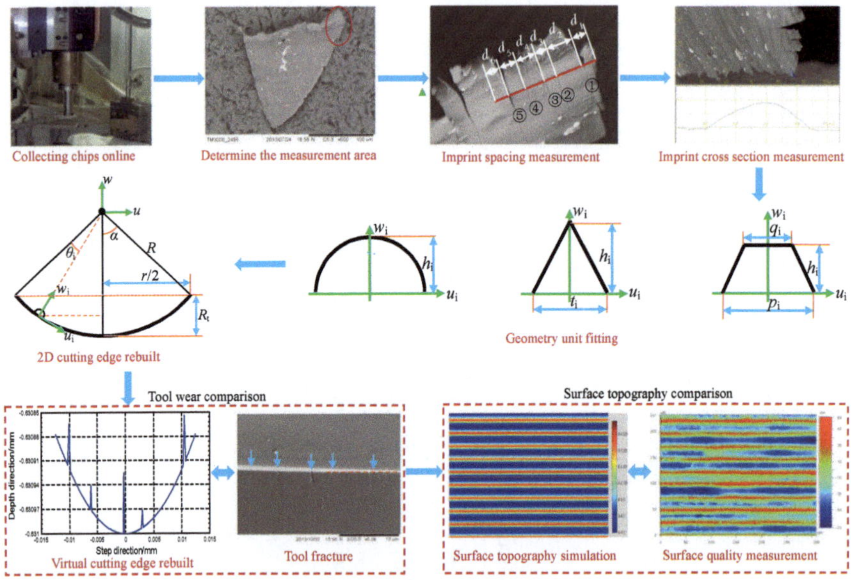

Fig. 18 Tool micro-fractures identification using cutting chips

wear patterns are also imprinted on the machined surface meantime, the 2D profile cutting edge could be used to simulate the topography of machined surface too. The simulated surface topography was then compared with the measured ones to prove the correctness of the simulation.

After obtaining the simulated topography of machined surface, the roughness of the machined surface cut by the fractured tools, e.g., R_a, R_z, and R_q, can be calculated from the 3D data cloud of the rebuilt surface. The calculation results are then used to evaluate the machined surface quality by considering the tool fractures' effect [38].

Similarly, tool flank wear level can also be identified by cutting chip morphology. From the cutting experiment, tool-entry area of a cutting chip was affected by tool flank wear and forms louver-like structures, which differs from a feather-like structure cut by a new cutting tool with sharp edge. According to the principle of minimum cutting depth, the formation of louver-like structure is caused by the comparable size of the width of tool flank wear and cutting chips thickness in this area.

It is found from prior cutting experiment that tool flank wear creates the truncation of cutting chips at both their tool-entry and exit areas. The truncation position of cutting chips depends on the cutting parameters, e.g., cutting depth, feed rate, and swing distance. Under the identical tool wear level, alternation of these parameters changes the chip morphology as well. This is because alteration of these parameters changes the position of comparable chip thickness to the width of flank wear land. For a worn tool, when the radius of its cutting edge is comparable to the chip thickness of a chip area, the chip will be truncated around this chip thickness-related position. Therefore, the thickness of cutting chip at its truncation position could be used to

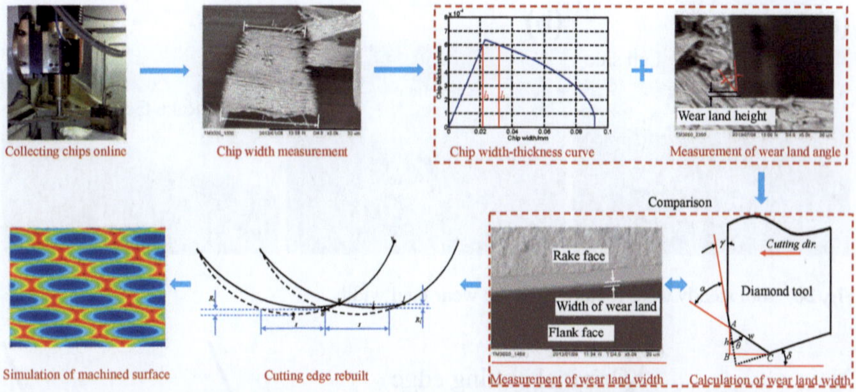

Fig. 19 Identification procedures for tool flank wear and machined surface topography

identify the width of flank wear land. In UPRFC, according to the minimum cutting depth principle, the chip thickness being cut must be bigger than the cutting-edge radius of a cutting tool to properly generate a chip. Therefore, to enhance the tool wear identification accuracy and improve the sensitivity, cutting parameters, e.g., cutting depth and feed rate, should be optimized.

According to the identified chip thickness at the chip truncation position, and the relation curve between chip thickness and wear land width, tool flank wear could be identified. First, measuring chip width w_1 and w_2 at chip truncation place. Secondly, calculating the chip thickness value and their mean value based on the measured chip width w_1 and w_2. Thirdly, according to the wear land angle and the mean value of chip thickness, calculating the width of flank wear land by:

$$w = c_t / \cos(\delta + \gamma) \tag{15}$$

where w is the width of flank wear land; c_t is the mean value of the thickness of cutting chip at its truncation position; δ is the wear land angle, 40.5° for copper material here; and $\gamma = -2.5°$ is the tool rake angle. The procedures for the identification of tool flank wear are shown in Fig. 19.

Moreover, the length of truncated chips could be applied to identify the width of flank wear land too. Because the chip length is seriously extruded by cutting tools during the cutting, tool flank wear identification using the length of truncated chips is not accurate. Therefore, the identified results using truncated chip length should be corrected carefully to eliminate the tool extrusion effect.

Tool flank wear will form a 'crescent-like' wear land, which changes the tool geometry and retracts the cutting edge, as is shown in Fig. 20. The tool flank wear occurrence directly influences the form accuracy and roughness of machined surface.

Tool flank wear results in the receding of cutting edge and therefore forms a worn cutting edge with different tool geometry parameters. Figure 21 shows a schematic illustration of the cutting edge receding duet to tool flank wear and its effects on the

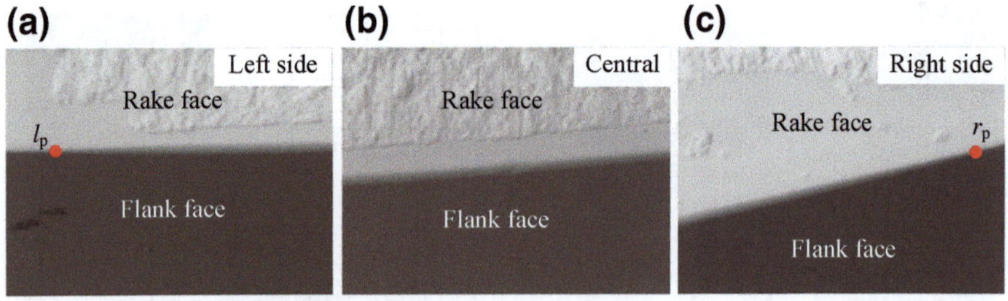

Fig. 20 Tool steady wear features and the wear land width

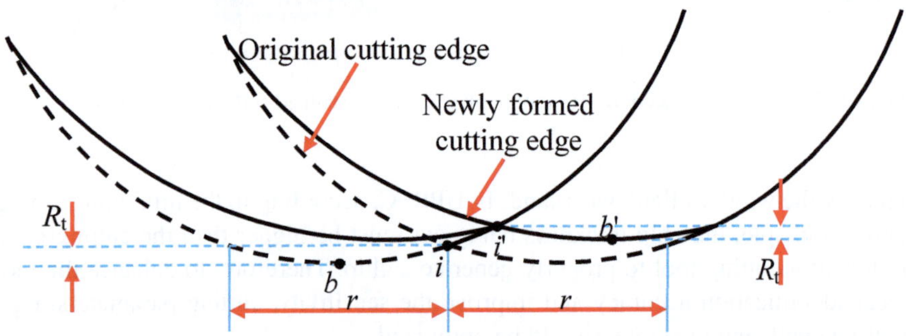

Fig. 21 Schematic of tool wear effect on surface roughness

surface profile. Because the wear land formed on the cutting edge is quite smooth (see Fig. 20), the profile of newly formed worn cutting edge is simplified to an arc with a larger radius. According to the geometric relation, relation between the nose radius of original tool and worn tool is derived as [39]:

$$F(R') = \arcsin\left(\frac{R\sin\theta}{R'}\right)R' - \theta R = 0 \tag{16}$$

where R is the nose radius of the original tool, R' is the nose radius of worn tool, and θ is the half angle of the wear land.

From Eq. (16), the nose radius of worn tool can be calculated and obtained. Therefore, the tool-flank-wear-effect-considered surface roughness, e.g., the root-mean-square roughness (R_q) and the maximum peak-to-valley height (R_t), could be calculated by:

$$R_q = \sqrt{\frac{f^4}{720s_w^2} + \frac{r^4}{720R'^2}} \tag{17}$$

$$R_t = s_w + R' - \frac{1}{2}\left(\sqrt{4s_w^2 - f^2} + \sqrt{4R'^2 - r^2}\right) \tag{18}$$

Based on Eqs. (17) and (18), the roughness of machined surface under tool wear could be calculated. However, some factors exist affecting the identification accuracy of tool flank wear and machined surface quality: First, material rebound during the cutting influences the measurement accuracy. Secondly, difference between the chip thickness at its truncation and the width of flank wear land limits the identification accuracy. The influence of these factors could be reduced by using correct coefficient.

5 Conclusion

This chapter mainly talks about the cutting mechanism and surface generation of ultra-precision raster fly cutting (UPRFC) process. The influence of material-sliding-caused vibration, size-effect-caused surface rough pattern, tool wear effects on the machined surface quality was studied. The present studies found that:

Microwaves were found accumulated in groups on the machined surface with regular distance and high frequency. The microwaves was caused by material sliding in chip formation, owing to the inconsistency of chip thickness, the microwaves tends to accumulate at the surface-exit of each tool feed imprint. The length of microwaves increases with tool wear progresses, while the group length of microwaves decreases to become more concentrated. Reduction in chip thickness and tool wear level could suppress the appearance of microwaves.

Tool wear features and their influences on the quality of machined surface in UPRFC are evaluated by using cutting chips. Tool fractures are imprinted on both the chip surface and machined surface as a group of ridge-like signatures, both the tool fractures and surface topography under that could be rebuilt and evaluated. Tool flank wear causes cutting chips truncated at the comparable chip thickness to the width of flank wear land. Through the identification of the thickness of cutting chips at their truncation position, the width of flank wear land and the surface roughness changes due to retraction of the cutting edge could be identified, respectively. The chip-morphology-based tool wear evaluation approach can effectively evaluate tool wear features and even their effects on machined surface quality.

Acknowledgements The work described in this chapter was supported by the grants from the National Natural Science Foundation of China (Grant No. 51505297), the Natural Science Foundation of Guangdong Province (Grant No. 2017A030313295), the Shenzhen Science and Technology Program (Grant No. JCYJ20160422170026058), and the Shenzhen Peacock Technology Innovation Project (Grant No. KQJSCX20170727101318462).

References

1. Zhang G, To S, Xiao G (2014) A novel spindle inclination error identification and compensation method in ultra-precision raster milling. Int J Mach Tools Manuf 78:8–17
2. Zhang G, To S (2017) Relation between tool wear and workpiece modal vibration in ultra-precision raster fly cutting. Int J Adv Manuf Technol 93(9–12):3505–3515

3. Kong LB, Cheung CF, To S et al (2009) An investigation into surface generation in ultra-precision raster milling. J Mater Process Technol 209(8):4178–4185
4. Kong LB, Cheung CF (2012) Prediction of surface generation in ultra-precision raster milling of optical freeform surfaces using an integrated kinematics error model. Adv Eng Softw 45(1):124–136
5. Cheung CF, Lee WB, To S (2004) A framework of a model-based simulation system for predicting surface generation in ultra-precision raster milling of freeform surfaces. In: Annual of American society for precision engineering proceedings, Orlando, Florida, pp 1149–1156
6. Cheung CF, Kong LB, Lee WB, To S (2006) Modelling and simulation of freeform surface generation in ultra-precision raster milling. Proc Inst Mech Eng Part B J Eng Manuf 220(11):1787–1801
7. To S, Zhang G (2014) Study of cutting force in ultra-precision raster milling of V-groove. Int J Adv Manuf Technol 75(5–8):967–978
8. Zhang SJ, To S, Zhang GQ, Zhu ZW (2015) A review of machine-tool vibration and its influence upon surface generation in ultra-precision machining. Int J Mach Tools Manuf 91:34–42
9. Wang H, To S, Chan CY, Cheung CF, Lee WB (2010) A study of regularly spaced shear bands and morphology of serrated chip formation in micro-cutting process. Scripta Mater 63:227–230
10. Zhang SJ, To S (2013) A theoretical and experimental investigation into multimode tool vibration with surface generation in ultra-precision diamond turning. Int J Mach Tools Manuf 72:32–36
11. Zhang SJ, To S, Wang HT (2013) A theoretical and experimental investigation into five-DOF dynamic characteristics of an aerostatic bearing spindle in ultra-precision diamond turning. Int J Mach Tools Manuf 71:1–10
12. Zhang SJ, To S (2013) A theoretical and experimental study of surface generation under spindle vibration in ultra-precision raster milling. Int J Mach Tools Manuf 75:36–45
13. Cheung CF, To S, Lee WB (2002) Anisotropy of surface roughness in diamond turning of brittle single crystals. Mater Manuf Process 17(2):251–267
14. Lee WB, Cheung CF (2001) A dynamic surface topography model for the prediction of nano-surface generation in ultra-precision machining. Int J Mech Sci 43:961–991
15. Su G, Liu Z, Li L, Wang B (2015) Influences of chip serration on micro-topography of machined surface in high-speed cutting. Int J Mach Tools Manuf 89:202–207
16. Özel T, Karpat Y (2005) Predictive modeling of surface roughness and tool wear in hard turning using regression and neural networks. Int J Mach Tools Manuf 45(4):467–479
17. Oraby SE, Hayhurst DR (1991) Development of models for tool wear force relationships in metal cutting. Int J Mach Sci 33(2):125–138
18. Sortino M (2003) Application of statistical filtering for optical detection of tool wear. Int J Mach Tools Manuf 43(5):493–497
19. Wang WH, Wong YS, Hong GS (2006) 3D measurement of crater wear by phase shifting method. Wear 261(2):164–171
20. Emel E, Kannatey-Asibu E (1989) Acoustic emission and force sensor fusion for monitoring the cutting process. Int J Mach Sci 31(11–12):795–809
21. Salgado DR, Alonso FJ (2007) An approach based on current and sound signals for in-process tool wear monitoring. Int J Mach Tools Manuf 47(14):2140–2152
22. Alonso FJ, Salgado DR (2008) Analysis of the structure of vibration signals for tool wear detection. Mech Syst Signal Process 22(3):735–748
23. Cheung CF, Lee WB (2001) Characterization of nanosurface generation in single-point diamond turning. Int J Mach Tools Manuf 41(6):851–875
24. Quinsat Y, Sabourin L, Lartigue C (2008) Surface topography in ball end milling process description of a 3D surface roughness parameter. J Mater Process Technol 195(1–3):135–143
25. Singh SK, Srinivasan K, Chakraborty D (2004) Acoustic characterization and prediction of surface roughness. J Mater Process Technol 152(2):127–130
26. Grzesik W, Brol S (2009) Wavelet and fractal approach to surface roughness characterization after finish turning of different workpiece materials. J Mater Process Technol 209(5):2522–2531

27. Cheng MN, Cheung CF, Lee WB, To S, Kong LB (2008) Theoretical and experimental analysis of nano-surface generation in ultra-precision raster milling. Int J Mach Tools Manuf 48:1090–1102
28. Wang SJ, To S, Cheung CF (2013) An investigation into material-induced surface roughness in ultra-precision milling. Int J Adv Manuf Technol 68:607–616
29. An CH, Zhang Y, Xu Q et al (2010) Modeling of dynamic characteristic of the aerostatic bearing spindle in an ultra-precision fly cutting machine. Int J Mach Tools Manuf 50(4):374–385
30. Zhang G, To S, Xiao G (2014) The relation between chip morphology and tool wear in ultra-precision raster milling. Int J Mach Tools Manuf 80:11–17
31. Wang S, To S, Chan CY et al (2010) A study of the cutting-induced heating effect on the machined surface in ultra-precision raster milling of 6061 Al alloy. Int J Adv Manuf Technol 51(1–4):69–78
32. Zhu YH, To S, Lee WB et al (2010) Ultra-precision raster milling-induced phase decomposition and plastic deformation at the surface of a Zn–Al-based alloy. Scripta Mater 62(2):101–104
33. Zhang SJ, To S (2016) Deformation-induced phase changes of Zn–Al alloy during ultra-precision raster milling. Int J Adv Manuf Technol 88(5–8):1–7
34. Zhang G, To S, Zhang S, Zhu Z (2016) Case study of surface micro-waves in ultra-precision raster fly cutting. Precis Eng 46:393–398
35. Astakhov VP (2006) Tribology of metal cutting, vol 52. Elsevier, Amsterdam
36. Zhang G, To S, Zhang S (2016) Relationships of tool wear characteristics to cutting mechanics, chip formation, and surface quality in ultra-precision fly cutting. Int J Adv Manuf Technol 83(1–4):133–144
37. Zhang G, To S, Xiao G (2014) Novel tool wear monitoring method in ultra-precision raster milling using cutting chips. Precis Eng 38(3):555–560
38. Zhang G, To S (2015) A novel surface quality evaluation method in ultra-precision raster milling using cutting chips. J Mater Process Technol 219:328–338
39. Zhang G, To S, Zhang S (2016) Evaluation for tool flank wear and its influences on surface roughness in ultra-precision raster fly cutting. Int J Mech Sci 118:125–134

Modeling and Experimental Study of Surfaces Optoelectronic Elements from Crystal Materials in Polishing

Yu. D. Filatov

Abstract Based on investigations of the mechanism of precision surface formation in workpieces of anisotropic monocrystalline materials for optoelectronics, a generalized model of material removal in polishing with suspensions of polishing powders has been constructed. The removal rate in polishing sapphire planes of different crystallographic orientations has been found to grow in the series $m < c < a < r$ with increasing volume, surface area, and most probable size of debris particles as well as with energy of dispersion of material from the face being polished. A study of the mechanism of formation of monocrystal planes of different crystallographic orientations has revealed that in polishing of sapphire the surface roughness parameters *Ra*, *Rq*, *R*max decrease in the series $c > r > m > a$ with decreasing dielectric permittivity and thermal conductivity coefficient of the workpiece material, debris particle height, and Lifshitz constant. As a result, studies of regularities mechanical polishing optoelectronic components of crystalline materials found that polishing efficiency decreases with an increase of the binding energy and the transfer energy. It is shown that the polishing efficiency increases with increasing heat conduction coefficient of the material being processed, a processed surface road friction element by lapping and Lifshitz force. It is found that the ratio of the volume wear coefficient to the temperature conductivity coefficient of the material being processed depends on the specific heat and the transfer energy. The relative roughness of the treated surfaces of silicon carbide crystals, gallium nitride, aluminum nitride, and sapphire is characterized by the ratio: 0.68:0.67:0.63:1.00.

1 Introduction

Finishing of monocrystalline materials is performed through the operations of fine grinding using diamond and conventional abrasive powders, either loose or bound, and ultrafine grinding with suspensions of diamond micron powders and powders of

---

Yu. D. Filatov (✉)
V. Bakul Institute for Superhard Materials, National Academy of Science of Ukraine, Kiev, Ukraine
e-mail: filatov@ism.kiev.ua

J. Zhang et al. (eds.), *Simulation and Experiments of Material-Oriented Ultra-Precision Machining*, Springer Tracts in Mechanical Engineering,
https://doi.org/10.1007/978-981-13-3335-4_6

metal and rare-earth element oxides [1, 2]. Polishing is also carried out through several operations—coarse polishing, prepolishing, polishing, and nanopolishing—using suspensions of diamond powders and polishing powders [2, 3]. In diamond–abrasive finishing of precision surfaces of components and substrates of monocrystalline sapphire, silicon carbide, semiconductor nitrides, optical ceramics, and other crystalline materials the high form accuracy and very low surface roughness should be achieved in a time-stable manner and with a high removal rate [4–6]. The most important performance parameter of polishing precision surfaces of optoelectronic components made of monocrystalline materials is the machining quality which is represented by surface roughness, reflecting and scattering properties of the machined surface as well as by the formation of a deposit of debris particles and tool material wear particles [7–12]. An increase in removal rate in polishing the above-mentioned precision surfaces with mandatory satisfaction of requirements for machined surface quality can be achieved only by elaborating and implementing new polishing suspensions, and optimization of composition and characteristics of such suspensions is a task of vital importance [13–15]. Monocrystalline materials including sapphire, silicon carbide, aluminum, and gallium nitrides are finding ever-widening applications in the manufacture of optoelectronic components (light-emitting diodes, Schottky diodes, solid-state lasers, microwave transistors, energy-saving devices). Finishing of monocrystalline materials is carried out using bound diamond and abrasive powders as well as polishing powder suspensions [4–6]. The polishing operation is performed in several steps (coarse polishing, prepolishing, polishing, nanopolishing, chemical-mechanical polishing) using polishing tools [16–20], suspensions of diamond and abrasive powders [1, 3], and colloid systems of nano-sized particles of polishing powders [13–15].

Polished surfaces of optical components must meet the requirements for geometrical accuracy: 3–5 interference rings; for surface roughness: $Rz = 0.05$ μm (as per GOST 2789–73); for optical purity: PIII–PV (according to GOST 11141–84); or for machined surface quality: 60–40 to 20–10 (as per U.S. Military Surface Quality Specification, MIL-PRF-13830B). The epi-ready substrate surfaces must satisfy the requirements for the crystal plane orientation accuracy (±0.2 to 0.5°), dislocation density ($(1–4) \times 10^3$ cm^{-2}), micropipe density (less than 30 cm^{-2}), structural perfection (FWHM (full width at half maximum) = 30 to 50 arcsecond), as well as "optical" requirements for out-of-flatness (less than 10 μm on a diameter of 76.2 mm), and roughness ($Ra = 0.3$ to 1.0 nm). The newly developed tools with bound polishing powders provide a fairly high removal rate but, unfortunately, are in principle incapable of meeting the surface quality requirements for components and substrates of monocrystalline materials for optoelectronics, in particularly because they might cause defects (scratches, deposits) [11, 12].

In the manufacture of precision elements and substrates of sapphire, silicon carbide, gallium nitride, aluminum nitride, and other crystalline materials for optoelectronic applications, the most difficult final machining operation is polishing [15, 21–23]. The currently available technologies of mechanical polishing are based on the use of bound-abrasive tools or polishing suspensions and polishing pads with abrasive grains localized on their surfaces. Tools with a bound polishing powder, such as ceria or diamond micron powder, in a resin or metal bond are used for mechanical polishing (MP) of optical glass, quartz, and monocrystalline silicon carbide [16, 20, 24]. However, the bound-abrasive tools for MP of elements and substrates for optoelectronics fail to find wide application for they are not able to ensure stable quality of machined surfaces. The MP operation during the manufacture of components for optical and electronic systems most often relies on the use of polishing suspensions of abrasive powders. Suspensions with diamond micron powders are employed for polishing sapphire, silicon carbide, and gallium nitride [3, 25, 26]. Mechanical polishing of monocrystalline silicon carbide is carried out with suspensions of ceria powders with a grain size less than 1 μm and cubic boron nitride and Ti_3AlC_2 MAX-phase powders [3, 27–29]. MP removes the major machining stock and produces the surface with roughness values from a few to dozens of nanometers.

Precision machining of optoelectronic components made of optical and semiconductor crystals (sapphire, silicon carbide, gallium nitride, aluminum nitride, etc.) is usually performed in several mechanical polishing (MP) and chemical-mechanical polishing (CMP) operations depending on specific requirements to surface roughness [1, 3, 14, 21, 22, 25, 29, 30]. The mechanical polishing of optoelectronic components can be carried out using bound or loose polishing powders. Tools of ceria polishing powers or diamond micron powders in resin or metal bonds are used for the mechanical polishing of optical glass, quartz, and single-crystal silicon carbide [20, 24, 31]. Such bound-abrasive tools fail to provide the required machined surface roughness with high repeatability during MP of components and substrates for optical and electronic systems. The MP operations on optical and electronic components usually involve the use of polishing suspensions and soft pads, where grains of polishing powders are localized. Sapphire, silicon carbide, and gallium nitride are polished with suspensions based on diamond micron powders [3, 14, 25, 26, 30, 32]. The mechanical polishing of silicon carbide is performed using suspensions of ceria powders with a grain size below 1 μm as well as cubic boron nitride and MAX-phase Ti_3AlC_2 powders [3, 27–29]. The MP provides the stock removal and workpiece surface formation with a rms deviation of profile ranging from a few to dozens of nanometers. The final machining of sapphire, silicon carbide, aluminum nitride components and substrates, composite AlN/substrates, etc., for optical and microelectronic applications is accomplished through CMP, where the removal rate is dozens, or even hundreds of times, lower than that in MP, while the *rms* deviation of profile $Rq \leq 1$ nm, and the CMP operations are mostly carried out using silica-based colloid systems [13, 14, 22, 33–35].

2 Material Removal Rate and Surface Roughness in Polishing Anisotropic Monocrystalline Materials for Optoelectronics

2.1 Removal Rate in Polishing Monocrystalline Materials

Nowadays, the removal rate in polishing precision surfaces of optoelectronic components using polishing suspensions is determined through the use of a generalized model of workpiece material removal in polishing. It is based on a cluster model of frictional wear of solids and a physical-statistical model of formation of debris particles and their removal from the workpiece surface [18, 19, 36, 37]. The calculation of workpiece material removal rate Q in polishing involves the use of machining process parameters (the workpiece-to-pad contact pressure, the velocities of the workpiece and pad relative motions, the contact area between the workpiece and the polishing pad, the contact temperatures) and characteristics of the workpiece material and polishing powder (the thermal conductivity coefficient of the workpiece material, the debris particle surface area, the mean size of polishing powder grains) and is carried out by the formula [1, 3]

$$Q = \eta L_{\mathrm{f}}, \tag{1}$$

where η is the volumetric wear coefficient [37] and L_{f} is the distance of friction of the workpiece surface over the polishing pad surface.

Depending on the debris particle size $d(i)$ and the duration of contact between the debris particles and the polishing grains t_c, the volumetric wear coefficient η in (1) is given by

$\eta = \sum_{i=1}^{m} \frac{d(i)^2}{4\beta(i)t_c}$, where the values of the dimensionless quantity $\beta(i)$ are defined as roots m of transcendental equations $\frac{\exp(-\beta(i)^2)}{\sqrt{\pi} erf(\beta(i))} = \frac{1}{N(i)\, S_i}\sqrt{L_f d S_c \vartheta}$, where $N(i)$ is the number of ith debris particles on the workpiece surface; S_i is the surface area of the ith debris particle; d is the mean size of polishing grains; S_c is the workpiece–pad contact area; $\vartheta = \frac{\lambda T L_f}{p_a u S_c}$ is the dimensionless parameter; λ is the thermal conductivity coefficient of the workpiece material; T is the contact temperature; p is the nominal workpiece-to-pad contact pressure; u is the velocity of the workpiece and pad relative motions [1, 3, 18, 36–38].

The number of the ith debris particles on the surface area S_c during the contact between the polishing grain and the workpiece surface $t_c = d/u$ was determined, in view of their distribution over the surface areas, by the formula

$$N(i) = \frac{A_L}{4\hbar\pi^3 l_0^2} S_c t_c P(i, \nu), \tag{2}$$

where $A_L = \hbar \int_{\omega_1}^{\omega_2} \frac{[\varepsilon_1(\omega_{lk}) - \varepsilon_3][\varepsilon_2(\omega_{lk}) - \varepsilon_3]}{[\varepsilon_1(\omega_{lk}) + \varepsilon_3][\varepsilon_2(\omega_{lk}) + \varepsilon_3]} \mathrm{d}\Omega$ is the Lifshitz constant; $\hbar = 1.054 \times 10^{-34}$ J s is the Plank constant; $\omega_{lk} = \frac{3\omega_{01l}\omega_{02k}}{2(\omega_{01l} + \omega_{02k})}$ is the characteristic frequency; $\Omega \in [\omega_1; \omega_2]$

is the frequency range; ω_{01l} and ω_{02k} are the natural frequencies of molecular fragments of the workpiece and tool materials; l, k are the ordinal numbers of frequencies; ω_1 and ω_2 are the minimum and maximum frequencies; ε_1 and ε_2 are the dielectric permittivities of the workpiece and tool materials; ε_3 is the dielectric permittivity of the polishing suspension; $P(i, \nu) = \frac{e^{-\nu}\nu^i}{i!}$ is the debris particles distribution over the surface areas (the Poisson distribution); $\nu = E_b/E_c$ is the distribution parameter; E_b is the bond energy in the workpiece material; $E_c = \hbar \sum_{lk} [\omega_{01l}\xi_{lk}]$ is the cluster energy; $\xi = \sum_{lk} \left[\frac{\omega_{01l}^2}{\omega_{01l}^2 - \omega_{02k}^2}\right]^2$ is the number of molecular fragments, l_0 is the thickness of the gap between the surfaces of the polishing pad and the workpiece, which is filled with the polishing suspension, and is governed by the maximum size of polishing grains.

Considering that the unit event of surface wear is represented as a transition of the cluster from the bound state to the free one, the outcome of transitions between equidistantly located energy levels is the formation of debris particles whose surface areas can take on some discrete values and the surface area of the ith debris particle is given by $S_i = S_0(i+1)$ [1, 3] ($i = \overline{1, N}$; N is the number of samples; S_0 is the minimum cluster surface area that depends on the workpiece material structure and the number of molecular fragments ξ). The most probable value of the debris particle surface area is determined, in view of the Poisson distribution, by the formula $S = \sum_i S_i P(i, \nu)$. A debris particle is represented in the form of a parallelepiped whose side length is given by the number of molecular fragments $\xi = k_1 k_2 k_3$ (k_1, k_2, k_3 are integers) that make up particle. The most probable size of a debris particle is found by the formula $a_\upsilon = \sum_i d(i)P(i, \nu)$. By analogy with the Feret diameter for planar figures, the size of the ith debris particle is defined as the diameter of a sphere whose surface area is equal to that of the debris particle according to the formula $d(i) = (S_i/\pi)^{1/2}$.

2.2 Polishing of Crystal Planes with Different Crystallographic Orientation

The removal rate in polishing surfaces of optoelectronic components made of monocrystalline materials greatly depends on the crystallographic orientation of the plane being polished [13, 14, 38–40]. Looking at hexagonally structured crystals, e.g., sapphire (density $\rho = 3.98$ g cm^{-3}, atomization energy = 731 kcal/mol, bond energy $E_b = 6.3$ eV, lattice parameters $a = 0.4758$ nm, $c = 1.2991$ nm, $c/a = 2.73$ [39–41]), we can define how clusters (which turn into debris particles when breaking away from the workpiece surface) are made up of individual molecular fragments.

In the case of polishing, the plane c [0001] the number of molecular fragments is dictated by the interplanar spacings $\Delta x = 1.5a$, $\Delta y = a\sqrt{3}$, and $\Delta z = c/6$ in the projections on the coordinate axes x, y, and z, respectively; the minimum surface area of the particles is given by $S_{0c} = a\xi\left(\frac{3a\sqrt{3}}{k_3} + \frac{c}{2k_2} + \frac{2c}{\sqrt{3}k_1}\right)$, and the distance between neighbor c planes $l_c = 0.2165$ nm. In polishing the plane a [11$\bar{2}$0] the number of molecular fragments is determined by the interplanar spacings $\Delta x = a\sqrt{3}$, $\Delta y = a/2$, and $\Delta z = c$, the minimum surface area of the particles $S_{0a} = a\xi\left(\frac{a\sqrt{3}}{k_3} + \frac{2\sqrt{3}c}{k_2} + \frac{c}{k_1}\right)$, and the distance between the neighbor a planes $l_a = 0.2379$ nm. In the case of polishing the plane m [10$\bar{1}$0] (the distance between the m planes is 0.4121 nm) the debris particles of minimum size are produced when the material located between the [30$\bar{3}$0] planes is removed; the number molecular fragments therein is governed by the interplanar spacings $\Delta x = 1.5a$, $\Delta y = a/2\sqrt{3} = 0.1374$ nm, and $\Delta z = c$, and the minimum surface area is given by $S_{0m} = a\xi\left(\frac{\sqrt{3}a}{2k_3} + \frac{2c}{3k_2} + \frac{7c}{3k_1}\right)$. In polishing the plane r [10$\bar{1}$2] the number of molecular fragments is determined by the interplanar spacings $\Delta x = \sqrt{3a^2 + c^2}$, $\Delta y = 3a/2$, and $\Delta z = \frac{ac\sqrt{3}}{2\sqrt{3a^2+c^2}}$, the minimum surface area of the particles $S_{0r} = a\xi\left(\frac{3\sqrt{3a^2+c^2}}{k_3} + \frac{c\sqrt{3}}{k_2}\sqrt{\frac{c^2+4a^2}{c^2+3a^2}} + \frac{ac\sqrt{3}}{\sqrt{3a^2+c^2}}\frac{1}{k_1}\right)$, and the distance between the neighbor r planes $l_r = 0.3479$ nm.

According to the Gibbs–Curie principle, the equilibrium (most stable) form of a crystal is the one having the smallest surface energy for a given volume. Put in other words, the debris particle that has formed in polishing a crystal features the minimum surface energy $\sigma = \sum_n \sigma_{0n} S_{gn}$ (σ_0 and S_g are the surface energy and the surface area of the nth face, respectively) [42]. Based on the values of the surface energy for the faces c, a, m, and r of sapphire—σ_{0c}, σ_{0a}, σ_{0m}, and σ_{0r} [39], we can determine the surface energy of debris particles in polishing a given crystal face using the respective formulas:

$$\sigma_c = a\xi\left(\frac{3a\sigma_{0c}\sqrt{3}}{k_3} + \frac{c\sigma_{0m}}{2k_2} + \frac{2c\sigma_{0a}}{\sqrt{3}k_1}\right); \tag{3}$$

$$\sigma_a = a\xi\left(\frac{a\sigma_{0c}\sqrt{3}}{k_3} + \frac{2\sqrt{3}c\sigma_{0a}}{k_2} + \frac{c\sigma_{0m}}{k_1}\right); \tag{4}$$

$$\sigma_m = a\xi\left(\frac{\sqrt{3}a\sigma_{0c}}{2k_3} + \frac{2c\sigma_{0a}}{3k_2} + \frac{7c\sigma_{0m}}{3k_1}\right); \tag{5}$$

$$\sigma_r = a\xi\left(\frac{3\sqrt{3a^2+c^2}}{k_3}\sigma_{0r} + \frac{c\sqrt{3}}{k_2}\sqrt{\frac{c^2+4a^2}{c^2+3a^2}}\sigma_{0a} + \frac{ac\sqrt{3}}{\sqrt{3a^2+c^2}}\frac{1}{k_1}\sigma_{0m}\right). \tag{6}$$

The surface energy of debris particles being removed in polishing of planes c, a, m, and r, which is determined by formulas (3) through (6), is a function of the integers k_1, k_2, k_3 that govern the dimensions of the particles in their projections on the coordinate axes x, y, and z.

Proceeding from the condition of the minimum value of σ and writing $k_3 = \xi/k_1k_2$, we can reduce the determination of numerical values of k_1, k_2, k_3 to solving the set of equations

$$\begin{cases} \frac{\partial \sigma}{\partial k_1} = 0 \\ \frac{\partial \sigma}{\partial k_2} = 0 \end{cases}. \tag{7}$$

To find the size of the debris particles produced in polishing the planes c, a, m, and r, we solve the set of Eq. (7) to have the values of k_1, k_2, k_3 for each plane: for plane c

$$k_1 = \frac{2}{3}\sqrt[3]{\sqrt{3}\xi\left(\frac{\sigma_a^2}{\sigma_m\sigma_c}\right)\left(\frac{c}{a}\right)}; \quad k_2 = \frac{1}{2}\sqrt[3]{\frac{\xi}{3}\left(\frac{\sigma_m^2}{\sigma_a\sigma_c}\right)\left(\frac{c}{a}\right)};$$

$$k_3 = 3\sqrt[3]{\sqrt{3}\xi\left(\frac{\sigma_c^2}{\sigma_a\sigma_m}\right)\left(\frac{a}{c}\right)^2},$$

for plane a

$$k_1 = \sqrt[3]{\frac{\xi}{6}\left(\frac{\sigma_m^2}{\sigma_a\sigma_c}\right)\left(\frac{c}{a}\right)}; \quad k_2 = \sqrt[3]{\frac{12\xi}{\sqrt{3}}\left(\frac{\sigma_a^2}{\sigma_c\sigma_m}\right)\left(\frac{c}{a}\right)};$$

$$k_3 = \sqrt[3]{\frac{\sqrt{3}}{2}\xi\left(\frac{\sigma_c^2}{\sigma_a\sigma_m}\right)\left(\frac{a}{c}\right)^2},$$

for plane m

$$k_1 = \sqrt[3]{\frac{49}{9}\sqrt{3}\xi\left(\frac{\sigma_m^2}{\sigma_c\sigma_a}\right)\left(\frac{c}{a}\right)}; \quad k_2 = \sqrt[3]{\frac{8}{21\sqrt{3}}\xi\left(\frac{\sigma_a^2}{\sigma_m\sigma_c}\right)\left(\frac{c}{a}\right)};$$

$$k_3 = \sqrt[3]{\frac{27}{56}\xi\left(\frac{\sigma_c^2}{\sigma_a\sigma_m}\right)\left(\frac{a}{c}\right)^2},$$

and for plane r

$$k_1 = \sqrt[3]{\frac{a^2\xi}{\sqrt{3}(c^2+3a^2)}\left(\frac{\sigma_m^2}{\sigma_a\sigma_r}\right)\left(\frac{c}{\sqrt{c^2+4a^2}}\right)};$$

$$k_2 = \sqrt[3]{\frac{\xi(c^2+4a^2)}{\sqrt{3}(c^2+3a^2)}\left(\frac{\sigma_a^2}{\sigma_r\sigma_m}\right)\left(\frac{c}{a}\right)};$$

$$k_3 = \sqrt[3]{\frac{3\xi}{a\sqrt{c^2+4a^2}}\left(\frac{\sigma_r^2}{\sigma_a\sigma_m}\right)\left(\frac{c^2+3a^2}{c}\right)^2}.$$

Given below are the findings of the investigation of polishing sapphire wafers with different crystallographic orientation using a cBN powder suspension. The polishing was performed by the conventional technology whereby flat surfaces of

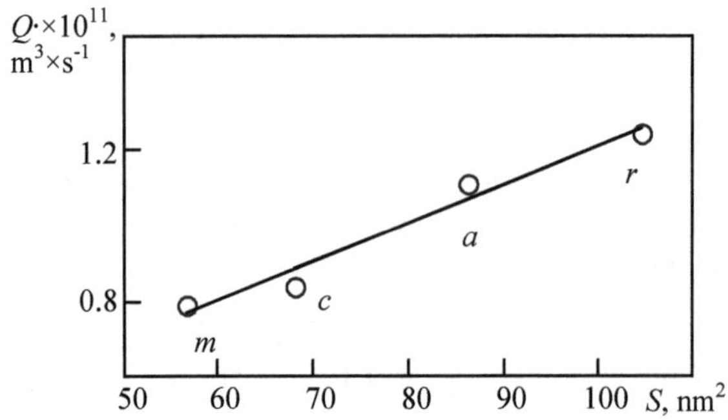

Fig. 1 Removal rate in polishing sapphire planes with different crystallographic orientations versus the debris particle surface area

25-mm-diameter cylindrical windows (21 pieces) discretely fixed on a 230-mm-diameter block were polished on a 350-mm-diameter pad under the following conditions: the pad speed = 40 rpm, the block speed = 36 rpm, the frequency of oscillating motion = 45 double pass/min, the workpiece-to-pad contact pressure = 26.2 kPa, the displacement = 30 mm, the hatch asymmetry = 5 mm, the hatch length = 160 mm, and mean contact temperature = 300 K. We considered the characteristics of sapphire and cBN (cubic boron nitride) as given below. Natural frequencies of sapphire molecular fragments, which were found by Raman spectra and IR absorption spectra, have the following values: $\omega_{01} = 10.8 \times 10^{13}$ s^{-1} (573 cm^{-1}), 12.1 × 10^{13} s^{-1} (642 cm^{-1}), and 14.1 × 10^{13} s^{-1} (748 cm^{-1}) [40, 43]. Natural frequencies of molecular fragments of the cBN polishing powder: $\omega_{02} = 19.9 \times 10^{13}$ s^{-1} (1056 cm^{-1}), 20.7 × 10^{13} s^{-1} (1100 cm^{-1}), and 24.6 × 10^{13} s^{-1} (1304 cm^{-1}) [3, 44].

The static permittivity ε_r and the thermal conductivity coefficient λ_r для for the sapphire plane r were calculated by the formula $\begin{pmatrix} \varepsilon_r \\ \lambda_r \end{pmatrix} = \begin{pmatrix} \varepsilon_c - \varepsilon_a \\ \lambda_c - \lambda_a \end{pmatrix} \cos\alpha$ ($\alpha = 57.6°$ is the angle between the planes r and c [39, 41]). The values of the static permittivity of the sapphire planes c, m, a, and r are 11.5, 9.3, 9.3, and 10.5, respectively. The static permittivity of cBN is 2.5 [3]. The thermal conductivity coefficients of the sapphire planes c, m, a, r are 32.5, 30.3, 30.3, 31.5 W m^{-1} K^{-1}, respectively.

The results of calculation of parameters of the interaction between the sapphire wafer flat surfaces of different crystallographic orientation and the polishing powder KM 05/0 are summarized in Table 1.

Based on the findings of computer modeling of the process of polishing sapphire planes with different orientations, it has been found out that the removal rate grows with increasing the most probable size, surface area, and volume of debris particles (see Table 1) [18, 36]. The dependence of the removal rate for the sapphire planes m, c, a, r on the debris particle surface area (Fig. 1) can be approximated by a linear function, $Q = k_1 S + b_1$ ($k_1 = 1.0 \times 10^5$ m s^{-1} and $b_1 = 2.1 \times 10^{-12}$ m^3 s^{-1} are the coefficients determined by the least-squares technique; the approximation error is 3%).

Table 1 Results of calculation of removal rate in polishing sapphire workpieces

Parameters	Sapphire plane being polished			
	m	*c*	*a*	*r*
Surface area of the plane S_g (nm^2)	3.5	7.1	8.0	11.0
Surface energy of the face σ_0 ($J\ m^{-2}$) [39]	7.126	3.357	3.858	3.456
Surface energy of a debris particle Σ ($J\ m^{-2}$)	4.34	4.36	4.31	4.30
Number of molecular fragments in the cluster $\xi = k_1k_2k_3 = 72$	(19–0) × (2–3) × (1–2)	4 × 3 × 6	5 × (9–10) × (1–2)	(2–3) × 4 × (7–8)
The most probable debris particle size a_v (nm)	4.0	4.5	5.1	5.6
Surface area of a debris particle *S* (nm^2)	53.4	68.0	86.0	104.6
The minimum surface area of a debris particle S_0 (nm^2)	27.0	34.4	43.5	52.9
Volume of a debris particle *V* (nm^3)	9.0	9.2	18.1	28.7
Volumetric wear coefficient η ($m^2\ s^{-1}$)	1.7×10^{-11}	1.9×10^{-11}	2.5×10^{-11}	2.8×10^{-11}
Energy of transfer *W* ($J\ kg^{-1}$)	13.0×10^{10}	13.0×10^{10}	9.1×10^{10}	8.5×10^{10}
Removal rate ($m^3\ s^{-1}$ $\mu m\ h^{-1}$)	0.79×10^{-11} 2.7	0.84×10^{-11} 3.0	1.11×10^{-11} 3.9	1.24×10^{-11} 4.3

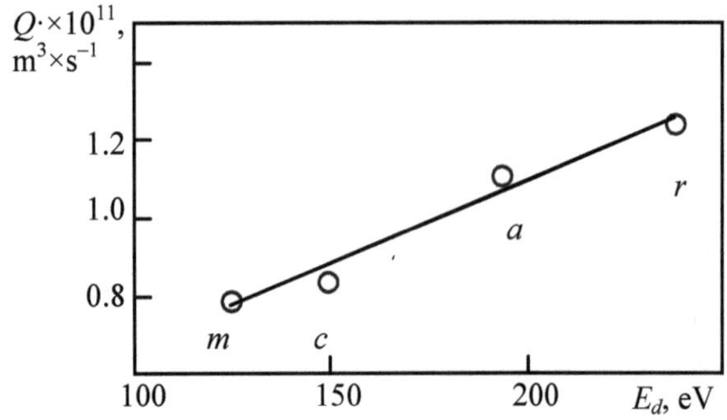

Fig. 2 Removal rate in polishing sapphire planes with different crystallographic orientations versus the energy of dispersion of material of the face being polished

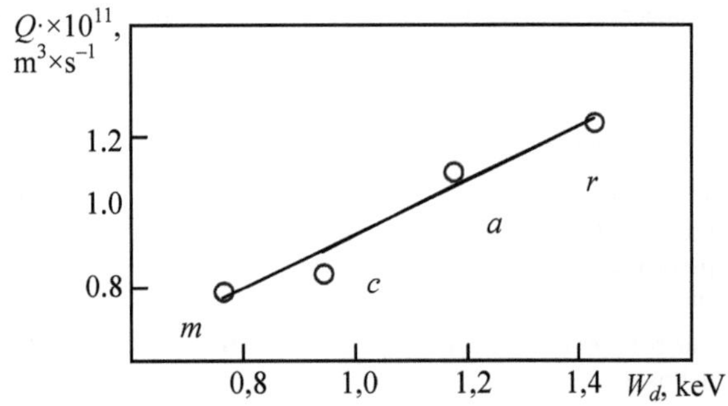

Fig. 3 Removal rate in polishing sapphire planes with different crystallographic orientations versus the energy of dispersion of debris particles

The anisotropy of properties of the monocrystalline workpiece material dictates the dependence of the volumetric wear coefficient, and hence of the removal rate, on the crystallographic orientation of the plane (crystal face) being polished. A relationship between the removal rate in polishing the sapphire planes *m*, *c*, *a*, *r* and the energy of dispersion $E_d = \sigma_0 S_g$ of material of the face being polished (Fig. 2) is also described by a linear function, $Q = k_2 E_d + b_2$ ($k_2 = 5.1 \times 10^{-14}$ m^2/(N^{-1} s^{-1}) and $b_2 = 5.7 \times 10^{-13}$ m^3 s^{-1} are the coefficients determined by the least-squares technique; the approximation error is 3%), which demonstrates that the more energy is spent for the process of dispersion of the workpiece material, the higher the removal rate in polishing.

The surface energy Σ of the debris particles produced in the course of polishing crystal planes with different crystallographic orientation, when multiplied by the minimum surface area of a debris particle S_0, gives the dispersion energy $W_d = \Sigma S_0$ which is spent for producing one debris particle and removing it from the respective crystal face. The higher this energy, the more intensive the workpiece material removal in polishing. The dependence of the removal rate in polishing the sapphire planes *m*, *c*, *a*, *r* with different crystallographic orientation on the dispersion energy of debris particles (Fig. 3) can be approximated by a linearly ascending function of the form $Q = k_3 W_d + b_3$ ($k_3 = 7.1 \times 10^{-15}$ m^3 eV^{-1} s^{-1} and $b_3 = 2.4 \times 10^{-12}$ m^3 s^{-1} are the coefficients determined by the least-squares technique; the approximation error is 4%).

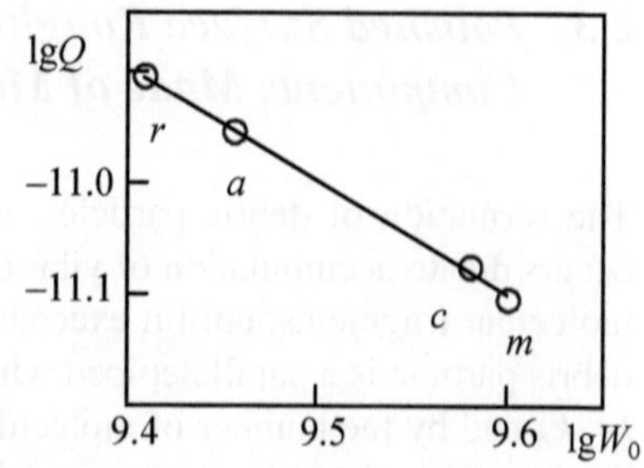

Fig. 4 Removal rate in polishing sapphire planes with different crystallographic orientations versus the reduced energy of transfer

The removal rate in polishing the sapphire planes *m*, *c*, *a*, *r* depends on the energy of transfer $W = \lambda T/\eta\rho$ (J kg^{-1}), which represents the energy transferred by debris particles from the workpiece surface to the contact zone. Considering that the energy of transfer *W* depends on the conditions and kinematic parameters of the polishing process, rheological characteristics of polishing powder suspensions, and on the thermal conductivity coefficient of the workpiece material, it is conventional to use the value of the reduced energy of transfer $W_0 = (\lambda_{K8}/\lambda)W$, which represents the contribution of debris particles to the transfer of energy in polishing sapphire relative to polishing of a reference material—K8 grade optical glass that has $\lambda_{K8} =$ 0.95 W m^{-1} K^{-1}) [1, 3, 37]. The dependence of the reduced energy of transfer W_0 in the logarithmic scale is written as the linear function $\log Q = -k_3 \log W_0 + b_3$ ($k_3 =$ 0.95 and $b_3 = -1.97$ are the coefficients determined by the least-squares technique), which permits assessing the efficiency of the polishing process using the energy of transfer as a criterion [1, 3] (Fig. 4).

The values of the reduced energy of transfer, which are typical of polishing sapphire planes with different crystallographic orientation using KM 05/0 polishing powder suspensions, fall into the interval $9.4 < \log W_0 < 9.6$ that corresponds to the classical process of polishing optical surfaces [1, 3, 37]. The relative removal rate in polishing sapphire planes with different crystallographic orientations (in relation to the plane *m*) $q = Q/Q_m$ is characterized by the ratio $q_m : q_c : q_a : q_r = 1.00 : 1.06 :$ $1.41 : 1.57$ and grows in the series $m < c < a < r$.

A study of the influence of crystallographic orientation of the workpiece plane on the removal rate in polishing sapphire [45] revealed a trend of increasing removal rate in the series $c < a < r$, which agrees with the ratio given above. The experimental measurement of the removal rate by the weight method has demonstrated that in polishing the sapphire planes *r*, *a*, and *c* using suspensions of diamond powders *RDDM* 4–8, *RDDM* 2–4, *RDDM* 0.25–0.50 polishing pads of a rubber material and pitch–rosin compound [45] the ratios between average values of removal rate for these planes are $q_r/q_a = 1.15$, $q_r/q_c = 1.53$, $q_a/q_c = 1.33$, respectively, and practically coincide with the theoretical ratios $(q_r/q_a)_t = 1.11$, $(q_r/q_c)_t = 1.48$, $(q_a/q_c)_t = 1.33$ (the discrepancy between the calculated and the experimental data is within 4%).

2.3 Polished Surface Roughness of Optoelectronic Components Made of Monocrystalline Materials

The formation of debris particles and their removal from the workpiece surface occurs due to accumulation of vibrational energy E_c in the cluster that consists of ξ molecular fragments, until it exceeds the bond energy E_b of workpiece material. A debris particle is a parallelepiped whose side length is governed by the numbers k_1, k_2, k_3 and by the number of molecular fragments $\xi = k_1 k_2 k_3$ in the debris particle. To determine the volume removed in polishing, the most probable size of debris particles is found from the minimum surface area S_0 that depends on the workpiece material structure and the number of molecular fragments ξ and is described by the Poisson distribution $P(i, \nu)$ with the parameter $\nu = E_b/E_c$ [3, 9, 38]. The size of the ith debris particle is defined as the Feret diameter, based on the condition of equality of surface areas of the sphere and the debris particle in accordance with the formula $d(i) = [S_0(i+1)/\pi]^{1/2}$ [38].

In the study of nanoprofile of polished surfaces, the above approach to the assessment of debris particle size is absolutely unacceptable for the height of irregularities depends not on the three numbers k_1, k_2, k_3, but only on the number that governs the number of material layers removed in polishing of planes with different crystallographic orientation and different interplanar spacings. Therefore, the height of the profile irregularities $z(x)$ of the workpiece surface is dictated not by the debris particle sizes $d(i)$ but by the height of the corresponding parallelepipeds $h(i)$, which is perpendicular to the plane being machined, and by its most probable value $h_\upsilon = \sum_i h(i)P(i, \nu)$. The sampling length for the determination of surface roughness is determined by lengths of projections of debris particles on the plane being machined $l(i)$. These parameters can be calculated based on the unit cell sizes of the lattice (the parallelepiped of the smallest surface area) and the numbers k_1, k_2, k_3 (Table 2). The height of the profile irregularities $z(x)$, depending on the coordinate, is described by the formulas proposed in [3, 46], after $d(i)$ is replaced with the debris particle height $h(i)$, and is given by the equation

$$z(x) = h(k_0)\Phi_{k_0}^2\left(\frac{x}{h(k_0)}\right) + \sum_{i\geq 1}^{G-1}\left[\frac{\Phi_{k_0}(\sqrt{2k_0+1})}{\Phi_{k_i}(\sqrt{2k_i+1})}\sqrt{\frac{h(k_i)}{h(k_0)}}\right]^2 h(k_i)$$
$$\times\, \Phi_{k_i}^2\left(\frac{x - \frac{1}{2}\left[h(k_0) + \sum_i h(k_i)\sqrt{2k_i+1}\right]}{h(k_i)}\right) \quad (8)$$

where $x \in [0, L]$ is the coordinate; $L = Gl_\nu$ is the sampling length of a model profile chart; G is the number of debris particles; $l_\upsilon = \sum_i l(i)P(i, \nu)$ is the most probable size of the debris particle projection; k_i is the random sequence of numbers, which is generated in accordance with the Poisson distribution and dictates the order of formation and sequence of their positioning on workpiece plane along the sampling length; Φ_k is the parabolic cylinder function. The derived Eq. (1) describes a nano-sized relief in the form of a stepped-terrace structure arising due to removal of

Table 2 Characteristics of workpiece surface and debris particles, parameters of interaction between sapphire and cBN polishing powder, and polished surface roughness parameters

Characteristic, parameter	Sapphire plane being polished			
	m	*c*	*a*	*r*
Number of molecular fragments in the cluster $\xi = k_1 k_2 k_3 = 72$	$k_1 = 16$	$k_1 = 4$	$k_1 = 5$	$k_1 = 2$–3
	$k_2 = 4$–5	$k_2 = 3$	$k_2 = 9$–10	$k_2 = 4$
	$k_3 = 1$	$k_3 = 6$	$k_3 = 1$–2	$k_3 = 7$–8
Unit cell dimensions (Å)	$Lx = 1.374$	$Lx = 7.137$	$Lx = 8.241$	$Lx = 15.384$
	$Ly = 7.137$	$Ly = 8.241$	$Ly = 2.379$	$Ly = 7.137$
	$Lz = 12.991$	$Lz = 2.165$	$Lz = 12.991$	$Lz = 3.479$
Debris particle height h_v (nm)	2.997	3.543	3.082	3.559
Projection of a debris particle on the workpiece plane l_v (nm)	4.724	5.150	6.216	6.532
Lifshitz constant A_L (10^{-21} J (meV))	5.6 (35)	9.3 (58)	5.6 (35)	7.7 (48)
Surface area of a debris particle *S* (nm^2)	53.4	68.0	86.0	104.6
Volumetric wear coefficient η (m^2/s)	1.7×10^{-11}	1.9×10^{-11}	2.5×10^{-11}	2.8×10^{-11}
Density of vibrational energy $\gamma_k \times 10^5$ (J/m^2)	10.4	13.7	6.5	7.4
Polished surface roughness				
Ra (nm)	6.8 ± 0.2	8.2 ± 0.1	6.7 ± 0.2	8.0 ± 0.2
Rq (nm)	7.2 ± 0.3	8.8 ± 0.1	7.3 ± 0.3	8.6 ± 0.3
*R*max (nm)	12.1 ± 1.5	14.8 ± 1.4	12.9 ± 1.0	14.4 ± 1.7
Minimum allowable surface roughness parameters				
Ra_0 (Å)	0.687	1.083	1.190	1.740
Rq_0 (Å)	0.971	1.531	1.682	2.460
$R\text{max}_0$ (Å)	1.374	2.165	2.379	3.479

parallelepiped-shaped debris particles from the crystallographic plane being polished, provided that the surface roughness parameters are close to the distance between neighbor sublattices in the lattice cell [47, 48]. Using the computer modeling of nanoprofile of a polished surface, the following roughness parameters are determined: arithmetic average deviation of the profile $Ra = \frac{1}{L}\int_0^L z(x)\mathrm{d}x$, root mean squared deviation of the profile $Rq = \sqrt{\frac{1}{L}\int_0^L [z(x)]^2\mathrm{d}x}(rms)$, and maximum height of the profile $R\max$.

In the course of polishing sapphire monocrystals (density $\rho = 3.98$ g/cm^3; bond energy $E_b = 6.3$ eV, lattice constants $a = 0.4758$ nm, $c = 1.2991$ nm [41]), some debris particles consisted of $\xi = k_1k_2k_3$ molecular fragments break away from the workpiece surface. In polishing planes with different crystallographic orientation, the surface area of the debris particles being formed depends on the number of molecular fragments and interplanar spacings. In the case of polishing, the planes c, m, a, and r the minimum surface area of the debris particles is given by the formulas [38, 49]

$$S_{0c} = a\xi\left(\frac{3a\sqrt{3}}{k_3} + \frac{c}{2k_2} + \frac{2c}{\sqrt{3}k_1}\right); \tag{9}$$

$$S_{0m} = a\xi\left(\frac{a\sqrt{3}}{2k_3} + \frac{2c}{3k_2} + \frac{7c}{3k_1}\right); \tag{10}$$

$$S_{0a} = a\xi\left(\frac{a\sqrt{3}}{k_3} + \frac{2\sqrt{3}c}{k_2} + \frac{c}{k_1}\right); \tag{11}$$

$$S_{0r} = a\xi\left(\frac{3\sqrt{3a^2+c^2}}{k_3} + \frac{c\sqrt{3}}{k_2}\sqrt{\frac{c^2+4a^2}{c^2+3a^2}} + \frac{ac\sqrt{3}}{k_1\sqrt{3a^2+c^2}}\right), \tag{12}$$

and the respective interplanar spacings are $l_c = 0.2165$ nm, $l_m = 0.1374$ nm, $l_a = 0.2379$ nm, and $l_r = 0.3479$ nm [38, 49]. Based on the minimum surface area according to the Gibbs–Curie principle of equilibrium crystal form [38, 42], for faces c, m, a, and r we found the values of k_1, k_2, k_3 and, using formulas (9) through (12), calculated the minimum surface areas S_0 of debris particles and their most probable values $S = \sum_i S_0(i+1)P(i,\nu)$ (see Table 2).

The computer modeling of the interaction between clusters on the polishing grains and on the workpiece surface during the polishing of sapphire planes of different crystallographic orientations has revealed that the polished surface roughness parameters Ra, Rq, $R\max$ grow linearly with increasing permittivity ε and thermal conductivity coefficient λ. The dependence of the roughness parameters Ra, Rq, $R\max$ on the crystal permittivity along the direction perpendicular to the respective plane has been approximated by linear functions of the form $\begin{pmatrix} Ra \\ Rq \\ R\max \end{pmatrix} = \alpha_1\varepsilon + \beta_1$, where

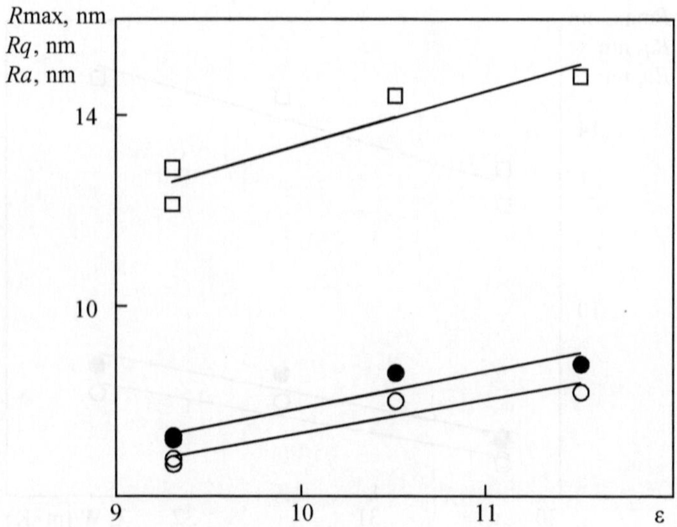

Fig. 5 Surface roughness parameters *Ra* (s), *Rq* (d), *R*max (u) of the sapphire planes with different crystallographic orientations versus dielectric permittivity

$\alpha_1 = \begin{pmatrix} 0.71 \\ 0.76 \\ 1.11 \end{pmatrix}$ nm, $\beta_1 = \begin{pmatrix} 0.25 \\ 0.30 \\ 0.27 \end{pmatrix}$ nm are the coefficients found by the least-squares method; the approximation error is 2–3% (Fig. 5).

The dependence of the roughness parameters of the sapphire planes *m*, *c*, *a*, and *r* on the thermal conductivity coefficient along the direction perpendicular to the planes has been approximated by linear functions of the form $\begin{pmatrix} Ra \\ Rq \\ R\text{max} \end{pmatrix} = \alpha_2\lambda - \beta_2$, where $\alpha_2 = \begin{pmatrix} 0.7 \\ 0.8 \\ 1.1 \end{pmatrix} \times 10^{-9}\ \text{W}^{-1}\ \text{m}^2\ \text{K}$, $\beta_2 = \begin{pmatrix} 14.6 \\ 15.6 \\ 21.1 \end{pmatrix}$ nm are the coefficients determined by the least-squares method; the approximation error is within 3% (Fig. 6).

In polishing monocrystal planes of different crystallographic orientations using a polishing suspension, the formation and removal of parallelepiped-shaped debris particles with a height multiple of the respective interplanar spacings occur due to the interaction between the workpiece surface and the pad surface with fixed polishing powder grains; this interaction is characterized by Lifshitz fluctuation forces [50]. During the interaction between the sapphire planes and the polishing powder (KM 0.5/0) layer on the pad, the thickness of the gap between them is $l_0 \approx 0.5$ μm. The wavelength corresponding to the range of natural vibrations of molecular fragments of sapphire and cBN polishing powder—from 573 to 1304 cm^{-1} [15, 43]—is $\lambda_0 =$ 7.7–17.5 μm. On the condition that $l_0 << \lambda_0$ the Lifshitz force $F_L = \frac{A_L}{8\pi^2 l_0^3}$ (A_L is

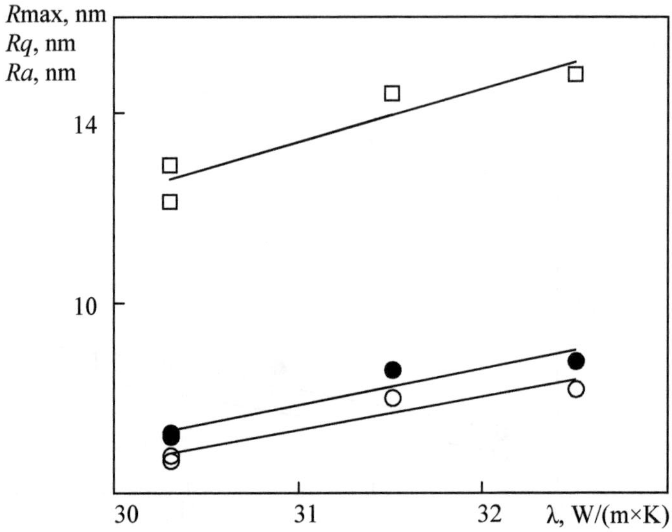

Fig. 6 Surface roughness parameters *Ra* (s), *Rq* (d), *R*max (u) of the sapphire planes with different crystallographic orientations versus the thermal conductivity coefficient

the Lifshitz constant [3, 50, 51]) is about 0.5 mN/m^2. The Lifshitz constant A_L (J) represents the energy of interaction between clusters on the polishing powder grains and on the sapphire workpiece surface and is assessed to be dozens of meV (see Table 2).

Figure 7 shows the dependence of the roughness parameters *Ra*, *Rq*, *R*max of the sapphire polished planes *m*, *c*, *a*, and *r* on the Lifshitz constant A_L, which can be approximated by linear functions of the form $\begin{pmatrix} Ra \\ Rq \\ R\max \end{pmatrix} = \alpha_3 A_L + \beta_3$ where ($\alpha_3 = \begin{pmatrix} 4.2 \\ 4.5 \\ 6.6 \end{pmatrix} \times 10^{11}\ \mathrm{N}^{-1}$, $\beta_3 = \begin{pmatrix} 4.5 \\ 4.8 \\ 8.9 \end{pmatrix}$ nm are the coefficients; the approximation error is 2–3%.

The dependence of the roughness parameters of the sapphire planes *m*, *c*, *a*, and *r* on the most probable value of the debris particle height h_v (the projection of the particles along the direction perpendicular to the planes) has been approximated by linear functions of the form $\begin{pmatrix} Ra \\ Rq \\ R\max \end{pmatrix} = \alpha_4 h_v - \beta_4$ where $\alpha_4 = \begin{pmatrix} 2.6 \\ 2.8 \\ 4.2 \end{pmatrix}$ nm, $\beta_4 =$

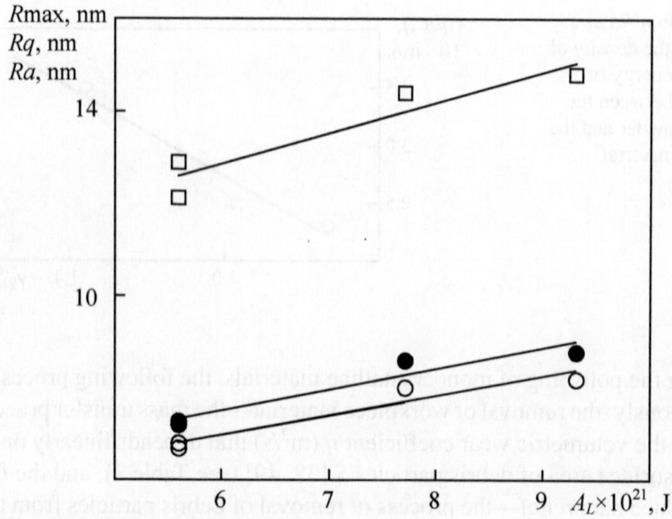

Fig. 7 Surface roughness parameters *Ra* (s), *Rq* (d), *R*max (u) of the sapphire planes with different crystallographic orientations versus the Lifshitz constant

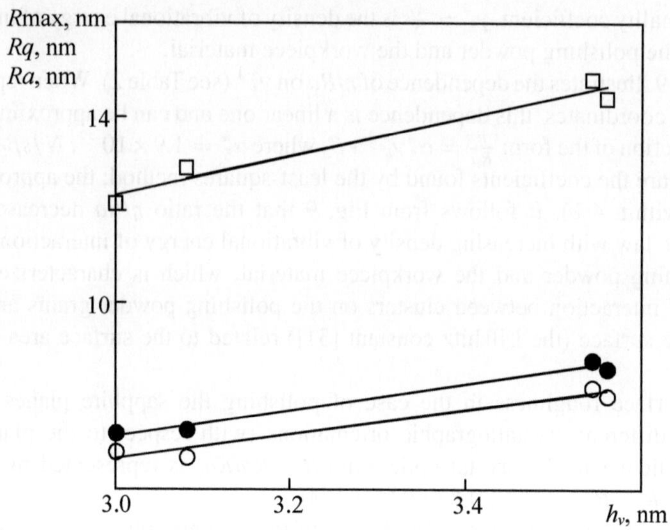

Fig. 8 Surface roughness parameters *Ra* (s), *Rq* (d), *R*max (u) of the sapphire planes *m*, *c*, *a*, and *r* versus the debris particle height

$\begin{pmatrix} 1.1 \\ 1.3 \\ 0.2 \end{pmatrix}$ nm are the coefficients found by the least-squares method; the approximation error is within 2% (Fig. 8).

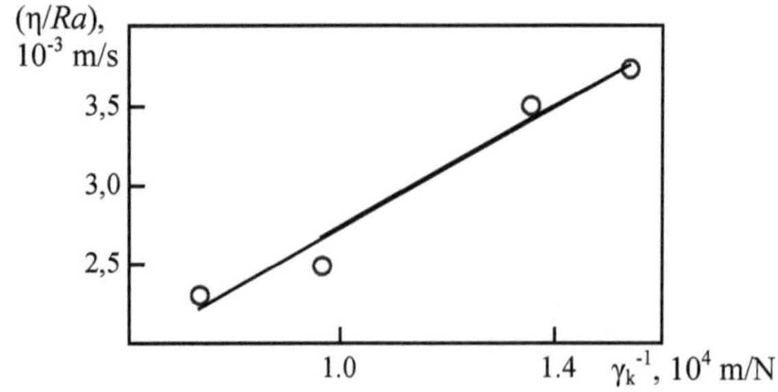

Fig. 9 Ratio η/Ra as a function of the density of vibrational energy of interaction between the polishing powder and the workpiece material

During the polishing of monocrystalline materials, the following processes occur simultaneously: the removal of workpiece material—the mass transfer process represented by the volumetric wear coefficient η (m^2/s) that depends linearly on the most probable surface area of debris particles S [38, 49] (see Table 2); and the formation of the surface nanorelief—the process of removal of debris particles from the workpiece surface, which is characterized by the roughness parameters Ra, Rq, Rmax that depend linearly on the Lifshitz constant (see Fig. 7). The ratio of the parameters (η/Ra) that represent these processes is given by $\frac{\eta}{Ra} = \alpha_5 \frac{S}{A_L} = \frac{\alpha_5}{\gamma_k}$, where α_5 is the proportionality coefficient, $\gamma_k = \frac{A_L}{S}$ is the density of vibrational energy of interaction between the polishing powder and the workpiece material.

Figure 9 illustrates the dependence of η/Ra on γ_k^{-1} (see Table 2). When represented in inverse coordinates, this dependence is a linear one and can be approximated by a linear function of the form $\frac{\eta}{Ra} = \alpha_5^* \gamma_k^{-1} + \beta_5$ where $\alpha_5^* = 1.9 \times 10^{-7}$, $N/s\beta_5 = 8.4 \times 10^{-4}$ m/s are the coefficients found by the least-squares method; the approximation error is within 4%). It follows from Fig. 9 that the ratio η/Ra decreases by the hyperbolic law with increasing density of vibrational energy of interaction between the polishing powder and the workpiece material, which is characterized by the energy of interaction between clusters on the polishing powder grains and on the workpiece surface (the Lifshitz constant [51]) related to the surface area of debris particles.

The surface roughness in the case of polishing the sapphire planes a, m, r, and c of different crystallographic orientations (with respect to the plane c that is perpendicular to the crystal optic axis) $R = Ra/Ra_c$ is represented by the ratio $R_a : R_m : R_r : R_c = 0.81 : 0.83 : 0.97 : 1.00$.

A study of the mechanism of polishing of the sapphire planes a, r, and c using *RDDM* 0.25–0.50 diamond powders on polishing pads made of a rubber material [45] revealed that the Ra values of these planes were 5.4, 6.2, and 7.5 nm, and their ratios were $R_a : R_r : R_c = 0.72 : 0.83 : 1.00$ and $Ra_a/Ra_r = 0.87$, which confirm the obtained ratios $R_a : R_r : R_c = 0.81 : 0.97 : 1.00$ and $Ra_a/Ra_r = 0.84$ (the respective deviations of the experimental and calculated data are 16, 12, and 4%). When investigating the influence of crystallographic orientation of the sapphire plane being machined on the removal rate and surface roughness in chemical-mechanical polishing, Budnikov et al. [14] revealed a general pattern of decreasing removal rate

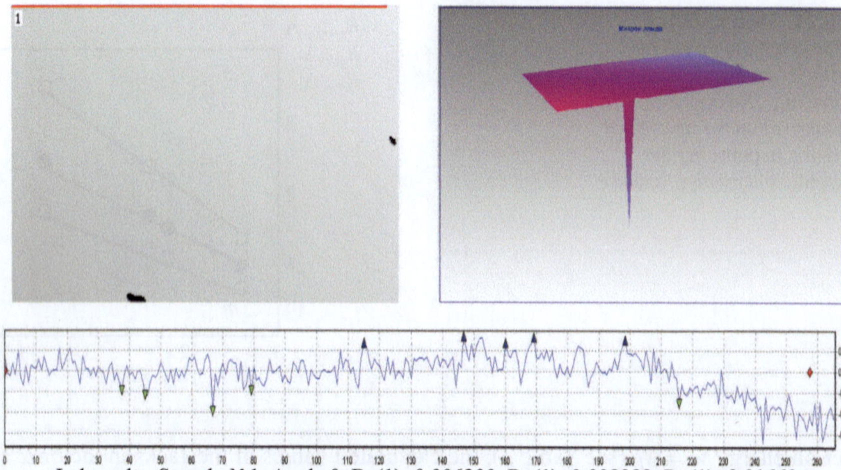

Ind.results: Sample №1, Angle 0, Ra(1)=0.006200; Rq(1)=0.008382; Rp(1)=0.01669

Fig. 10 Polished surface sapphire (plane a) condition: Mod. Micron-Alpha non-contact interference 3D roughness measuring instrument

and surface roughness parameters in the series $c > r > a$, which is consistent with the above-mentioned ratio.

In the experimental assessment of surface roughness of the planes a and c of the sapphire windows polished under the above-mentioned conditions, which was performed using a MarSurf PS1 roughness measuring instrument (Mahr GmbH, Germany) and a Micron-Alpha non-contact interference 3D surface contour recorder, we have obtained the following Ra values: 6.5 ± 1.4 nm for the plane a (Fig. 10) and 8.6 ± 0.7 nm for the plane c; their ratio $Ra_a/Ra_c = 0.76$ slightly differs from the calculated value 0.81 (the deviation being 6–7%).

As the debris particles whose height h_v is 3.0–3.6 nm (see Table 2) and which consist of dozens of elementary fragments are formed and removed from the surface of the sapphire planes a, m, r, and c, a polished surface is produced whose roughness is assessed by the parameters $Ra = 6.7$–8.2 nm, $Rq = 7.2$–8.8 nm, and $R\text{max} = 12.1$–14.8 nm (see Table 2). A different situation is observed when the debris particles formed on the workpiece surface consist of one or a few elementary fragments. The material removal in polishing, which is essentially an opposite process to the crystal growth, may result in producing either atomically rough or atomically smooth surfaces in as per the Jackson criterion [52] according to which the surface entropy factor is defined as $\alpha = \frac{zw}{k_B T}$, where z is the total number of possible bonds a cluster has with its nearest neighbors in the crystal; w is the energy of interaction between the clusters on the polishing powder grains and on the workpiece surface; T is the mean temperature in the workpiece–tool contact zone; $k_B = 1.38 \times 10^{-23}$ J/K is Boltzmann's constant. The machined surfaces for which $\alpha > 5.3$ are atomically smooth, while the surfaces for which $\alpha < 2$ are atomically rough. In polishing the sapphire planes m, c, a, and r (the total number of bonds $z = 4$), with $T = 300$ K and inter-

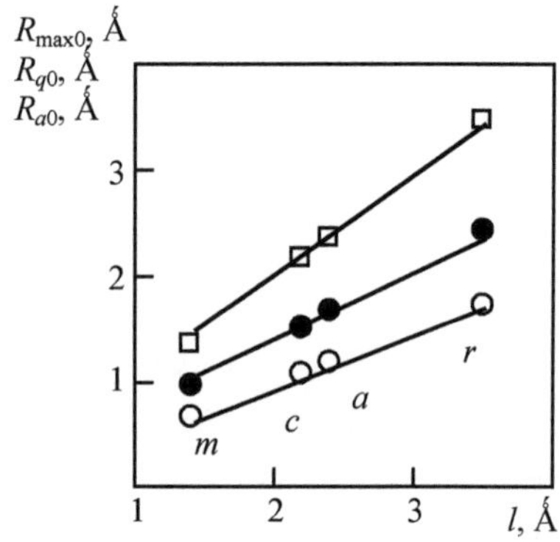

Fig. 11 Dependence of allowable limit values of surface roughness parameters Ra_0 (s), Rq_0 (d), $R\text{max}_0$ (u) on the interplanar spacing in polishing the sapphire planes *m*, *c*, *a*, and *r*

action energy $w = A_L$ (see Table 2), the calculated values of the Jackson factor $\alpha =$ 5.4, 9.0, 5.4, 7.4 are higher than 5.3. This means that atomically smooth surface can be produced in polishing the sapphire planes *m*, *c*, *a*, and *r*.

In the case of chemical-mechanical polishing of sapphire using colloid systems of nanoparticles of polishing powders, it is possible to achieved a surface with angstrom-sized roughness: the polished planes *m*, *a*, *c* have the roughness parameters Rq = 3 Å [13]; Ra < 3 Å [14] and Rq = 2.3 Å [15]; Ra = 2.9 Å and Rq = 3.6 Å [53]. To define the minimum allowable values of surface roughness parameters of atomically smooth surfaces, we should make an evident assumption that min{*Ra*, *Rq*, *R*max} corresponds to the minimum number of steps in the plane (one), and that the height of such step *h* is equal to (multiple of) the interplanar distance *l* which is 2.165, 1.374, 2.379, and 3.479 Å for the sapphire planes *c*, *m*, *a*, and *r*, respectively. In this context, the profile of such a surface is represented by the function $z(x) = \begin{cases} h, x < L/2 \\ 0, x \geq L/2 \end{cases}$, and the surface roughness parameters are $Ra_0 = h/2$, $Rq_0 = h/\sqrt{2}$, and $R\text{max}_0 = h$, respectively. The minimum allowable values of roughness parameters for the sapphire planes *m*, *c*, *a*, and *r* are given in Table 2. Noteworthy is that in polishing the sapphire planes of different crystallographic orientations their roughness depends on the Lifshitz constant and decreases in the series $c > r > m > a$. On the other hand, this trend is not valid for the minimum allowable values of the roughness parameters for atomically smooth surfaces for which we observe a linear dependence on the value of the interplanar spacing and a decrease in the series $r > a > c > m$ (Fig. 11).

Nanopolishing provides the following minimum allowable surface roughness parameters: Ra_0 = 0.69–1.74 Å, Rq_0 = 0.97–2.46 Å, and $R\text{max}_0$ = 1.374–3.479 Å, which are consistent with the physical meaning and agree with the crystal lattice parameters of sapphire. The values of surface roughness parameters of various sapphire planes [13–15, 41, 45, 53], which are achieved at various stages of polishing, are in good agreement with the values given above.

3 Mechanical Polishing Optoelectronic Components of Crystal Materials

According to the cluster model, the concentration of debris particles (DP) on the workpiece surface is governed by the fluctuation forces of interaction (Lifshitz forces) that arise within a unit area of contact between the workpiece and the polishing pad and are represented by frequency-dependent permittivities [30, 36, 38, 54]:

$$n(i) = \frac{F_L l_0}{\pi \hbar} P(i, \nu),\ (\mathrm{m}^{-2}\ \mathrm{s}^{-1}), \tag{13}$$

where $F_L = \frac{\hbar \overline{\omega}}{8\pi^2 l_0^3}$ is the Lifshitz force (N/m^2) [1]; $\overline{\omega} = \int_{\omega_1}^{\omega_2} \frac{[\varepsilon_1(\omega_{lk})-\varepsilon_3][\varepsilon_2(\omega_{lk})-\varepsilon_3]}{[\varepsilon_1(\omega_{lk})+\varepsilon_3][\varepsilon_2(\omega_{lk})+\varepsilon_3]} \mathrm{d}\Omega$ is the dispersion function of permittivities [3, 50, 54, 55].

The investigation of the process of mechanical polishing of crystalline materials using polishing suspensions was carried out on the following assumptions:

- polishing grains are localized on the pad surface at the level of maximum height of profile irregularities (the active particles) or inside the surface microvoids (the inactive particles) [56];
- the formation of debris particles results from the action of Lifshitz forces (the intermolecular interaction between the workpiece and pad surfaces);
- debris particles are formed in case the cluster energy exceeds the bond energy of the workpiece material ($E_{cl} \geq E_b$ [54]);
- the thickness of the polishing suspension layer between the pad and the workpiece is determined by the polishing grain size, is almost independent of the suspension flow characteristics, and meets the condition $l_0 << \lambda_0$ [50] (λ_0 is the wavelength whereby the substantial absorption is observed).

Mechanical polishing of flat surfaces of optoelectronic components was performed by the conventional technology. Wafers of monocrystalline silicon carbide SiC (the surface area to be machined $S_n = 10.2$ cm^2), gallium nitride GaN ($S_n = 8.1$ cm^2), aluminum nitride AlN ($S_n = 26.0$ cm^2), and sapphire Al_2O_3 ($S_n = 16.8$ cm^2), which were discretely fixed on a block of diameter 60 mm, were polished using a suspension with ACM 2/1 diamond micron powders on a polishing pad of diameter 100 mm with a contact pressure of 50.5 N, speed 90 rpm, stroke displacement 30 mm, stroke length 50 mm, and average contact temperature 300 K.

The main physical properties of silicon carbide, gallium nitride, aluminum nitride, and sapphire are summarized in Table 3 [1, 3, 29, 38, 57–60]. We also took into consideration the following physical characteristics of the diamond powder: natural frequencies of molecular fragments ω_{02} 1135 and 1332 cm^{-1}, static permittivity of diamond 2.5 [3, 30].

In polishing of plane *c* [0001] of crystalline materials with a hexagonal structure (SiC, GaN, AlN, and Al_2O_3), the number of molecular fragments in a debris particle is denoted by ξ and its surface area is given by $S(i) = S_0(i+1)$ (where $i = \overline{1, N}$, N is the number of samples, S_0 is the minimum surface area) [3, 29, 38]. Table 4 gives the

Table 3 Physical properties of workpiece materials

Parameter	Material			
	SiC	GaN	AlN	Al_2O_3
Density ρ (g/cm^3)	3.21	6.15	3.26	3.97
Bond energy E_b (eV)	3.2	5.4	6.2	6.3
Static permittivity ε_1	6.5	9.5	8.5	11.5
Thermal conductivity λ (W/(m K))	490.0	130.0	200.0	32.5
Specific heat capacity c_p (J/(kg K))	690.0	491.0	748.0	750.0
Thermal diffusivity $\chi \times 10^5$ (m^2/s)	22.1	4.3	8.2	1.1
Natural frequency ω_{01} (cm^{-1})	794.0	144.0	248.6	573.0
	1000.0	531.8	611.0	642.0
		558.8	657.4	748.0
		567.6	670.8	
		734.0	890.0	
		741.0	912.0	

calculated values of the dispersion function of permittivities $\overline{\omega}$, which represents the interaction forces F_L (the Lifshitz forces) that act within a unit contact area between the crystalline material surface being polished and the pad surface, as well as the dimensionless parameter θ. The data of calculations and experimental determination of removal rate and machined surface roughness in mechanical polishing of SiC, GaN, AlN, and Al_2O_3 wafers were obtained by following the procedures described in [1, 3, 5, 8, 29, 38, 46].

Analysis of the data given in Tables 3 and 4 has revealed some regularities of mechanical polishing of crystalline materials. Figure 12 shows, in the semi-log scale, the removal rate versus the bond energy of the workpiece materials. This relation has been approximated by a linearly decreasing function $\ln Q = -k_1 E_b + b_1$ (where $k_1 = 0.26$ eV^{-1}, $b_1 = -25.3$ are the coefficients found by the least-squares method; the approximation error is 1%). It confirms the activation nature of the stock removal process and the exponential dependence of the removal rate on the bond energy of the workpiece material in line with the cluster model [36, 54].

The relation between the removal rate in polishing of the materials at hand on their thermal conductivity coefficient λ can be approximated by a linearly increasing

Table 4 Results of calculation of removal rate in polishing of crystalline materials

Parameters of interaction with ACM 2/1 polishing powder	Workpiece material			
	SiC	GaN	AlN	Al_2O_3
Dispersion function $\overline{\omega} \times 10^{-13}$ (s^{-1})	1.9	2.5	1.4	1.5
Lifshitz force F_L (μPa)	5.2	3.4	3.3	3.0
Parameter θ	1939.0	572.0	881.0	143.0
Number of molecular fragments ξ	27.0	62.0	57.0	73.0
Volumetric wear coefficient $\eta \times 10^{11}$ (m^2/s)	2.5	2.6	1.3	1.1
Energy of transfer $W \times 10^{-11}$ (J/kg)	18.5	2.4	15.0	2.2
Removal rate in polishing Q				
Calculated (10^{-12} m^3/s (μm/h))	4.2 (14.8)	2.7 (12.0)	2.3 (3.2)	1.6 (3.4)
Experimental (mg/min (μm/h))	0.8 (15.5)	1.0 (12.4)	0.4 (3.0)	0.4 (3.9)
Calculation error (%)	4.0	7.0	3.0	11.0

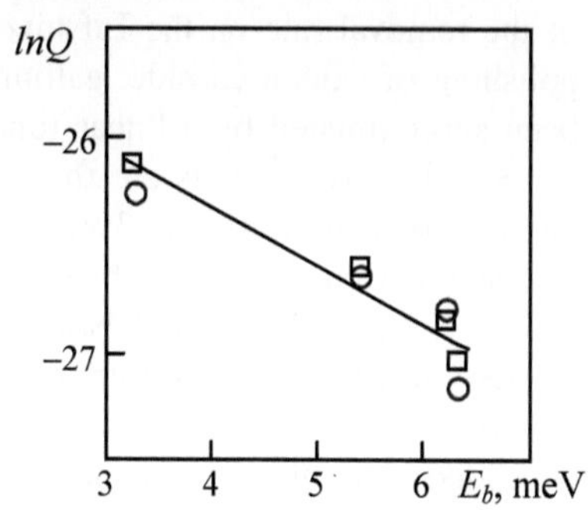

Fig. 12 Removal rate in polishing as a function of the workpiece material bond energy: calculation (s), experiment (u), approximation (—)

function $Q = k_2\lambda + b_2$ (where $k_2 = 5.3 \times 10^{-15}$ J^{-1} K m^4, $b_2 = 1.6 \times 10^{-12}$ m^3/s are the coefficients determined by the least-squares method; the approximation error is 10%). Figure 13 illustrates a more general relation between the removal rate and

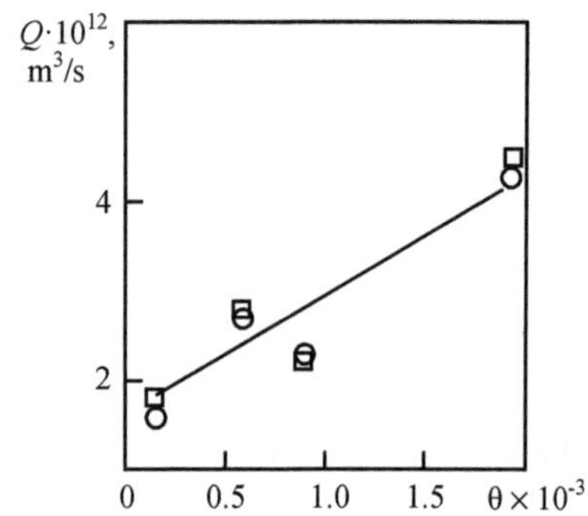

Fig. 13 Removal rate in polishing of crystalline materials versus the parameter θ: calculation (s), experiment (u), approximation (—)

the dimensionless parameter θ. It can also be approximated by a linearly increasing function of the form $Q = k_3\theta + b_3$ (where $k_3 = 1.4 \times 10^{-15}$ m^3/s, $b_3 = 1.5 \times 10^{-12}$ m^3/s are the coefficients found by the least-squares method; the approximation error is 10%).

This relation shows that the removal rate in mechanical polishing of the materials under study depends on the thermal characteristics of the workpiece material, workpiece design parameters, machining conditions and kinematic parameters which are represented by the dimensionless parameter θ, in full agreement with the physical-statistical model of formation and removal of debris particles from the workpiece surface [5, 37, 38].

Its has been also found out that the removal rate in mechanical polishing essentially depends on the energy of interaction between clusters on the polishing grains and on the workpiece surface, which is represented by the Lifshitz forces [50] acting within a unit contact area between the interacting surfaces. These forces are expressed in terms of the Lifshitz constant A_L according to the formula $F_L = \frac{A_L}{8\pi^2 l_0^3}$ (where l_0 is the distance between the workpiece and pad surfaces) [29, 38]. The dependence of the removal rate on the Lifshitz force, which is given in Fig. 14 for diamond polishing of silicon carbide, gallium nitride, aluminum nitride, and sapphire, has been approximated by a linear function $Q = k_4F_L + b_4$ ($k_4 = 1.1 \times 10^{-6}$ W m^6, $b_4 = -1.3 \times 10^{-12}$ m^3/s are the coefficients found by the least-squares method; the approximation error is 3%). Using this formula, we can assess the minimum (threshold) value of the Lifshitz force whereby stock removal occurs in polishing: $F_L^l = -(b_4/k_4) \approx 1.2 \times 10^{-6}$ N/m^2.

Also, it has been demonstrated that the removal rate in polishing of crystalline materials depends on the transfer energy W (see Table 4) that represents energy expended to remove a unit mass of the workpiece material from the workpiece surface [37]. Figure 15 shows the removal rate versus the reduced transfer energy $W_0 = (\lambda_{K8}/\lambda)W$ that characterizes the contribution of debris particles to the transfer of heat energy during mechanical polishing of sapphire, silicon carbide, gallium nitride, and aluminum nitride against the case of polishing a reference material (optical glass K8 whose thermal conductivity λ_{K8} is 0.95 W/(m K) [1, 3, 38]. This relation is expressed, in the semi-log scale, by the linearly decreasing function: $\log Q = -k_5 \lg W_0$

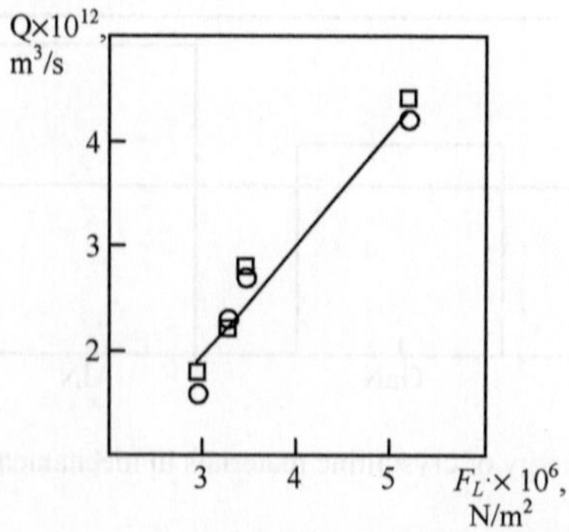

Fig. 14 Removal rate in polishing of crystalline materials as a function of the Lifshitz force: calculation (s), experiment (u), approximation (—)

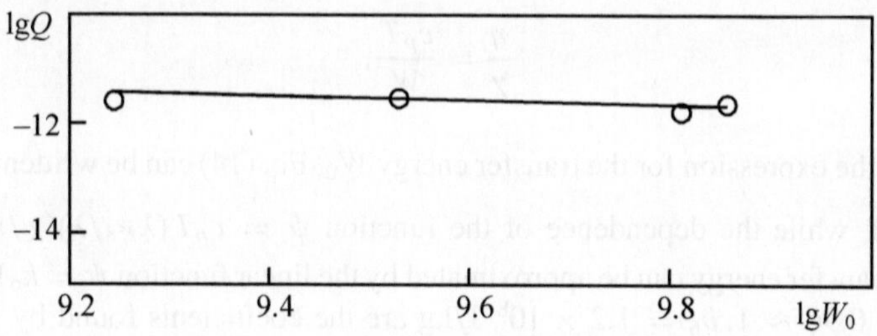

Fig. 15 Removal rate in polishing of polycrystalline materials versus the reduced transfer energy

$+b_5$ (where $k_5 = 0.3$, $b_5 = -8.7$ are the coefficients determined by the least-squares method; the approximation error is 1%), which permits assessing the efficiency of mechanical polishing of crystalline materials using the transfer energy as a criterion [3].

The study of the mechanism of polishing of nonmetallic materials as a transfer phenomenon, which was carried out by applying the physical-statistical model of formation and removal of debris particles from the workpiece surface [37], has enabled us to derive an equation that relates the volumetric wear coefficient η to the thermal conductivity coefficient λ, contact temperature, and transfer energy: $\eta = \lambda T/(\rho W)$. This equation suggests that the volumetric wear coefficient η (and thus the removal rate in mechanical polishing) is inversely proportional to the transfer energy W, which is further supported by the data given in Fig. 15. In addition, the equation shows that between the coefficients η and λ there is a relation which is also confirmed by the linear dependence of the removal rate on the thermal conductivity coefficient.

The analysis of the well-known expressions that relate the transfer coefficients—thermal diffusivity, diffusion constant, and viscosity/density ratio, whose dimension is the same as that of the volumetric wear coefficient (m^2/s)—has revealed that the ratio between the volumetric wear coefficient η and diffusivity coefficient χ

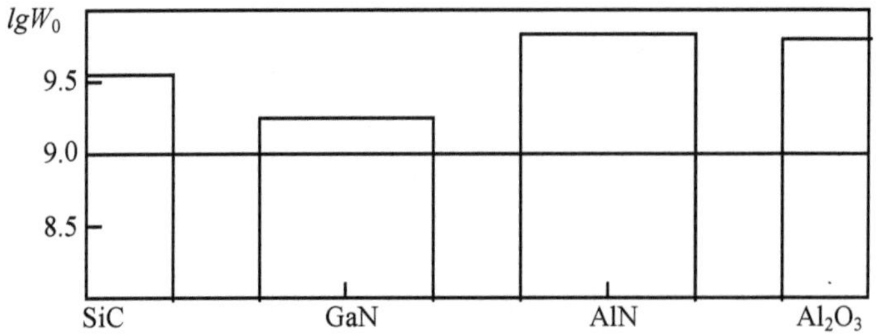

Fig. 16 Reduced transfer energy of crystalline materials in mechanical polishing

$= \lambda/(\rho c_p)$ (where c_p is the specific heat capacity; ρ is the workpiece material density) depends on the transfer energy W and is given by

$$\frac{\eta}{\chi} = \frac{c_p T}{W}. \tag{14}$$

Using the expression for the transfer energy W_0, Eq. (14) can be written as $\frac{W_0}{c_p T} = \left(\frac{\lambda_{K8}}{\lambda}\right)\left(\frac{\chi}{\eta}\right)$, while the dependence of the function $\psi = c_p T(\lambda_{K8}/\lambda)(\chi/\eta)$ on the reduced transfer energy can be approximated by the linear function $\psi = k_6 W_0 + b_6 \approx W_0 (k_6 = 0.96 \approx 1, b_6 = 1.2 \times 10^8$ J/kg are the coefficients found by the least-squares method; the approximation error is 3%). The resulting expression $\psi = W_0$ suggests the validity of Eq. (14) that defines a relation between the volumetric wear coefficient and thermal conductivity coefficient as a function of specific heat capacity and transfer energy of the workpiece material and temperature in the contact zone between the workpiece and the polishing pad.

The diagram in Fig. 16 shows, in the log scale, the transfer energy values typical of the process of mechanical polishing of silicon carbide, gallium nitride, aluminum nitride, and sapphire crystals as well as the horizontal lines $\log W_0 = 9$ and $\log W_0 = 10$, which are characteristic of the respective processes of ultrafine grinding and polishing of precision optical surfaces [1, 3, 38].

From the values of the reduced transfer energy (see Fig. 16), we can infer that when the suspension of ACM 2/1 diamond micron powder acts on the workpiece surface of the crystalline materials at hand by the conventional technology, the stock removal occurs through the classic polishing mechanism ($9 < \log W_0 < 10$). The relative removal rate in mechanical polishing of silicon carbide, gallium nitride, and aluminum nitride crystals, $q = Q/Q_0$ (where Q_0 is the removal rate in polishing of sapphire), is represented by the ratio $q_{SiC} : q_{GaN} : q_{AlN} : q_{sapphire} = 2.6 : 1.7 : 1.4 : 1.0$ (while the experimentally found ratio is 2.4:1.6:1.2:1.0; the mean deviation of the calculated data from the experimental ones is 8%).

4 Surface Roughness of Optoelectronic Components in Mechanical Polishing

Mechanical polishing is a transfer phenomenon whereby both the material removal and surface roughness generation result from the formation of debris particles and their removal from the workpiece surface. The removal rate is determined by the volumetric wear coefficient [29, 30, 37, 38], while the machined surface roughness parameters depend on the size of debris particles being removed from the workpiece surface in the process of polishing [3, 5, 8, 46, 49].

We studied the mechanism of the nanoprofile generation in optoelectronic crystalline materials under the action of a polishing suspension on the assumptions that the abrasive powder grains are localized on the surface of a polishing pad or at the level of maximum height of profile irregularities or in the micropores of the surface [56], and that the debris particles are formed through intermolecular interaction between the workpiece surface and the pad surface. A debris particle consists of molecular fragments and has the shape of a parallelepiped whose side lengths are given by k_1, k_2, k_3, [54, 61]. The height of irregularities of the workpiece surface is governed by the debris particle size $d(i)$ or by the dimensions of the corresponding parallelepipeds $h(i)$, while roughness width cut off is given by the lengths of the debris particle projections on the workpiece surface plane $l(i)$. The surface roughness parameters Ra, Rq, and $R\max$ were found by calculations using model profile charts plotted by means of computer modeling [3, 49, 61].

In the case of polishing the plane c [0001] of the crystalline materials SiC, GaN, AlN, and Al_2O_3 that have a hexagonal structure with interplanar spacings $L_x = 1.5a$, $L_y = a\sqrt{3}$ and $L_z = c/6$, the debris particle dimensions are found by the formulas: $l_x = k_1L_x$, $l_y = k_2L_y$, $l_z = k_3L_z$. Since the surface area of debris particles $S_i = S_0(i+1)$, $i = \overline{1, N}$, N is the number of samplings [3, 29, 30, 38, 49], their dimensions can be represented by the most probable size $a_\upsilon = \sum_i d(i)P(i, \nu)$, the height of parallelepipeds $h(i) \approx h_0(i+1)^{1/2}$, and its most probable value $h_\upsilon = \sum_i h(i)P(i, \nu)$, as well as the length of particles projections on the workpiece surface plane $l(i) \approx [(l_x^2 + l_y^2)(i+1)]^{1/2}$ and their most probable value $l_\upsilon = \sum_i l(i)P(i, \nu)$. The calculated values of the number of molecular fragments ξ, unit cell dimensions, most probable values of the debris particle size are summarized in Table 5. The surfaces of samples of crystalline materials SiC, GaN, AlN and Al2O3 (plane c), treated by mechanical polishing with a polishing suspension of diamond powder and powders of chromium oxide and cubic boron nitride, are shown in Figs. 17, 18, 19 and 20 (Optical microscope Mod. "LOMO-METAM R-1" with Vision camera "STD-Res Series"). Analysis of the data given in the table has revealed the following regular patterns of the workpiece surface nanorelief generation during the mechanical polishing of crystalline materials. Figure 21 shows the surface roughness parameters Ra, Rq, and $R\max$ versus the most probable debris particle size a_ν. These plots can be approximated by the linear functions: $\begin{pmatrix} Ra \\ Rq \\ R\max \end{pmatrix} = \alpha_1 a_\upsilon + \beta_1$, where $\alpha_1 =$

Fig. 17 Polished surface SiC condition: **a** diamond powder, **b** chromium oxide powder

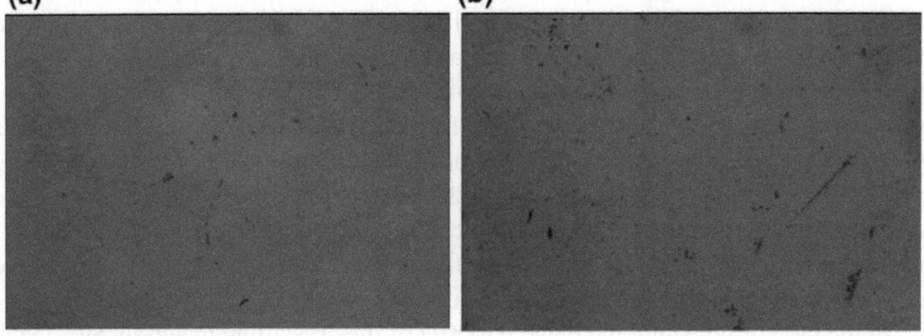

Fig. 18 Polished surface GaN condition: **a** diamond powder, **b** chromium oxide powder

$\begin{pmatrix}2.0\\2.1\\2.8\end{pmatrix}$, $\beta_1 = \begin{pmatrix}3.2\\3.3\\7.3\end{pmatrix}$ nm are the coefficients found by the least-squares method; the approximation error is within 2%. Hence, the smaller the debris particles being removed from the workpiece surface of the crystalline materials SiC, GaN, AlN, and Al_2O_3, the lower the surface roughness height parameters. The MP-machined surface of these materials is represented by the minimum roughness parameters: $Ra \approx 3.2$ nm, $Rq \approx 3.3$ nm, and $R\max \approx 7.3$ nm. The linear dependence of the roughness parameters on the most probable debris particle size is the fundamental factor for the mechanical polishing processes and is fully consistent with the experimental findings in polishing single-crystal silicon carbide using suspensions based on diamond micron powders, cBN powders, and MAX-phase Ti_3AlC_2 powders with various grain size distributions [1, 3].

Figure 22 shows the surface roughness parameters Ra, Rq, and $R\max$ versus the most probable debris particle height h_v. These relations can also be approximated by the linear functions: $\begin{pmatrix}Ra\\Rq\end{pmatrix} = \alpha_2 h_v$ and $R\max = \alpha_{2m} h_v + \beta_2$, where $\alpha_2 = \begin{pmatrix}2.3\\2.5\end{pmatrix}$,

Table 5 Results of calculation of workpiece surface roughness

Parameter	Workpiece material			
	SiC	GaN	AlN	Al_2O_3
Number of molecular fragments in a debris particle $\xi = k_1k_2k_3$	27	62	57	73
	$k_1 = 4$	$k_1 = 4$	$k_1 = 3–4$	$k_1 = 4$
	$k_2 = 2$	$k_2 = 1–2$	$k_2 = 1–2$	$k_2 = 3$
	$k_3 = 3–4$	$k_3 = 10$	$k_3 = 10$	$k_3 = 6$
Unit cell dimension (Å)	$L_x = 4.610$	$L_x = 4.784$	$L_x = 4.668$	$L_x = 7.137$
	$L_y = 5.323$	$L_y = 5.524$	$L_y = 5.390$	$L_y = 8.241$
	$L_z = 2.518$	$L_z = 0.864$	$L_z = 0.830$	$L_z = 2.165$
Surface area of the plane (Å^2)	24.5	26.4	25.2	58.8
Cell volume V_0 (Å^3)	62	23	21	127
The most probable debris particle size a_v (nm)	2.5	2.4	2.3	4.5
Particle height h_v (nm)	2.40	2.37	2.27	3.56
Particle projection on the workpiece surface plane l_v (nm)	2.90	2.86	2.49	5.18
Surface area of a debris particle S (nm^2)	21.6	19.3	17.3	68.6
Lifshitz constant A_L (10^{-21} J (meV))	2.0 (13)	2.7 (17)	1.5 (9)	1.6 (10)
Density of vibrational energy γ (μJ/m^2)	93	140	87	23
Polished surface roughness				
Ra (nm)	8.1 ± 0.2	7.9 ± 0.4	7.5 ± 0.2	12.0 ± 0.4
Rq (nm)	8.7 ± 0.3	8.4 ± 0.4	8.0 ± 0.1	12.8 ± 0.4
*R*max (nm)	14.6 ± 1.8	13.9 ± 1.1	13.1 ± 0.4	19.6 ± 2.1

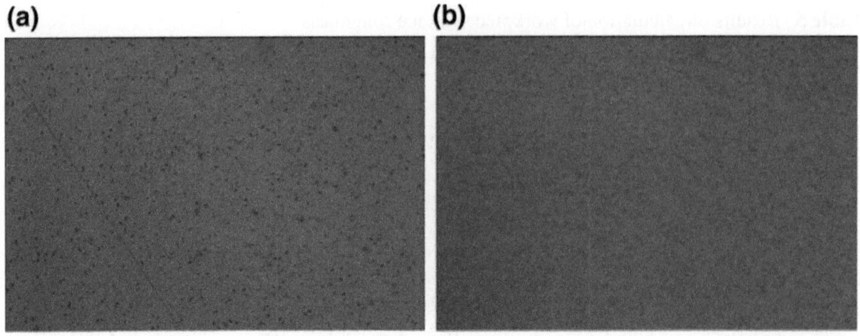

Fig. 19 Polished surface AlN condition: **a** diamond powder, **b** powder of cubic boron nitride

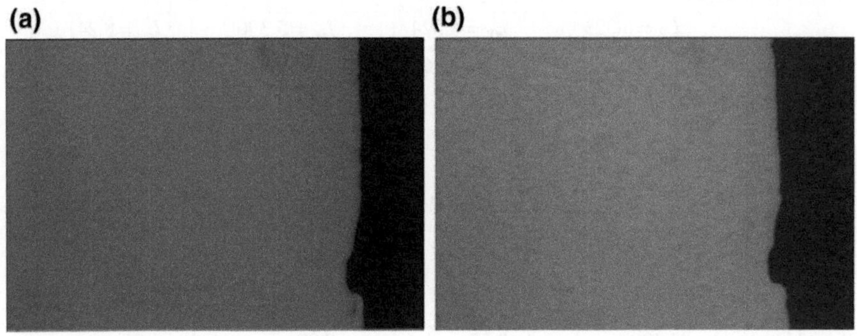

Fig. 20 Polished surface Al_2O_3 condition: **a** diamond powder, **b** powder of cubic boron nitride

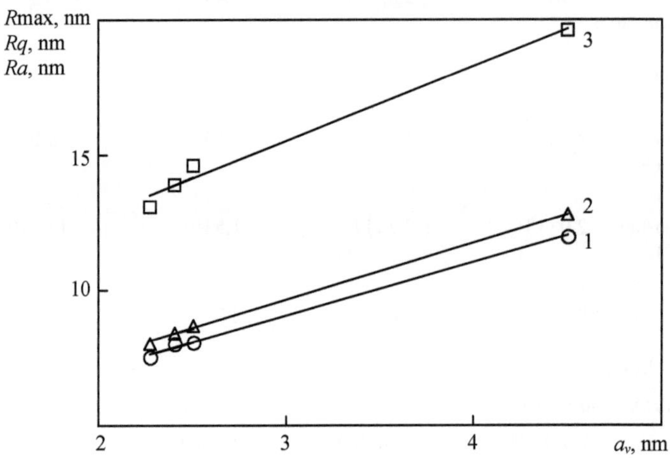

Fig. 21 Surface roughness parameters *Ra* (1), *Rq* (2), *R*max (3) versus the most probable debris particle size

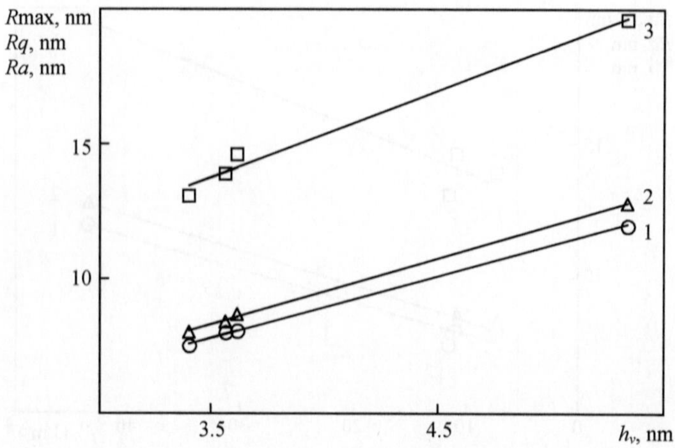

Fig. 22 Surface roughness parameters *Ra* (1), *Rq* (2), *R*max (3) versus the most probable debris particle height

$\alpha_{2m} = 3.2$, $\beta_2 = 2.6$ nm are the coefficients found by the least-squares method; the approximation error is within 2%.

The linear dependence of the roughness parameters on the most probable debris particle height agrees well with the results obtained in the polishing of sapphire planes with different crystallographic orientations [49], where the workpiece surface roughness is dictated not by debris particle dimensions but by the height of the corresponding parallelepipeds, which is normal to the workpiece surface and depends on the interplanar spacing of the crystal. The fact that there are only some insignificant discrepancies in the results of computer modeling of polished surface nanoprofile for crystalline materials of the same structure (in this case, the hexagonal one) and in the same plane (in this case, [0001]), which was performed based on the calculations of the most probable size a_v [3] and most probable height h_v of debris particle [49], suggests that when analyzing roughness of surfaces with different crystallographic orientations one can apply the Poisson distribution of debris particles in height.

Also, we observed the linear dependence of the surface roughness parameters *Ra*, *Rq*, and *R*max on the surface area of the crystal face to be polished, unit cell volume, and debris particle surface area (see Table 5), which, along with the functions shown in Figs. 21 and 22, supports the validity of the results of the computer modeling of the polished surface nanoprofile. During the mechanical polishing, the workpiece surface nanorelief is generated through removal of wear particles whose most probable size essentially depends on the energy of intermolecular interactions between clusters of the abrasive grain surfaces and workpiece surface elements, which is represented by the Lifshitz constant A_L [29, 38, 49] (see Table 5).

When looking into the influence of the density of vibrational energy $\gamma = A_L/S$ (J/m^2), which shows the energy consumptions for the formation of one debris particle with a surface area *S*, we have found out that the most probable debris size is inversely

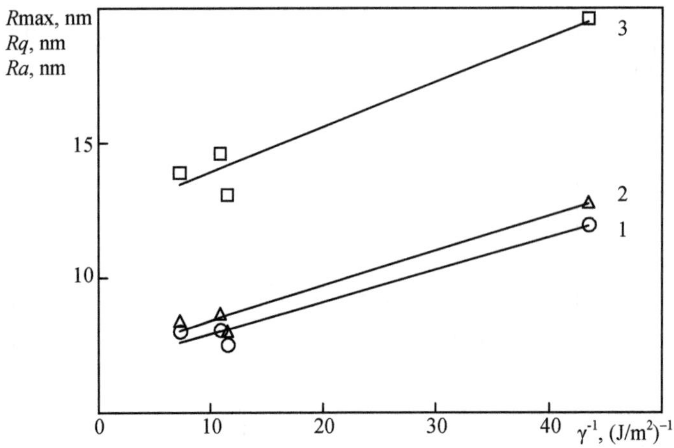

Fig. 23 Surface roughness parameters *Ra* (1), *Rq* (2), *R*max (3) versus the density of vibrational energy γ

proportional to this quantity [3]. Figure 23 shows the surface roughness parameters *Ra*, *Rq*, and *R*max versus the energy consumed to form a debris particle γ.

These relations can be approximated by the linear functions (see Fig. 23):

$$\begin{pmatrix} Ra \\ Rq \\ R\max \end{pmatrix} = \alpha_3 \gamma^{-1} + \beta_3, \text{ where } \alpha_3 = \begin{pmatrix} 1.2 \\ 1.3 \\ 1.7 \end{pmatrix} \times 10^{-13}\ \text{J/m}, \beta_3 = \begin{pmatrix} 6.7 \\ 7.1 \\ 12.3 \end{pmatrix} \text{nm}$$

are the coefficients found by the least-squares method; the approximation error is within 3–4%.

The hyperbolically decreasing surface roughness parameters *Ra*, *Rq*, and *R*max with increasing density of vibrational energy γ demonstrate that the mechanical polishing to provide the best surface finish should be performed using those abrasive powders which would ensure the maximum value of the Lifshitz constant, i.e., the largest energy of intermolecular interaction between the abrasive frains and the workpiece material, for a particular material to be polished.

The relative surface roughness upon polishing of the crystalline materials SiC, GaN, AlN, and Al_2O_3 with a diamond micron powder based suspension, i.e., with respect to surface roughness of sapphire, $R = Ra/Ra_{\text{sapphire}}$, is represented by the ratio: $R_{\text{SiC}} : R_{\text{GaN}} : R_{\text{AlN}} : R_{\text{Sapphire}} = 0.68 : 0.67 : 0.63 : 1.00$.

5 Conclusions

Based on the theoretical and experimental investigations of the mechanism of precision surface formation in workpieces of anisotropic monocrystalline materials for optoelectronics, we have come up with a generalized model of material removal in polishing using suspensions of polishing powders. Applying this model to sapphire single crystals we have studied the influence of the crystallographic orientation of the plane being polished on the removal rate. The removal rate has been found to grow with increasing volume, surface area, and most probable size of debris particles, and energy of dispersion of material from the face being polished. It has been shown that the dependence of the removal rate in polishing the sapphire planes *m*, *c*, *a*, and *r* with different crystallographic orientation on the energy of transfer in the logarithmic scale is represented by a linear function. Using the value of the reduced energy of transfer as criterion of efficiency of removal rate, we have demonstrated that in polishing sapphire planes *m*, *c*, *a*, and *r* with a cBN powder suspension by the conventional technology corresponds to the classical process of polishing optical surfaces. The relative removal rate in polishing the sapphire planes *m*, *c*, *a*, and *r* with different crystallographic orientations is characterized by the ratio 1.00:1.06:1.41:1.57.

The present study of the mechanism of polishing of surfaces of optoelectronic components made of monocrystalline materials, with sapphire taken as an example, and the computer modeling of the process of generation of planes with different crystallographic orientations have demonstrated that the polished surface roughness parameters *Ra*, *Rq*, *R*max grow linearly with increasing permittivity, thermal conductivity coefficient, height of debris particles, and the Lifshitz constant that represents the energy of interaction between the polishing powder grains and the workpiece surface. The ratio of the volumetric wear coefficient to the room mean squared deviation of the profile η/Ra in polishing the sapphire planes of different crystallographic orientation has been found to decrease by the hyperbolic law with increasing density γ_k of the vibrational energy of interaction between the polishing powder and the workpiece material. The investigation of the mechanism of polishing of the sapphire planes *a*, *m*, *r*, and *c* with different crystallographic orientations has revealed that the polished surface roughness parameters are characterized by the ratio 0.81:0.83:0.97:1.00 and has found the general trend of decreasing roughness in the series $c > r > m > a$. Also, the Jackson entropy factor for these sapphire planes has been shown exceed 5.3, which confirms the feasibility of producing atomically smooth surfaces in the process of polishing. We have defined the minimum allowable values of surface roughness parameters for atomically smooth surfaces that are generated in nanopolishing of the sapphire planes with different crystallographic orientations; these values linearly depend on interplanar spacings and decrease in the series $r > a > c > m$.

The present theoretical and experimental investigations of the mechanism of mechanical polishing of flat surfaces of optoelectronic components of crystalline materials using a suspension of diamond micron powders have demonstrated that the removal rate in polishing of silicon carbide, gallium nitride, aluminum nitride, and sapphire crystals decreases exponentially with increasing bond energy of the workpiece materials. The removal rate in mechanical polishing of the materials under consideration is proportional to the dimensionless parameter θ that depends on the sliding distance of a workpiece material element against the pad surface, on the thermal conductivity coefficient of the workpiece material, contact temperature, nominal contact pressure, velocity of the workpiece and pad relative motion, and their contact area.

The cluster interaction energy on the polishing grains and on the workpiece surface governs the intensity of stock removal, and the removal rate in polishing of crystalline materials is proportional to the Lifshitz force acting within a unit contact area between the interacting workpiece and pad surface. The dependence of the removal rate in polishing of crystalline materials on the reduced transfer energy is a linearly decreasing function, which permits using the transfer energy as a criterion of effectiveness of mechanical polishing. A study of the relationship between the process of a workpiece surface wear in mechanical polishing and the process of heat energy propagation has enabled us to determine a ratio between the volumetric wear coefficient and the thermal conductivity coefficient of the workpiece material, which depends on specific heat capacity, transfer energy of the workpiece material, and temperature in the workpiece–pad contact zone, and has revealed that in mechanical polishing of crystalline materials using ACM 2/1 diamond micron powder suspension the relative removal rate goes down in the series of SiC, GaN, AlN, Al_2O_3 and is represented by the ratio 2.6:1.7:1.4:1.0.

The present investigation of the mechanism of the workpiece surface nanoprofile generation during the mechanical polishing of optoelectronic components made of crystalline materials, using a diamond micron powder suspension, on the basis of the model of debris particle formation and removal from the workpiece surface, along with the computer modeling has revealed that the surface roughness parameters of single-crystal silicon carbide, gallium nitride, aluminum nitride, and sapphire linearly grow with increasing most probable values of the debris particle size, height, and surface area as well as unit cell volume and surface area of the crystal face being polished. The study of the influence of the energy of intermolecular interaction between the polishing powder grains and the workpiece surface, which is determined by the Lifshitz constant, on the machined surface roughness has demonstrated that the parameters *Ra*, *Rq*, and *R*max of the crystalline materials are inversely proportional to the energy consumed to form a debris particle with a unit surface area. It has been found out that in the case of mechanical polishing of the plane [0001] of silicon carbide, gallium nitride, aluminum nitride, and sapphire crystals with a suspension based on ACM 2/1 diamond micron powder the relative roughness of the machined surfaces is represented by the ratio: 0.68:0.67:0.63:1.00.

References

1. Filatov YuD, Vetrov AG, Sidorko VI, Filatov OYu, Kovalev SV (2013) A mechanism of diamond-abrasive finishing of monocrystalline silicon carbide. J Superhard Mater 35(5):303–308
2. Prikhna TA, Starostina AV, Lizkendorf D et al (2014) Studies of the oxidation stability, mechanical characteristics of materials based on max phases of the Ti–Al–(C, N) systems, and of the possibility of their use as tool bonds and materials for polishing. J Superhard Mater 36(1):9–17
3. Filatov YuD, Vetrov AG, Sidorko VI, Filatov OYu et al (2015) Polishing of optoelectronic components made of monocrystalline silicon carbide. J Superhard Mater 37(1):48–56

4. Filatov OYu, Sidorko VI, Filatov YuD (2007) Special features of macro- and microrelief formation in nonmetallic flat surfaces in diamond grinding. J Superhard Mater 29(6):361–368
5. Filatov YuD, Sidorko VI, Filatov OYu et al (2009) Surface roughness in diamond abrasive finishing. J Superhard Mater 31(3):191–195
6. Filatov YD, Sidorko VI, Filatov OYu, Yaschuk VP, Heisel U, Storchak M (2009) Surface quality control in diamond abrasive finishing. Measurement Systems for Industrial Inspection VI, Proceedings of SPIE, 2009, vol 7389, art. 73892O (9)
7. Filatov OYu, Poperenko LV (2006) In situ ellipsometry of surface layer of nonmetallic transparent materials during its finish processing. Appl Surf Sci 253(1):163–166
8. Filatov YuD, Yashchuk VPFilatov OYu, Heisel U, Storchak M, Monteil G (2009) Assessment of surface roughness and reflectance of nonmetallic products upon diamond abrasive finishing. J Superhard Mater 31(5), art. 338
9. Filatov YuD, Filatov OYu, Heisel U, Storchak M, Monteil G (2010) In situ control of roughness of processed surfaces by reflectometric method. Optical Micro- and Nanometrology III, Proceedings of SPIE, vol 7718, art. 77181 J (9)
10. Filatov YuD, Filatov AYu, Syrota OO et al (2010) The influence of tool wear particles scattering in the contact zone on the workpiece surface microprofile formation in polishing quartz. J Superhard Mater 32(6):415–422
11. Filatov YuD, Sidorko VI (2011) Localization of fragments of a deposit on the workpiece surface in polishing nonmetallic materials. J Superhard Mater 33(5), art. 340
12. Filatov YuD, Monteil G, Sidorko VI, Filatov OYu (2013) Formation of a deposit on workpiece surface in polishing nonmetallic materials. Smart Sensors, Actuators, and MEMS VI. Proceedings of SPIE, vol 8763, art. 876336 (15)
13. Zhu H, Tessaroto LA, Sabia R, Greenhut VA, Smith M, Niesz DE (2004) Chemical mechanical polishing (CMP) anisotropy in sapphire. Appl Surf Sci 236(1–4):120–130
14. Budnikov AT, Vovk EA, Krivonogov SI, Danko AYa, Lukiyenko OA (2010) Anisotropy of sapphire properties associated with chemical-mechanical polishing with silica. Funct Mater 17(4):488–494
15. Wang Y, Liuu S, Peng G, Zhou S, Xu J (2005) Effects of surface treatment on sapphire substrates. J Crystal Growth 274:241–245
16. Li Y, Wu Y, Wang J, Yang W, Guo Y, Xu Q (2012) Tentative investigation towards precision polishing of optical components with ultrasonically vibrating bound-abrasive pellets. Opt Express 20(1):568–575
17. Zhou C, Zhang Q, He C, Li Y (2014) Function of liquid and tool wear in ultrasonic bound-abrasive polishing of fused silica with different polishing tools. Opt-Int J Light Electron Opt 125(15):4064–4068
18. Filatov YuD (2008) Bound-abrasive polishing of precision surfaces of nonmetallic materials. J Superhard Mater 30(1), art. 44
19. Filatov YuD, Filatov OYu, Monteil G, Heisel U, Storchak MG (2011) Bound-abrasive grinding and polishing of surfaces of optical materials. Opt Eng 50(6), art. 063401
20. Filatov YuD, Filatov OYu, Monteil G, Heisel U, Storchak M (2010) Bound-abrasive grinding and polishing of surfaces of optical materials. Current developments in lens design and optical engineering XI; and advances in thin film coatings VI, Proceedings of SPIE, 2010, vol 7786, art. 778613 (9)
21. Chen XF, Siche D, Albrecht M, Hartmann C, Wollweber J, Xu XG (2008) Surface preparation of AlN substrates. Crystal Res Technol 43(6):651–655
22. Vovk EA, Budnikov AT, Nizhankovskyi SV et al (2013) Polishing of AlN/sapphire substrate obtained by thermochemical nitridation of sapphire. Funct Mater 20(2):253–258
23. Satoh I, Arakawa S, Tanizaki K, Miyanaga M, Sakurada T, Yamamoto Y, Nakahata H (2010) Development of aluminum nitride single-crystal substrates. SEI Techn Rev 71:78–82
24. Ling Y, Low-damage grinding. In: Polishing of silicon carbide surfaces: SIMTech Technical Report (PT/01/001/PM)
25. Aida H, Takeda H, Kim S-W, Aota N, Koyama K, Yamazaki T, Doi T (2014) Evaluation of subsurface damage in GaN substrate induced by mechanical polishing with diamond abrasives. Appl Surf Sci 292:531–536

26. Nowak G, Xia XH, Kelly JJ, Weyher JL, Porowski S (2001) Electrochemical etching of highly conductive GaN single crystals. J Cryst Growth 222:735–740
27. Golabczak M (2011) Polishing of hard machining semiconductor materials made of silicon carbide. Mech Mech Eng 15(1):81–93
28. Deng Hui, Hosoya Kenji, Imanishi Yusuke, Endo Katsuyoshi, Yamamura Kazuya (2015) Electro-chemical mechanical polishing of single-crystal SiC using CeO_2 slurry. Electrochem Commun 52:5–8
29. Filatov OYu, Sidorko VI, Kovalev SV et al (2016) Polishing substrates of single-crystal silicon carbide and sapphire for optoelectronics. Funct Mater 23(1):104–110
30. Filatov YuD (2017) Diamond polishing of crystalline materials for optoelectronics. J Superhard Mater 39(6):427–433
31. Li Yaguo, Yongbo Wu, Zhou Libo, Fujimoto Masakazu (2014) Vibration-assisted dry polishing of fused silica using a fixed-abrasive polisher. Int J Machine Tools Manufact 77:93–102
32. Budnikov AT, Vovk EA, Kanishchev VN, Krivonogov SI (2012) Investigation of residual stresses in sapphire plates after grinding and polishing. Funct Mater 19(4):478–482
33. Yan W, Zhang Z, Guo X, Liu W, Song Z (2015) The effect of pH on sapphire chemical mechanical polishing. ECS J Solid State Sci Technol 4(3):108–111
34. Vovk EA (2015) Chemical-mechanical polishing of sapphire by polishing suspension based on aerosol. Funct Mater 22(2):252–257
35. Vovk EA (2015) Deagglomeration of aerosil in polishing suspension for chemical-mechanical polishing of sapphire. Funct Mater 22(1):110–115
36. Filatov YuD (2001) Polishing of aluminosilicate materials with bound-abrasive tools. Sverkhtverdye Materialy 23(3):36–49 (J Superhard Mater 2001, no. 3)
37. Filatov YuD, Sidorko VI (2005) Statistical approach to wear of nonmetallic workpiece surfaces in polishing. Sverkhtverdye Materialy 27(1):58–66 (J Superhard Mater 2005, no. 1)
38. Filatov OYu, Sidorko VI, Kovalev SV, Filatov YuD, Vetrov AG (2016) Material removal rate in polishing anisotropic monocrystalline materials for optoelectronics. J Superhard Mater 38(2):123–131
39. Bakholdin SI, Maslov VN (2015) Modeling of surface energies of sapphire crystals. Fiz Tverd Tela 57(6):1213–1219
40. Vodenitcharova T, Zhang LC, Zarudi I, Yin Y, Domyo H, Ho T, Sato M (2007) The effect of anisotropy on the deformation and fracture of sapphire wafers subjected to thermal shocks. J Mater Proc Tech 194:52–62
41. Dobrovinskaya ER, Litvinov LA, Pishchik VV (2004) Entsiklopediya sapfira (Encyclopedia of Sapphire). Institute for Monocrystals, National Academy of Sciences of Ukraine, Kharkov
42. Barybin AA, Tomilin VI, Shapovalov VI (2011) Fiziko-tehnologicheskie osnovy makro-, mikro, i nanoelektroniki (Physical-Technological Fundamentals of Macro-, Micro-, and Nano-electronics). Fizmatlit, Moscow
43. Kadleikova M, Breza J, Vesely M (2001) Raman spectra of synthetic sapphire. Microelectron J 32:955–958
44. Fainer NI, Kosinova ML, Rumyantsev YuM (2001) Thin films of silicon and boron carbonitrides. Synthesis, investigation of composition and structure, Ros. Khim. Zhurn. Obshch. im. D. I. Mendeleeva vol XLV, no 3, pp 101–108
45. Gerasimov SA, Dmitrieva NM, Kosareva TA (2013) A study of machining of leucosapphire. Nauch-Tekhn Zhurn Kontenant 12(2):84–89
46. Filatov YuD (1991) The mechanism of the surface microrelief formation in glass polishing. Sverkhtverdye Materialy 5:61–65 (J Superhard Mater 1991, no. 5)
47. Cuccureddu F, Murphy S, Shvets IV, Porcu M, Zandberger HW, Sidorov NS, Bozhko SI (2010) Surface morphology of *c*-plane sapphire (α-alumina) produced by high temperature anneal. Surf Sci 604:1294–1299
48. Butashin AV, Vlasov VP, Kanevskii VM, Muslimov AE, Fedorov VA (2012) Specific features of the formation of terrace-step nanostructures on the (0001) surface of sapphire crystals. Crystallogr Rep 57(6):824–830

49. Filatov OYu, Sidorko VI, Kovalev SV, Filatov YuD, Vetrov AG (2016) Polished surface roughness of optoelectronic components made of monocrystalline materials. J Superhard Mater 38(3):197–206
50. Kats EI (2015) Van der Waals, Casimir, and Lifshitz forces in soft substances. Usp Fiz Nauk 185(9):964–969
51. Rekhviashvili SSh (2007) The application of fractional integro-differentiation for calculating thermodynamical properties of surfaces. Fiz Tverd Tela 49(4):756–759
52. Voloshin AE, Rashkovich LN, Rudneva EB, Manomenova VL (2014) Growing the crystals. Priroda 10:62–72
53. Rogov VV, Filatov YD, Kottler W, Sobol VP (2001) New technology of precision polishing of glass optic. Opt Eng 40(8):1641–1645
54. Filatov YuD, Rogov VV (1994) A cluster model of fatigue wear mechanism of SiO_2-containing materials in polishing with tools containing bound ceria-based polishing powders. Part 1. Sverkhtverdye Materialy 3:40–43 (J Superhard Mater 1994, no. 3)
55. Deryagin BV, Abrikosova II, Lifshitz EM (2015) Molecular attraction of condensed bodies. Usp Fiz Nauk 185(9):981–1001
56. Lee HS, Jeong HD, Dornfeld DA (2013) Semi-empirical material removal rate distribution model for SiO_2 chemical mechanical polishing (CMP) processes. Precis Eng 37:483–490
57. Harima H (2002) Properties of GaN and related compounds studied by means of Raman scattering. J Phys: Condens Matter 14:967–993
58. Davydov VYu, Kitaev YuE, Goncharuk IN et al (1998) Phonon dispersion and Raman scattering in hexagonal GaN and AlN. Phys Rev B 58(19):12899–12907
59. Quay R (2008) Gallium nitride electronics. Springer, Berlin
60. Asghar K, Qasim M, Das D (2014) Effect of polishing parameters on chemical mechanical planarization of C-plane (0001) gallium nitride surface using SiO_2 and Al_2O_3 abrasives. ECS J Solid State Sci Technol 3(8):277–284
61. Filatov YuD (2018) Surface roughness of optoelectronic components in mechanical polishing. J Superhard Mater 40(1):52–57

Advanced Applications of Elliptical Vibration Cutting in Micro/Nanomachining of Difficult-to-Cut Materials

Jianguo Zhang, Norikazu Suzuki and Eiji Shamoto

Abstract Textured surfaces with sophisticated micro/nanostructures can provide advanced and useful functions. To promote widespread use of these textured surfaces with mass production, manufacturing technology of structured surfaces for ultra-precision dies and molds made of hardened steel and tungsten carbide becomes essential. Nowadays, elliptical vibration cutting (EVC) is attracting more and more attention due to its excellent machining performances in precision machining of difficult-to-cut materials. The emphasis on this chapter is the practical applications of EVC in micro/nanomanufacture process. The development of the EVC technology is introduced firstly, and then, the advantageousness of EVC in the micro/nanomachining process is explored in detail. Moreover, a unique amplitude control sculpturing method, where the depth of cut is arbitrary changed by controlling the vibration amplitude, is introduced. As following, a criterion to indicate how to obtain ductile machining of tungsten carbide is clarified by applying EVC. The feasibility of highly accurate micro/nanomachining of tungsten carbide and hardened steel is experimentally verified with a machining accuracy of about ±1 nm in the depth-of-cut direction. A series of functional surfaces with textured grooves, dimple patterns, and sinusoidal grids were successfully fabricated on tungsten carbide, hardened steel, and single-crystal silicon, which could be applied to the molding, encoder, optoelectronics, and semiconductor industries. The EVC technology is expected to promote the development of micro/nanomachining process in the actual industrial applications.

J. Zhang (✉)
School of Mechanical Science and Engineering, Huazhong University of Science and Technology, Wuhan 430074, China
e-mail: nuzhjg@gmail.com

J. Zhang · N. Suzuki · E. Shamoto
Department of Mechanical and Aerospace Engineering,
Nagoya University, Furo-cho, Chikusa-ku, Nagoya 464-8603, Japan

J. Zhang et al. (eds.), *Simulation and Experiments of Material-Oriented Ultra-Precision Machining*, Springer Tracts in Mechanical Engineering,
https://doi.org/10.1007/978-981-13-3335-4_7

1 Introduction

Textured surfaces with micro/nanostructures can provide advanced and useful functions. These functional textured surfaces can exhibit a number of novel and excellent features as compared with just simple smooth surfaces. They have been increasingly demanded in various applications including optics [1, 2], solar energy technology [3, 4], bioengineering [5, 6], self-cleaning [7, 8], and advanced manufacturing [9, 10] [11, 12]. To promote widespread use of these textured surfaces with mass production, manufacturing technology of structured surfaces for ultra-precision dies and molds made of hardened steel and tungsten carbide becomes essential. Hardened steel and tungsten carbide can be used in high-performance molding applications because of their unique and practical characteristics in physical, mechanical, thermal, and chemical properties. When the feature size of the designed structure is downscaled into micrometer or nanometer level, the fabrication technology becomes extremely challenging. To overcome this problem, several typical methods for the micro/nanostructure fabrication have been proposed so far, including the lithographic machining, the laser beam machining, the focused ion beam machining, the electron beam machining, and the diamond cutting.

Lithography, focused ion beam machining, and electron beam machining play a critical role in micro/nanostructure fabrication. They are advantageous to fabricate highly dense micro/nanostructures with a high aspect ratio and straight sidewalls. The feature size can be downscaled into tens of nanometers or even several nanometers. However, these manufacturing technologies are not available for a large structure height of several hundred micrometers due to their low removal rate and time-consuming nature. The machining efficiency is improved by laser machining; however, it is difficult to fabricate highly dense nanometer-scale structures due to the physical restriction in the minimum dimension of the focused laser beam. Considering the mechanical micromachining technologies, diamond cutting has a large dimension span in the micro/nanostructure fabrication, which is flexible and capable of many different designs. Ultra-precision diamond cutting is superior to produce ultra-precision and sophisticated structures in the feature size of several to hundreds of micrometers practically. It also has many advantages of high geometrical accuracy, good surface quality, and high machining efficiency. Diamond cutting allows a high degree of freedom for the structural design as compared with other methods, and thus, it has been widely used especially for plastic molding applications of a variety of optical elements. Combining with mass production process such as injection molding and compression molding, diamond cutting becomes available for manufacturing of high-quality and low-cost consumer products and, hence, it is quickly popularized among the related industries.

Ultra-precision diamond cutting is usually applied to the fabrication of precision parts on plastic materials, such as the soft metals including oxygen-free copper, brass, aluminum alloy, polymeric material such as PMMA, electroless nickel–phosphorus plating [13]. However, the conventional diamond cutting is not applicable directly to steel materials due to the extremely rapid thermochemical tool wear [14]. In addition, tungsten carbide is a typically hard and brittle material, and its ductile machining is extremely difficult by the conventional diamond cutting due to the generations of brittle fracture on machined workpiece surface and the excessive tool damage [15]. On the other hand, micro/nanostructured components are mass produced by glass molding and injection molding with complicated micro/nanostructured molds/dies made of hardened steel and tungsten carbide. Hence, advanced diamond cutting technology is highly required for the micro/nanostructure fabrication on hardened steel, tungsten carbide, and other difficult-to-cut materials.

2 Elliptical Vibration Cutting Process

2.1 *Principle of Elliptical Vibration Cutting*

For the last few decades, ultrasonic vibration cutting technology has been successfully applied to difficult-to-cut materials' machining [16]. In particular, Shamoto and Moriwaki [17] proposed a new cutting method, named elliptical vibration cutting (EVC). The feasibility of steel material machining and tungsten carbide machining was verified by applying EVC with single-crystal diamond (SCD) tools. Furthermore, Suzuki et al. [18] proposed a unique micro/nano-sculpturing method by controlling the vibration amplitude in EVC. It is expected to enable highly efficient fabrication of sophisticated micro/nanostructures on hardened steel, tungsten carbide, and other difficult-to-cut materials.

Figure 1 shows the schematic illustration of the EVC process presented by Shamoto et al. [19, 20]. The cutting tool is fed at a nominal cutting speed, and the tool tip is generally controlled to vibrate elliptically in the plane determined by the nominal cutting direction and the chip flow direction. In the EVC process, the tool is vibrated at an angular frequency ω and fed at a nominal cutting speed v_c in *x*-*z* plane. Then, the trajectory of the tool can be expressed as follows:

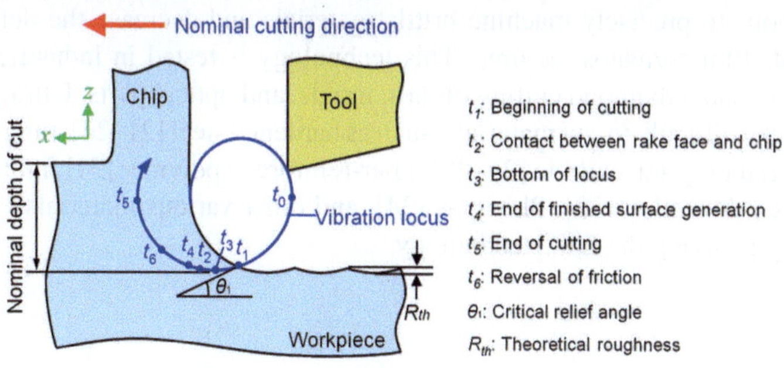

Fig. 1 Elliptical vibration cutting process [19, 20]

$$x_e = A_c \cos(\omega\tau) + v_c\tau, \quad z_e = A_d \cos(\omega\tau + \varphi), \tag{1}$$

where x_e and z_e denote x, z components of a relative position between the tool and the workpiece. x-axis and z-axis are defined to be parallel to the nominal cutting direction and the chip flow direction, respectively. A_c and A_d are mean-to-peak amplitudes in the both directions. φ is a phase shift of the vibration, which is typically set to be $-90°$. τ is the time during the elliptical vibration process. Based on the fundamental Eq. (1), the basic characteristics of the EVC process can be numerically clarified before carrying out the actual cutting, e.g., the theoretical roughness, the minimum clearance angle, and rake angle.

As shown in Fig. 1, in each cycle of the elliptical vibration, the tool starts to cut the workpiece at time t_1, and then, the workpiece material is removed in the form of a chip. After the tangential direction of the tool trajectory becomes parallel to the rake face, the cutting tool separates from the chip at t_5. In the general EVC process, the nominal cutting speed is set to be lower than the maximum vibration speed in the nominal cutting direction, ensuring that the tool is separated from the workpiece in each vibration cycle. Due to the separation in each vibration cycle, the contact surfaces of the cutting edge and the workpiece can be exposed into the surrounding gas and/or the cutting fluid, probably suppressing thermo-chemical activity of the newly developed surface of the workpiece. As a result, not only the thermochemical wear but also the adhesion and diffusion between the workpiece and the diamond tool can be suppressed efficiently. Furthermore, when the tangential direction of the tool trajectory exceeds the shear direction of the material removing, the friction direction between the tool rake face and the chip is reversed as compared with that in ordinary cutting (OC) process. This reverse of the friction leads to an increase in the nominal shear angle, resulting in significant decreases in the chip thickness and the average cutting forces [17, 19, 20]. Additionally, the tool cuts the surface that is cut in the previous vibration cycle in EVC process. The actual instantaneous uncut chip thickness, especially when the finished surface is generated in each vibration cycle, becomes generally smaller as compared with the nominal uncut chip thickness. As compared with the OC process, the average pushing stress and bending stress in workpiece deformation zone are effectively decreased in EVC process, which is advantageous to precisely machine brittle materials and decrease the deformation of fabricated micro/nanostructures. This technology is tested in industries mainly for ultra-precision diamond cutting of dies, molds, and optical parts. Ultra-precision machining of difficult-to-cut materials, such as hardened steel [21–24], tungsten alloy [25], sintered tungsten carbide [26–30], fiber-reinforced polymer [31], molybdenum [32], Co–Cr–Mo alloy [33], Plexiglas [34], and other various materials [35], were realized by applying the EVC technology.

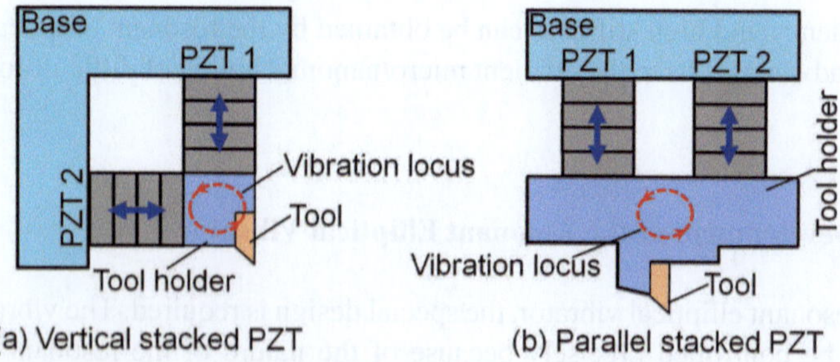

Fig. 2 Non-resonant elliptical vibrators [36, 37]

2.2 *Development of Elliptical Vibrators*

2.2.1 Development of Non-resonant Elliptical Vibrator

The initial elliptical vibrator is proposed in 1994 by Shamoto and Moriwaki [17]. The tool is vibrated by two piezoelectric actuators (PZT) arranged separately at a right angle. With sinusoidally controlling the motion of PZT, the vibration locus is generated on the tool tip by converting the linear expansion and contraction of the PZT. This elliptical vibrator is able to work at a continuous frequency range, which is quite different from the natural resonant vibration frequencies. This type of vibration device is named as non-resonant elliptical vibrator. Figure 2 illustrates two types of non-resonant elliptical vibrator used in the researches [36, 37]. Shamoto and Moriwaki [17] firstly applied the vertical stacked vibrator to investigate the EVC process directly inside a scanning electron microscope (SEM). Following that, Ahn et al. [38] and Kim et al. [39–41] developed the similar vertical stacked vibrators to fabricate the microstructures on various materials, such as nickel-plated mold steel, nickel alloy, nickel, brass, copper, and so on. Later on, another type of non-resonant elliptical vibrator named as parallel stacked PZT vibrator was developed at North Carolina State University [16]. Brehl et al. [16, 42–45] and Brocato et al. [46] applied this type of vibrator to the micro/nanostructure fabrication on the hard-plated copper and the stainless steel. Due to the decreasing of burr generation in EVC, Kim and Loh [35, 47] developed a similar parallel stacked vibrator to fabricate the micro-V-grooves on the brass and the copper.

Although the non-resonant vibrator has a relatively simpler design and their vibration parameters can be adjusted conveniently, it has a lower mechanical stiffness and a lower vibration speed. The machining efficiency is low due to the lower nominal cutting speed adopted. Moreover, the non-resonant elliptical vibrator is rarely applied to the micro/nanomachining of high-hardness molding materials, e.g., hardened steel and tungsten carbide, so far. In order to attain a higher machining efficiency in the industrial application, the tool needs to be vibrated at a much high frequency, for example, ultrasonic frequency by exciting the resonant modes of the vibrator. This type of vibrator is named as a resonant elliptical vibrator. The benefits of high vibration frequency and high stiffness can be obtained by the resonant elliptical vibrator, which is advantageous to the efficient micro/nanomachining of difficult-to-cut materials.

2.2.2 Development of the Resonant Elliptical Vibrator

For the resonant elliptical vibrator, the special design is required. The vibration locus needs to be controlled precisely because of the nature of the resonant vibrations and the phase lag between the excitation and the mechanical response. The first resonant ultrasonic elliptical vibration cutting device was developed in 1995 by Moriwaki and Shamoto [48]. It is vibrated at its first resonant mode of bending in two vertical directions by applying the sinusoidal voltage to the PZTs, whose resonant frequency is 20 kHz. The amplitude of the vibration is magnified by the step horns and maximized at the tooltip. The difficulties in the design of the resonant ultrasonic elliptical vibrator were clarified, such as the optimum shape of the horn, the position of the supporting points, heat generation. As following, a new resonant vibrator with the third resonant mode of bending was designed and manufactured by Shamoto and Moriwaki [21]. The actual application of micro/nanomachining of hardened steel by applying the SCD tool is experimentally clarified. Figure 3 shows a schematic illustration of the ultrasonic elliptical vibration tool and its third resonant mode of bending. The vibrator is resonated elliptically at about 20 kHz by exciting the two piezoelectric plates with some phase shift. In order to improve the shape accuracy by applying the EVC, a new control system is developed to stabilize the elliptical vibration locus. The above-mentioned vibrator in Fig. 3 is modified slightly by attaching two small PZT plates as feedback sensors [49]. The interference between the two directional vibrations is efficiently decreased.

In order to promote the industrial application of EVC technology, the tool vibration frequency and the vibration amplitudes need to be improved. A commercial ultrasonic elliptical vibrator [25, 50, 51] was developed for industrial use by the collaborative research between the Nagoya University and the Taga electric Co., Ltd. The vibrator can be actuated by exciting PZT actuators that re-sandwiched with metal cylindrical parts, as shown in Fig. 4 [25]. The sandwiched structure has many advantages, such as the high output power, durability, easy to be designed, and assembled. Since the vibrator is designed to have the same resonant frequencies in the second resonant mode of longitudinal vibration and the fifth resonant mode of bending vibration, it can generate large longitudinal and bending vibrations simultaneously at the same ultrasonic frequency of 39 kHz. Suzuki et al. [25] applied this vibrator to the micro/nanomachining of tungsten alloy, which is a kind of typical molding materials in the optical industry. With this commercial of vibrator, Nath et al. [28] also carried out several experimental investigations on elliptical vibration cutting of sintered tungsten carbide by applying the polycrystalline diamond (PCD) tool. Zhang et al. [24] conducted further investigations on machining performance of hardened steel using PCD tools. Moreover, some other similar vibrators are also reported in the published literature. For example, Song et al. [33] fabricated an ellip-

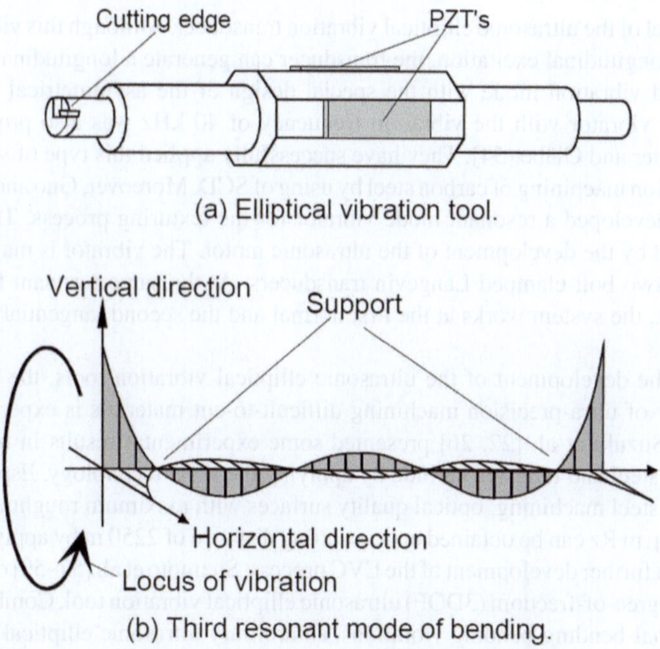

Fig. 3 Resonant elliptical vibrator with the third resonant mode of bending [21]

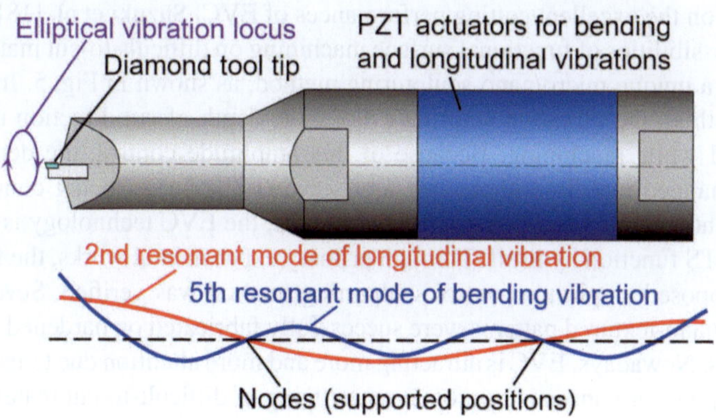

Fig. 4 Commercial elliptical vibrator with longitudinal and bending modes [25]

tical vibrator with a resonant frequency of approximately 15 kHz for mirror finishing of Co–Cr–Mo alloy by applying the SCD. This vibrator works at the first resonant mode of longitudinal vibration and the third resonant mode of bending vibration. With the same vibration modes in 20 kHz, Zhou and Hu [52] fabricated a similar vibrator for ultra-precision machining of hardened steel by applying the SCD.

Based on the resonant longitudinal-bending vibration mode, some other ultrasonic vibrators were also developed. Li and Zhang [53] presented an asymmetrical structural model of the ultrasonic elliptical vibration transducer. Although this vibrator has only the longitudinal excitation, the transducer can generate a longitudinal-bending compound vibration mode with the special design of the asymmetrical structure. A similar vibrator with the vibration frequency of 40 kHz was also proposed by Brinksmeier and Gläbe [54]. They have successfully applied this type of vibrator to the precision machining of carbon steel by using of SCD. Moreover, Guo and Ehmann [55, 56] developed a resonant mode vibrator for the texturing process. The design is inspired by the development of the ultrasonic motor. The vibrator is mainly composed of two bolt-clamped Langevin transducers. At the same resonant frequency of 28 kHz, the system works at the first normal and the second tangential vibration modes.

With the development of the ultrasonic elliptical vibration tools, the industrial feasibility of ultra-precision machining difficult-to-cut materials is experimentally verified. Suzuki et al. [22, 26] presented some experimental results in machining hardened steel and tungsten carbide by applying the EVC technology. Especially in hardened steel machining, optical quality surfaces with maximum roughness of less than 0.05 μm Rz can be obtained up to a cutting distance of 2250 m by applying EVC [21]. With further development of the EVC process, Shamoto et al. [57–59] developed a three-degree-of-freedom (3DOF) ultrasonic elliptical vibration tool. Combining the longitudinal-bending-bending vibrations, an arbitrary ultrasonic elliptical vibration locus in the 3D space can be generated. It becomes suitable to machine the 3D sculptured surfaces.

Based on the excellent cutting performances of EVC, Suzuki et al. [18] explored further possibilities of functional surface machining on difficult-to-cut materials and proposed a unique micro/nano-sculpturing method, as shown in Fig. 5. In this proposed method, the vibration amplitude along the depth-of-cut direction is actively controlled while machining. Because of this amplitude control, the depth of cut can be changed rapidly without fast tool servo (FTS) but as being controlled by the conventional FTS technology. In other words, the EVC technology is equipped with an FTS function by itself. Through primary experimental works, the feasibility of the proposed amplitude control sculpturing method was verified. Several kinds of micro/nano-textured patterns were successfully fabricated on hardened steel with SCD tools. Nowadays, EVC is attracting more and more attention due to its excellent machining performances in micro/nanomachining of difficult-to-cut materials.

3 Tungsten Carbide Machining by Applying EVC

With the rapidly developing of optoelectronics industry, demand for advanced manufacturing technology for sophisticated micro/nanostructures on optical systems is increasing drastically. For example, glass lenses with small radius and large curvature have been used in various devices to miniaturize their structures and/or to attain large storage capacity. Some micro/nano-optical gratings have been used in optical communication equipment, encoders, and digital cameras. In order to realize

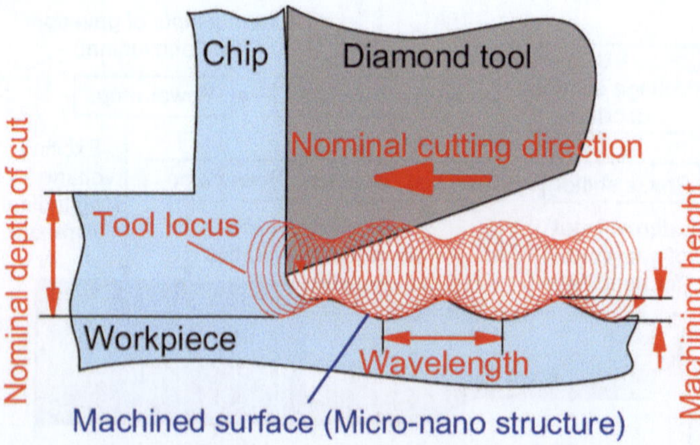

Fig. 5 Amplitude control sculpturing method in EVC [18]

mass production of those functional glass devices, tungsten carbide is heavily used in the glass molding industry because of its unique mechanical, thermal, and chemical properties. Ultra-precision diamond cutting equipped with FTS has been generally applied to fabricate sophisticated micro/nano-structures, and it has been widely used especially for a variety of plastic molding applications. However, tungsten carbide is a typically hard and brittle material, and its ductile machining is extremely difficult by the ordinary ultra-precision diamond cutting technology due to the generations of brittle fracture in the workpiece and excessive tool damage. In contrast to diamond cutting, ultra-precision diamond grinding is applicable to ductile mode machining of tungsten carbide [60–62]. Suzuki et al. [63, 64] additionally reported that the following lapping and/or polishing are also effective to achieve the higher-quality surface finish. However, it is extremely difficult to fabricate sophisticated micro/nanoscale structures especially with sharp edges by grinding and polishing. These facts impose significant restrictions in practical use of tungsten carbide and define the bottleneck of ultra-precision machining of tungsten carbide. In particular, elliptical vibration cutting technology has been the foreseen alternative to attain micro/nanomachining of hard/brittle materials.

3.1 Fundamental Grooving Experiments

3.1.1 Utilized Elliptical Vibrator

A two-degree-of-freedom (2DOF) elliptical vibrator is actuated by exciting PZT actuators that are sandwiched with metal cylindrical parts as shown in Fig. 4. The mean-to-peak vibration amplitude can be adjusted arbitrary within 2 $\mu m_{0\text{-}p}$ by controlling the amplifier gains as shown in Fig. 6 [18, 29, 36].

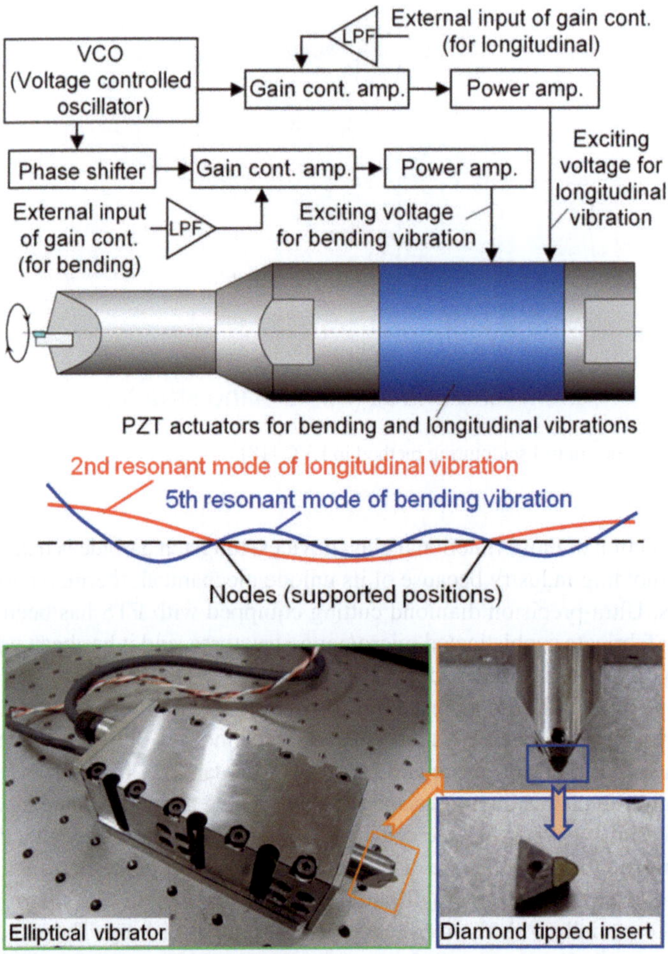

Fig. 6 2DOF elliptical vibration tool [18, 29, 36]

The vibration amplitudes along nominal cutting and depth-of-cut directions are measured by using the orthogonally aligned laser Doppler vibrometer. Amplitudes of the vibration can be obtained in both directions simultaneously. Then the elliptical vibration locus can be plotted as a Lissajous curve. For example, when the mean-to-peak amplitudes of elliptical vibration in nominal cutting and depth-of-cut directions are set to be 2 $\mu m_{0\text{-}p}$ and 1 $\mu m_{0\text{-}p}$, respectively, and a phase difference is 90°, then the measurement locus is shown in Fig. 7 [36].

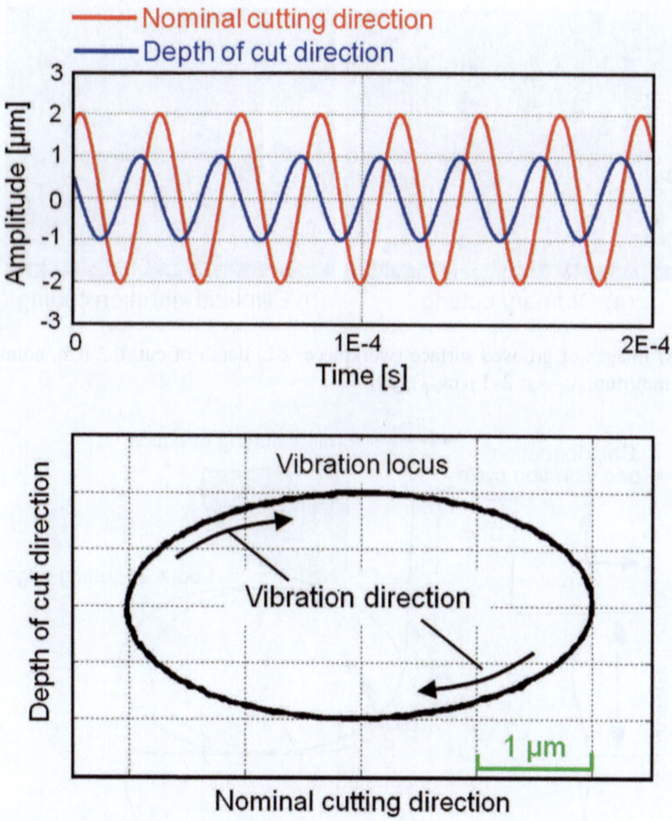

Fig. 7 Measured vibration locus (A_c–A_d: 2–1 µm$_{0\text{-p}}$) [36]

3.1.2 Comparison of EVC with OC

At first, experimental investigations are carried out to compare EVC performance against OC. A binderless tungsten carbide (average grain size: 0.3 µm; hardness: 25.48 GPa; elastic modulus: 675 GPa) labeled as BL is grooved by OC and EVC. In this experiment, the nominal cutting speed is set to be 150 mm/min. Mean-to-peak amplitudes of elliptical vibration in nominal cutting and depth-of-cut directions A_c and A_d are set to be 2 µm$_{0\text{-p}}$ and 1 µm$_{0\text{-p}}$ to realize an elliptical vibration locus. A SCD tool with a nose radius of 1 mm, a clearance angle of 10°, and a negative rake angle of −20° is used. Figure 8 shows the scanning electron microscope (SEM) images of the grooved surface at a depth of cut of 0.2 µm [29]. The groove formed by OC is filled with numerous brittle fractures. In contrast, a smooth surface without any brittle cracks and asperities can be obtained by EVC. Ductile mode machining was successfully obtained in EVC. From this fact, it can be expected that the inside of tungsten carbide grains existing at the topmost layer of the remained surface was

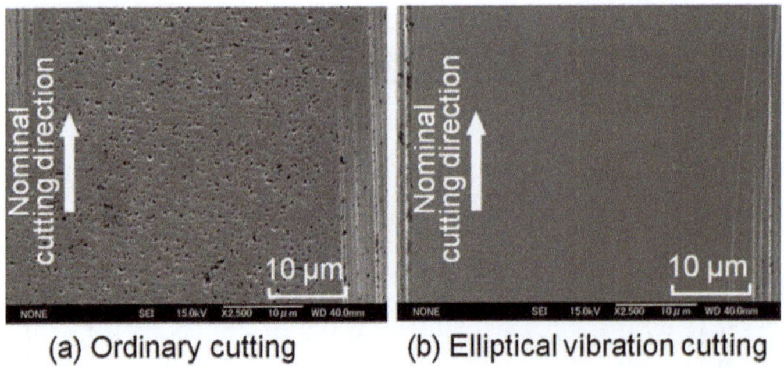

Fig. 8 SEM images of grooved surface (workpiece: BL, depth of cut: 0.2 μm, nominal cutting speed: 150 mm/min, A_c–A_d: 2–1 $\mu m_{0\text{-}p}$) [29]

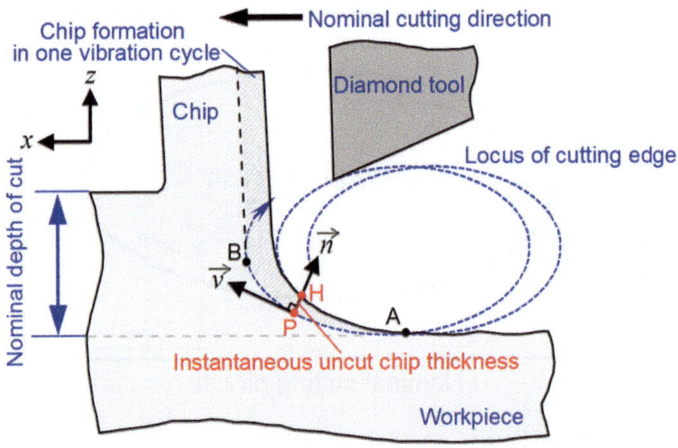

Fig. 9 Elliptical vibration cutting with thin instantaneous uncut chip thickness [65]

cut without fracture generation on the surface or pullout of the grains, which are observed in OC.

As following, the mechanism of the ductile mode machining of tungsten carbide in EVC is considered. In EVC process, the tool cuts the surface that is finished in the previous vibration cycle. Thus, the actual uncut chip, i.e., instantaneous uncut chip thickness as shown in Fig. 9, becomes extremely thin [65]. Note that this uncut chip thickness becomes generally much smaller as compared with the nominal depth of cut. In other words, the process is similar to that of micro-milling. This instantaneous uncut chip thickness in each vibration cycle becomes significantly small especially when the tool cuts the finished surface around the bottom of the elliptical vibration. Because of this process, the actual depth of cut becomes smaller than the critical value for ductile machining due to the size effect in fracture toughness, resulting in significant improvement of nominal critical depth of cut for ductile machining of tungsten carbide.

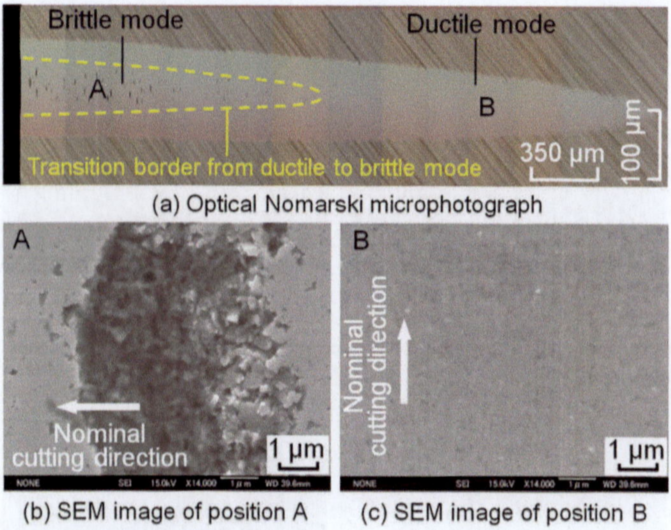

Fig. 10 Machined groove with the increasing of depth of cut [29]

Furthermore, the influence of the depth of cut on the surface quality is also investigated. Figure 10 demonstrates a groove surface of BL, which is machined with a circular vibration (1 $\mu m_{0\text{-}p}$) at a cutting speed of 150 mm/min [29]. The depth of cut is gradually increased from 0 to 2 μm. Surface quality changes depending on the depth of cut. The transition border from ductile to brittle modes is observed at the depth of cut of about 1.5 μm. In the brittle machining, agglomerated grains seem to be pulled out at one time from the machined surface. It is also considered that the defect morphology may relate to the chip pulling-up motion in EVC process.

And then, the influence of nominal cutting speed on machined surface quality is also examined by several grooving experiments. The machined surfaces of BL at a depth of cut of 0.5 μm are shown in Fig. 11 [29]. It is confirmed that higher-quality surfaces in the ductile mode can be attained with a lower nominal cutting speed of 250 mm/min. On the contrary, poor surfaces with numerous defects are observed distinctively at a higher nominal cutting speed of 1100 mm/min. The defect size is less than several hundred nanometers, which is almost identical to the dimension of grain size. This cutting speed dependency of surface quality is considered to be associated with the cut volume in each vibration cycle. It should be noted that the instantaneous uncut chip thickness is almost proportional to the nominal cutting speed. Thus, the cutting forces decrease and the strain energy at the deformed zone is reduced when the nominal cutting speed decreases. With a small nominal cutting speed, small forces and strain energy are generated in the cutting regime, which is advantageous to suppress crack propagations.

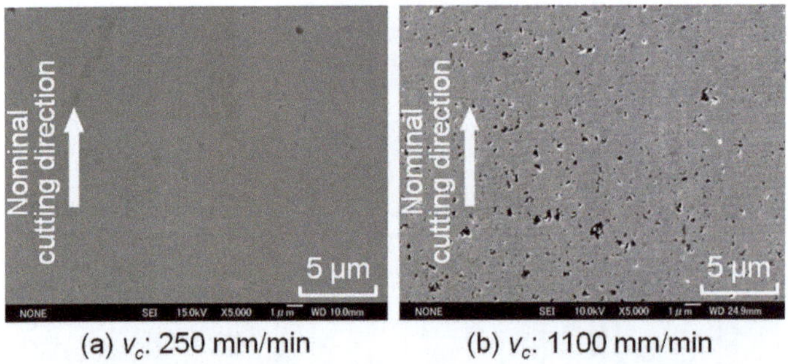

Fig. 11 Effect of nominal cutting speed on surface quality (workpiece: BL, amplitudes: 2–1 $\mu m_{0\text{-}p}$, depth of cut: 0.5 μm) [29]

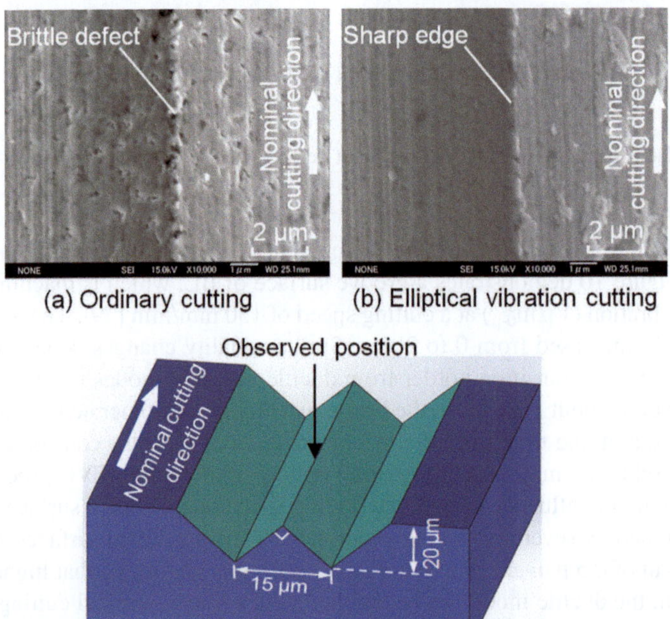

Fig. 12 V-grooved surface with sharp edge fabricated by elliptical vibration cutting [36]

3.2 *Feasibility of Micro/Nanostructure Fabrication on Tungsten Carbide*

Based on the critical findings in grooving investigations, optimal machining conditions are utilized to ductile cutting of tungsten carbide. Binderless tungsten carbide of BL (with grain size of 0.3 μm) and binder-containing tungsten carbide of Co (with binder phase of Cobalt and grain size of 0.5 μm) are applied to the following

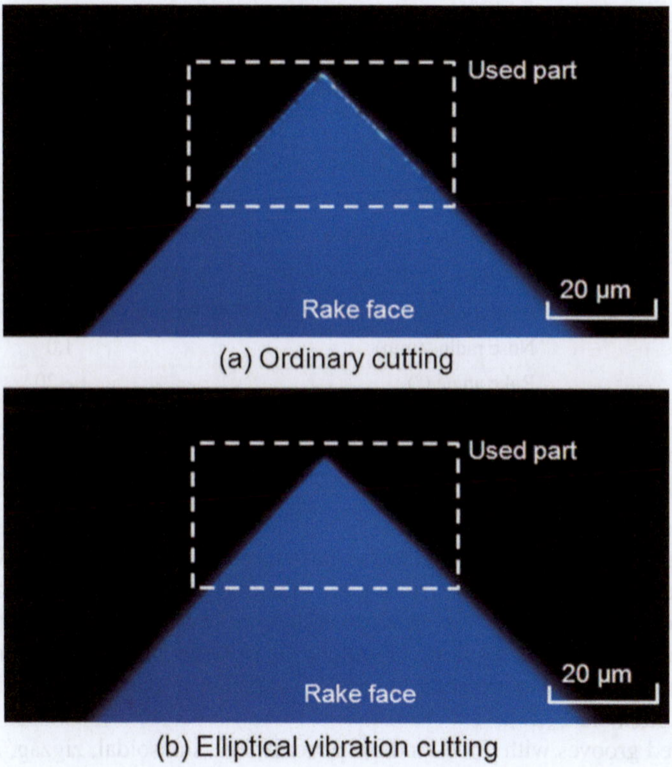

Fig. 13 Microphotographs of V-shaped tool edge in different cutting methods [36]

experimental investigations. In order to investigate the feasibility of microstructure fabrication with sharp edges, V-groove machining experiments are conducted on binder-containing tungsten carbide by using V-shaped diamond tool. Two V-shaped grooves with a depth of 20 μm are machined, where the depth of groove is increased by cumulating the grooving with a depth of cut of 0.5 μm 40 times. A constant nominal cutting speed of 250 mm/min is adopted. A V-shaped single-crystal diamond tool with a nose radius of less than 0.5 μm, an included angle of 90°, a rake angle of −20°, and a clearance angle of 10° is used. As shown in Fig. 12 [36], V-grooved surface with sharp edges can be obtained with no considerable defect generation by applying EVC. However, V-grooved surface formed by OC is filled with numerous brittle fractures. Hence, ultra-precision machining of sintered tungsten carbide is not feasible by means of OC technology.

The microphotographs of the cutting edges after the V-shape grooving experiments are shown in Fig. 13 [36]. No adhesion and no serious tool wear can be observed in the elliptical vibration cutting.

As following, the feasibility of micro/nanoscale sculpturing is clarified through micro-textured grooving and nano-dimple pattern machining experiments. The

Table 1 Experimental conditions for sculpturing of micro/nano-structures [29]

Conditions		A	B
Cutting conditions	Nominal cutting speed (mm/min)	150	284
	Pick feed (μm)	–	100
	Maximum depth of cut (μm)	1.2	0.2
Vibration conditions	Frequency of elliptical vibration (kHz)	36.2	36.2
	Frequency of amplitude oscillation (Hz)	42–84	60
	Amplitude in depth-of-cut direction ($\mu m_{0\text{-}p}$)	0.5–1	0.5–1
	Amplitude in nominal cutting direction ($\mu m_{0\text{-}p}$)	2	2
Tool	Nose radius (mm)	1.0	1.0
	Rake angle (°)	−20	−20
	Clearance angle (°)	10	10
Workpiece	Material	BL	Co

amplitude control sculpturing method as introduced in Fig. 5 is applied, where the depth of cut is regulated by controlling the vibration amplitude in the depth-of-cut direction during machining. The envelope of the cutting edge trajectory is transferred to the machined surface, resulting in sculpturing of target micro/nano-structures on the workpiece. The experimental conditions and cutting parameter selections are presented in Table 1 [29].

Textured grooves with three kinds of patterns, i.e., sinusoidal, zigzag, and ramp waves, are fabricated under Condition A. Figure 14 shows the microphotographs of micro-textured grooves and the measured surface profile along the nominal cutting direction [29]. The amplitude of these wave commands is set to be about 0.5 μm. The profiles of the textured grooves are measured by an optical surface profiler (ZYGO NewView6200). The experimental results indicate that the grooves with ultra-precision micro/nano-textures can be machined successfully in ductile mode on tungsten carbide, and a defect-free smooth surface can be successfully obtained by EVC.

Figure 15 shows the photograph of dimples sculptured under Condition B and measured profiles [29]. In this machining, the mean-to-peak amplitude in the depth-of-cut direction is controlled by sinusoidal wave pattern, which is changed from 0.5 $\mu m_{0\text{-}p}$ to 1 $\mu m_{0\text{-}p}$. The maximum depth of cut is, however, set to be about 0.2 μm, which is less than the mean-to-peak amplitude variation of 0.5 μm. Consequently, independent concave dimples with a maximum depth of about 0.2 μm were machined successfully in ductile mode on tungsten carbide. As shown in Fig. 16, the principle of independent dimples fabrication is also illustrated in detail by applying amplitude control sculpturing method in EVC [65]. Note that these micro/nano-sculpturing patterns cannot be attained by OC methods.

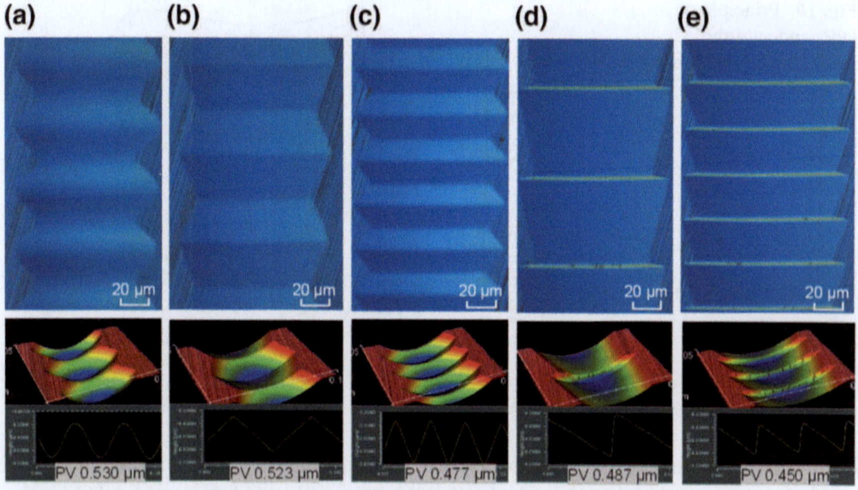

Fig. 14 Microphotographs of textured grooves: **a** sinusoidal (60 Hz), **b** ramp (42 Hz), **c** ramp (84 Hz), **d** zigzag (42 Hz), and **e** zigzag (84 Hz) [29]

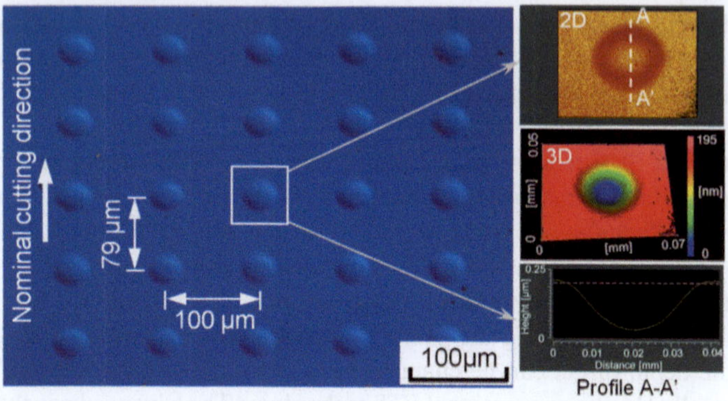

Fig. 15 Microphotograph of fabricated dimple structures and their measured profiles [29]

4 Hardened Steel Machining by Applying EVC

Considering widespread use of textured surface applications and their mass production, manufacturing technology of functional textured surfaces for a variety of materials, especially for steel materials, is greatly required. Steel is a typical material that is widely used for mechanical elements and molding. However, Paul et al. [14] found that in ferrous material machining chemical affinity of materials triggers extreme chemical reaction of SCD tools. Because of this chemical nature, conventional diamond cutting is not applicable to machining of steel due to rapid tool wear and surface deterioration. Many researchers have dedicated to attain the conventional

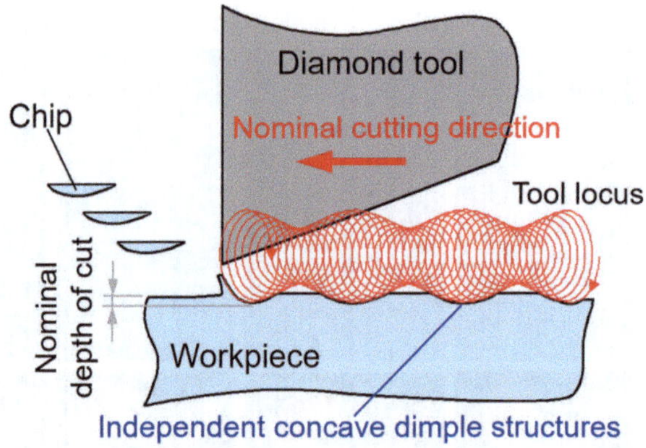

Fig. 16 Principle of independent dimple structures fabrication [65]

diamond machining of steel materials with suppressing tool wear propagation, such as cutting process with inert atmosphere and carbon-saturated atmosphere [66], cryogenic cutting by cooling the tool–workpiece system [67], nitriding process based on workpiece modifications for avoiding chemical reactions between the diamond tool and the workpiece material [68] and ultra-intermittent cutting method to reduce tool wear by decreasing the contact time [69]. However, none of these methods was found applicable to industrial applications. These restrictions in conventional machining methods in fact define the bottleneck of practical use of hardened steel with functional surfaces.

In particular, EVC technology is considered as a potential candidate for the functional surface fabrication on steel materials. With the development of ultrasonic EVC devices, Shamoto and Moriwaki [21] verified the feasibility of steel material machining by the use of SCD tools. Based on successful cutting performance of EVC, Suzuki et al. [18] explored further possibilities of functional surface machining on steel material by controlling the vibration amplitudes. Because of this amplitude control, the depth of cut can be changed precisely and rapidly without FTS but as being controlled by the conventional FTS technology. The advanced advantageous of EVC and FTS is efficiently combined by this amplitude control sculpturing method.

4.1 Applications of Elliptical Vibration Sculpturing Method in Hardened Steel Machining

4.1.1 Investigation of Machining Accuracy

In order to investigate machining accuracy on hardened steel (JIS: SUS420J2, UTS $\geq$ 540 MPa, HRC53) by the amplitude control sculpturing method in EVC, nano-stair-stepping structures were fabricated with a trapezoidal amplitude command wave at 100 Hz. The structure height of trapezoidal command is controlled to change a depth

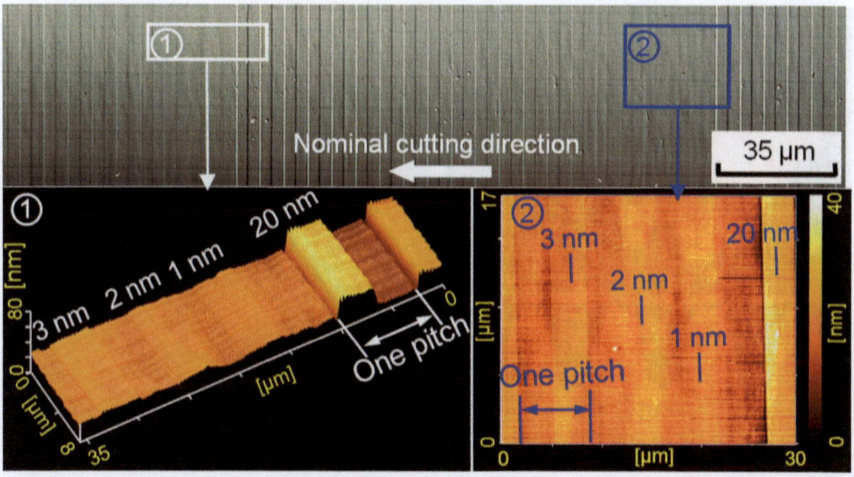

Fig. 17 Optical Nomarski microphotograph and AFM images of machined trapezoidal structures [36, 70]

of cut from 1 to 20 nm, and a pitch of the trapezoidal structures is set to be a constant value of 7 μm. The machine tool is simply controlled to machine a plane surface with a pick feed of 5 μm at a constant nominal cutting speed of 42 mm/min. The vibration amplitude in the depth-of-cut direction is controlled in synchronization with the cutting feed along the nominal cutting direction. Figure 17 shows the experimental results. Nano-structures with a designed structure height of 1 nm are marginally distinctive from the machined surface on hardened steel [36, 70]. From the measured results, it is confirmed that nano-structures can be machined with surprisingly high accuracy of about ±1 nm in the depth-of-cut direction by applying the amplitude control sculpturing method.

4.1.2 Feasibility of Nano-structure Fabrication

As following, several different nano-patterns are fabricated by applying the amplitude control sculpturing method in EVC. In order to avoid the flank face contact to the fabricated structures, the slope of amplitude change in EVC needs to be less than the clearance angle of the cutting edge at least. Figure 18 shows a trapezoidal structure machined by an amplitude corresponding to 25 nm structure height with a pitch value of 2 μm [36, 37]. The fabricated patterns are measured by using the atomic force microscope and the optical microscope, respectively. As the slope of the trapezoidal structure is kept being 30°, the tool with a clearance angle of 40° is utilized here.

In what follows, textured grooves with a 20 nm structure height are also sculptured on a hardened steel workpiece, and the feasibility of nano-structure sculpturing is investigated. Figure 19 shows several textured grooves with different patterns, i.e.,

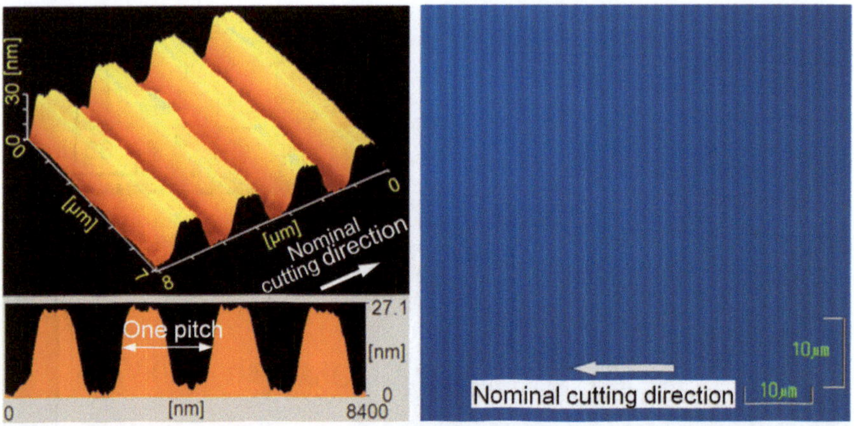

Fig. 18 Trapezoidal structures with the structural height of 25 nm and pitch of 2 μm [36, 37]

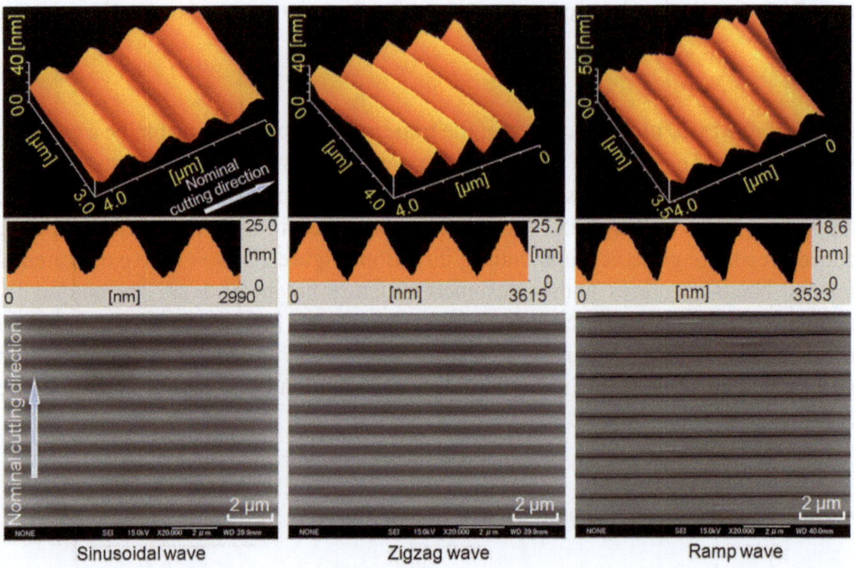

Fig. 19 Nano-structures with the structural height of 20 nm and wavelength of 1 μm [36, 37, 71]

sinusoidal, zigzag, and ramp waves, with a steady amplitude command at 100 Hz [36, 37, 71]. The fabricated patterns were measured by using the atomic force microscope and the scanning electron microscope, respectively. Experimental conditions are summarized in Table 2 [71].

Table 2 Experiment conditions for nano-textured grooves machining [71]

Cutting conditions	Nominal cutting speed (mm/min)	6
	Maximum depth of cut (μm)	3
	Cutting fluid	Oil mist (PS-FM-A)
Vibration conditions	Frequency (kHz)	36.2
	Phase shift (°)	−90
	Amplitude in depth-of-cut direction ($\mu m_{0\text{-}p}$)	0.48–0.5
	Amplitude in nominal cutting direction ($\mu m_{0\text{-}p}$)	0.5
Tool	Material	Single-crystal diamond
	Nose radius (mm)	1
	Rake angle (°)	0
	Clearance angle (°)	10
Workpiece	Hardened steel	SUS420J2 (HRC53)

4.1.3 Feasibility of Three-Dimensional Structure Machining

In the previous research, some micro/nano-structures have been successfully fabricated on hardened steel by applying the proposed amplitude control sculpturing method [18]. For example, sinusoidal commands to control the vibration amplitude were input to the elliptical vibration control system during machining and the phase of the sinusoidal commands was changed by 180° in every cutting feed, so that hexagonal dimple patterns can be sculptured based on vibration conditions, cutting conditions, and tool geometries. Table 3 summarizes the experimental conditions [36].

Figure 20 shows the optical microphotographs and a cross-section profile of the fabricated hexagonal dimple patterns [36, 37]. The hexagonal dimple patterns with a measured dimple depth of 0.65 μm and a side length of about 36.14 μm.

In order to investigate the feasibility of practical application, the proposed machining method is applied to a sophisticated three-dimensional micro/nano-structure sculpturing. Figure 21 shows the target machining geometry, i.e., an angle grid surface [36, 70, 71].

The height of the arbitrary position on the angle grid surface can be given by Eq. (2), which is presented by:

$$z = A_x \sin(2\pi x/\lambda_x) + A_y \sin\left(2\pi y/\lambda_y\right), \quad (2)$$

where A_x and A_y are amplitudes of the sine functions in the x-direction and the y-direction, respectively. λ_x and λ_y are the corresponding wavelengths.

A SCD tool with a nose radius of 1 mm, a rake angle of 0°, and a clearance angle of 10° is used. When the target amplitude is set to be $A_x = A_y = 0.5$ μm, the wavelength was set to be $\lambda_x = \lambda_x = 150$ μm from the restrictions of tool nose radius in the nominal

Table 3 Experiment conditions for hexagonal dimple pattern fabrication [36]

Cutting conditions	Nominal cutting speed (mm/min)	650
	Maximum depth of cut (μm)	5
	Pick feed (μm)	31
Vibration conditions	Frequency of elliptical vibration (kHz)	36.2
	Frequency of amplitude oscillation (Hz)	100
	Phase shift (°)	90
	Amplitude in depth-of-cut direction ($\mu m_{0\text{-}p}$)	0.88–1.64
	Amplitude in nominal cutting direction ($\mu m_{0\text{-}p}$)	2
Tool	Material	Single-crystal diamond
	Nose radius (mm)	1
	Rake angle (°)	0
	Clearance angle (°)	10
Workpiece	Hardened steel	SUS420J2 (HRC53)

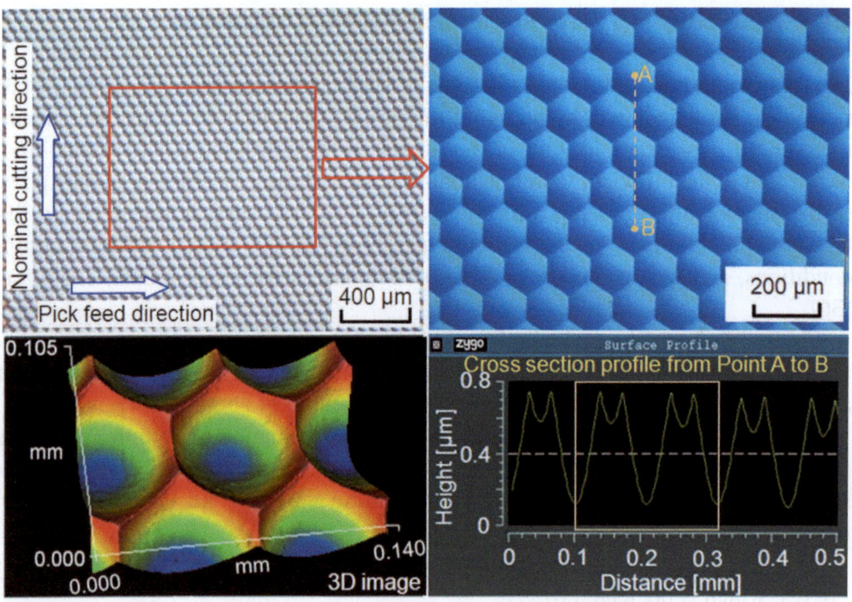

Fig. 20 Microphotographs of sculptured hexagonal dimples [36, 37]

pick feed direction. The sinusoidal structure is machined by the amplitude control sculpturing method with a sinusoidal amplitude command at 100 Hz. The nominal cutting speed is set to be 900 mm/min. On the other hand, the height position along the nominal pick feed direction is controlled by utilizing positioning of the ultra-precision machine tool itself. The pick feed is set to be a constant value of 3 μm.

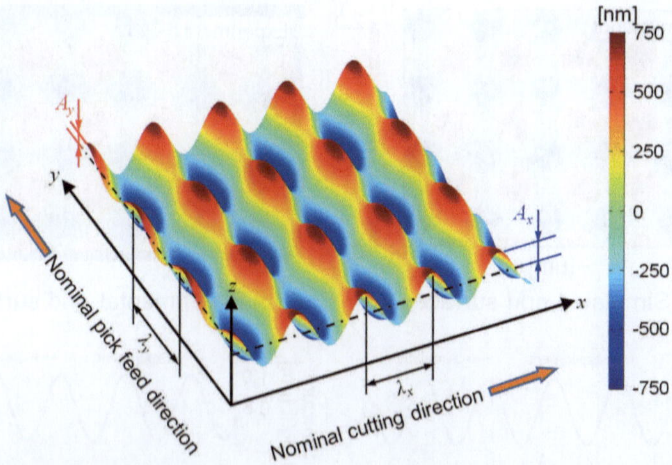

Fig. 21 Schematic of the angle grid surface [36, 70, 71]

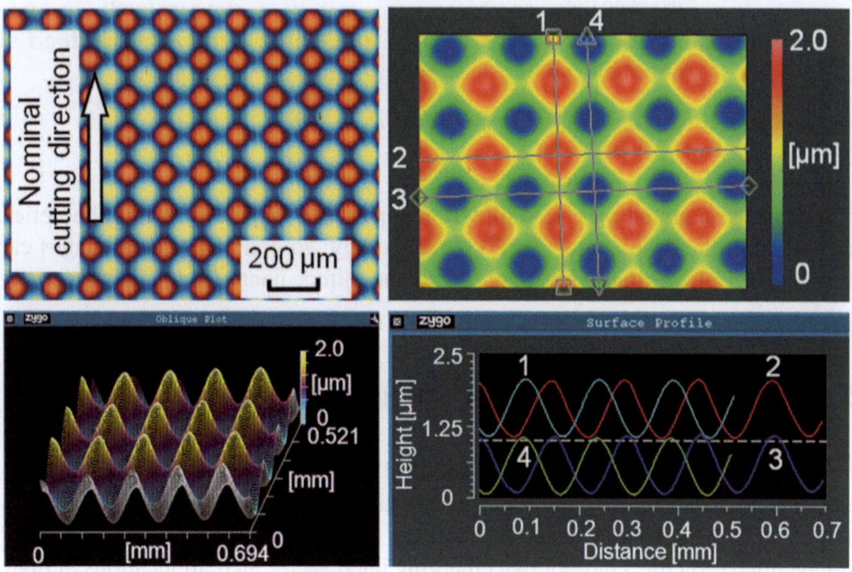

Fig. 22 Machined angle grid surface with the structure height of 1 μm and the wavelength of 150 μm [36, 70]

Figure 22 shows a microphotograph and a profile of the machined angle grid surface measured by optical microscopes (Nikon MM-40 and ZYGO NewView6200) [36, 70]. It was confirmed that the periodic micro/nano-structure with smooth surface is obtained on the hardened steel workpiece.

Figure 23 shows a comparison of the simulated and the experimental results [36]. It was confirmed that machined height and the wavelength along the nominal cutting

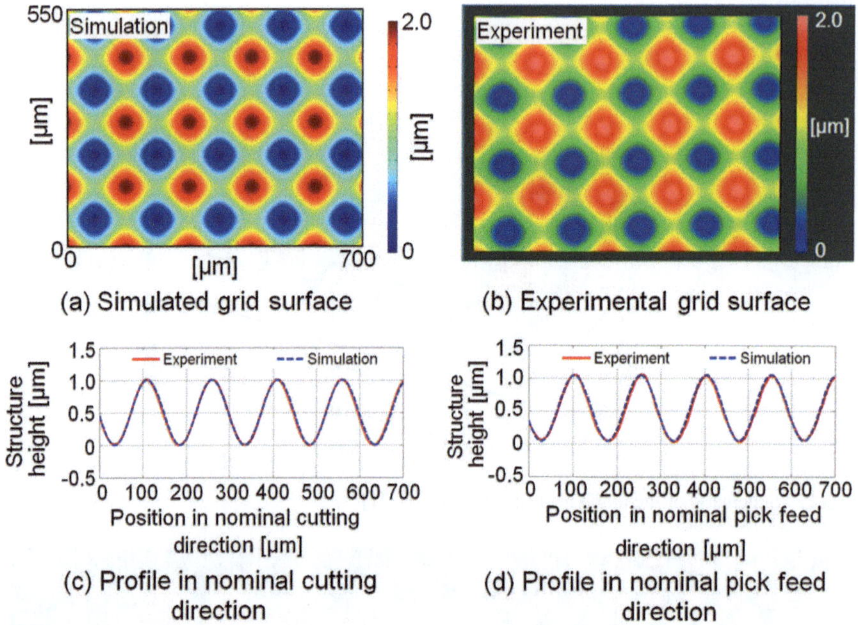

Fig. 23 Comparison of simulation and experimental result [36]

direction and nominal pick feed direction correspond accurately to the theoretical values. With the development of EVC technology, arbitrary grayscale images can also be precisely sculptured by applying the amplitude control sculpturing method.

5 Single-Crystal Silicon Machining by Applying EVC

Single-crystal silicon is a basic technological material in semiconductor and optoelectronics industries for its excellent properties of high hardness, high wear resistance, lightweight, excellent stability, and low oxides formability [72–75]. Recently, silicon microstructures have drawn emerging interests for their applications of high-performance imaging, concentration, and illumination in fields of semiconductors [76, 77] and infrared optics [78, 79]. However, the precision fabricating of silicon microstructures is challenging due to its brittle fracture. In particular, the feasibility of ductile cutting of silicon with SCD tool has been investigated experimentally and theoretically over the last two decades. It has been demonstrated that with a depth of cut (DOC) ranging from 50 to 250 nm, nanometric surface finish, sub-micrometric form accuracy, and minimum subsurface damages of silicon can be achieved by ductile cutting. However, the extremely small DOC utilized in ductile cutting not only requires extremely high rigidity of ultra-precision machine tools, but also introduces difficulties in decreasing undesired inclination of workpiece surface with respect to workpiece holder and alignment errors of workpiece. Therefore, improving the nominal critical DOC for ductile cutting is critical to simplify the setting up of ductile cutting of silicon, which also facilitates the industrial applications of low-cost machine tools.

5.1 Fundamental Grooving Experiments with EVC Versus OC

Fundamental grooving experiments with EVC and OC are firstly carried out to obtain the first impression of machinability of silicon. For each cutting process, the nominal DOC is increased from 0 to the maximum value of 2.5 μm, while the nominal cutting speed is fixed as 200 mm/min. The mean-to-peak amplitudes in the nominal cutting direction and the DOC direction are 2 $\mu m_{0\text{-}p}$ and 1 $\mu m_{0\text{-}p}$, respectively. Figure 24 presents microphotographs of grooves machined at a nominal depth of cut of 0.4 μm through the OC and the EVC, respectively [65]. While the groove formed by the OC is filled with brittle fractures, a smooth surface can be obtained successfully by applying the EVC in ductile cutting mode. The nominal critical DOC in the EVC is 475 nm, which is 12.5 times higher than the critical DOC of 38 nm in OC.

Due to the ultrasonic vibration frequency and the small nominal cutting velocity, the instantaneous uncut chip thickness is almost within several tens of nanometers by applying the EVC, as shown in Fig. 9. Fractures initiate before pronounced permanent plastic deformation under conventional machining of silicon, due to its lower fracture strength than yield strength. However, with the extremely small instantaneous uncut chip thickness, very large stress can be achieved at small volumes, which triggers

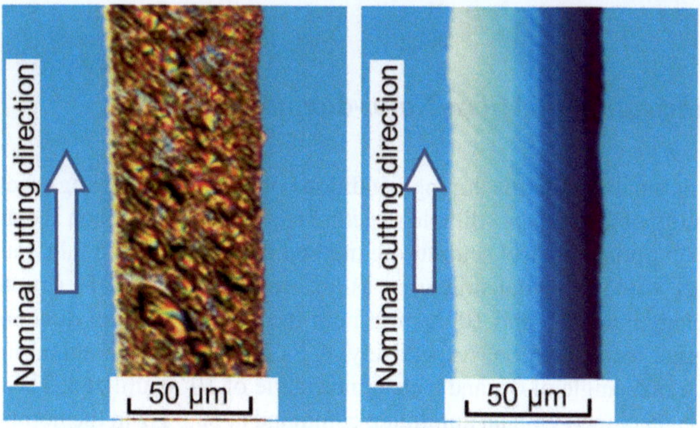

(a) Ordinary cutting (b) Elliptical vibration cutting

Fig. 24 Micro-grooving of silicon by applying OC and EVC [65] dislocation activities under high shear stress. Since the actual DOC is smaller than the critical value for ductile machining due to the size effect in fracture toughness of silicon, the EVC results achieve a significant improvement of the nominal critical DOC for ductile cutting of silicon. Moreover, the actual rake angle acting on cutting regime becomes negative due to the tool moving up from the bottom of elliptical vibration locus, as shown in Fig. 9. It was reported by Komanduri et al. [80] and Yan et al. [81] that the negative rake angle can produce hydrostatic compressive pressure that is advantageous to suppress crack propagation and obtain ductile cutting of silicon.

Additionally, the surfaces machined by the EVC and OC in brittle cutting modes are also examined. Figure 25 shows that the nominal DOC is gradually increased from 0.6 to 1.5 μm, which has a strong influence on machined surface quality for both types of cutting [65]. However, as compared with the brittle defects generation in OC process, the deterioration of machined surface quality is less pronounced in EVC process. Moreover, the defect geometry is quite different in EVC from that in OC. While there are cleavage defects caused by the formation of continuous brittle defects on the whole machine surface in OC, there are only discrete defects with smaller defect sizes generated in EVC. Even though at the largest DOC of 1.5 μm, the brittle defect propagation can also be efficiently suppressed and smooth side edges of grooves can be obtained by applying the EVC. For the OC process, the generated defects are mainly in the size of about tens of micrometers and fulfilled with the whole coarse machined surface, suggesting that huge brittle defects are mainly caused by the lateral crack system. On the other hand, small pits with the size of about several micrometers are generated on the machined surface in EVC, which are primarily caused by the median crack system. As the threshold load for median cracks is less than that for lateral cracks, the lateral cracks that cause huge brittle fracture can be successfully suppressed in the EVC process with the small cutting forces.

5.2 *Fabrication of Micro/Nano-dimple Structures*

With the optimal ductile machining conditions obtained in grooving investigations, the feasibility of micro/nano-dimple structures fabrication on silicon is clarified. The proposed amplitude control sculpturing method is first applied to fabricate an angle grid surface with sinusoidal structures, as illustrated in Fig. 21. The structure height and wavelength are 0.5 and 100 μm in both the nominal cutting direction and the nominal pick feed direction, respectively. A SCD tool with a nominal nose radius of 1 mm, a rake angle of 0°, and a clearance angle of 10° is utilized. The sinusoidal structure is machined with an amplitude command at 33.3 Hz in the nominal cutting direction. The mean-to-peak vibration amplitude is fixed to 2 $\mu m_{0\text{-}p}$ in the nominal cutting direction, while varied within 0.5 $\mu m_{0\text{-}p}$ to 1 $\mu m_{0\text{-}p}$ in the DOC direction. The nominal cutting speed is set to be 200 mm/min, and the maximum DOC is adopted as 2 μm. On the other hand, the height position along the pick feed direction

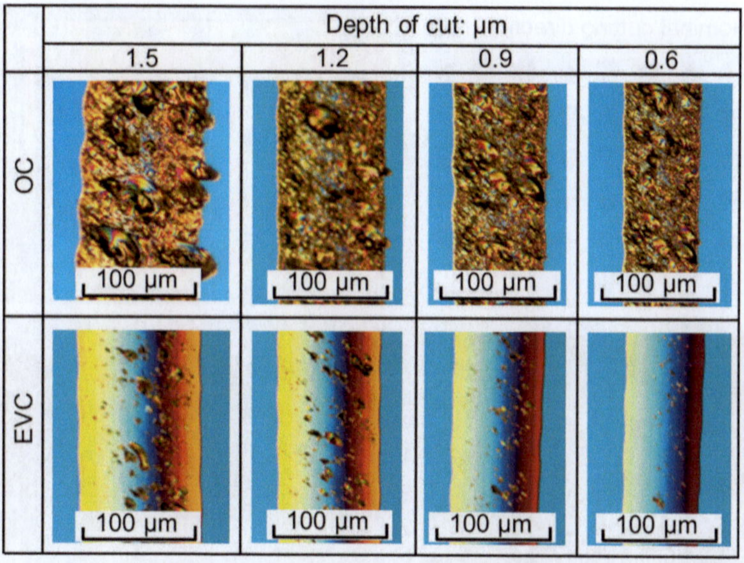

Fig. 25 Grooves formed by OC and EVC in brittle cutting with different depth of cut [65]

is controlled by utilizing positioning of the ultra-precision machine tool itself. The pick feed value is set to be 5 μm. As a round cutting tool with arc cutting edge is used, the maximum effective depth of cut d_e is calculated to be 304 nm by the following equation [82]:

$$d_e = R - \sqrt{(R - d)^2 + \left(\sqrt{(2R - d)d} - f\right)^2} \tag{3}$$

where R is the tool radius, d is the maximum depth of cut, and f is the pick feed value. Since the calculated maximum effective DOC is smaller than the critical DOC of 456 nm, the ductile sculpturing of silicon can be guaranteed by applying the above-mentioned machining conditions. Figure 26 shows a microphotograph and a profile of the machined angle grid surface measured by optical microscope [65]. It is confirmed that the periodic micro/nano-structures with smooth surface can be successfully obtained in ductile mode. The machined structural height (0.483 μm) and the wavelength (100 μm) along the cutting direction and the pick feed direction correspond accurately to the designed target values. It should be noted that even though the cutting distance is increased into 45 m, the tool wear deterioration is not serious and ductile machining can also be guaranteed.

In order to investigate the chip formation mechanism in elliptical vibration cutting of silicon, chips are also collected in ductile machining the angle grid surface. Figure 27 shows SEM photographs of the chips collected during sculpting of silicon at the nominal cutting distance of 45 m. The ductile needle-type-like segments are

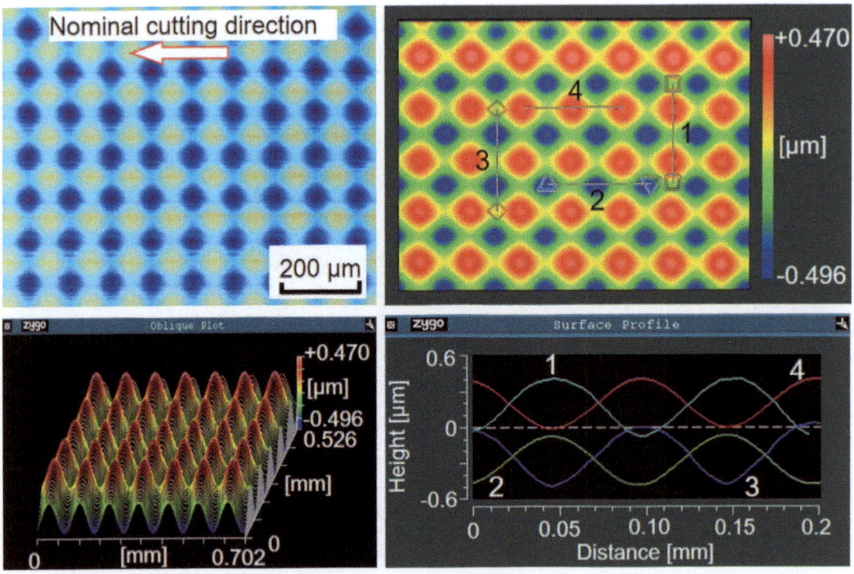

Fig. 26 Angle grid surface fabricated on silicon workpiece in ductile mode [65]

Fig. 27 Ductile needle-type-like chips of silicon in EVC

observed, while thin chip segments are almost discontinuous. This fact indicates that ductile mode machining is attained successfully in each vibration cycle.

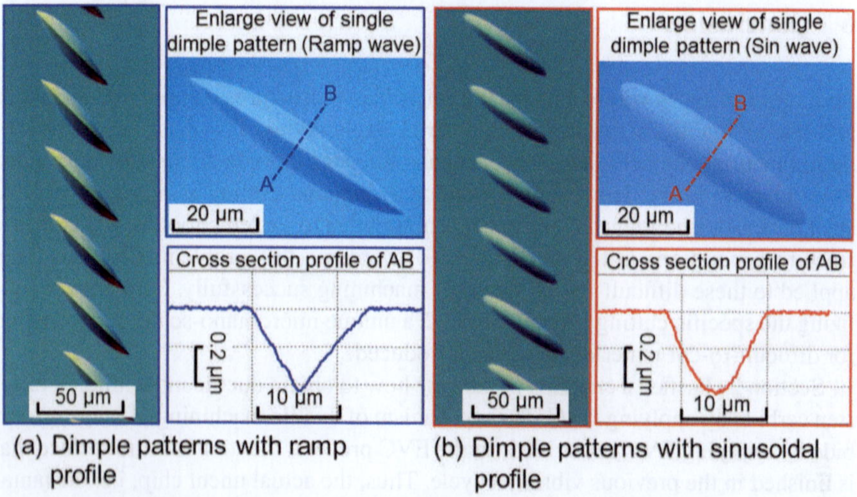

Fig. 28 Microphotograph and profiles of fabricated silicon dimple structures [65]

As following, the independent concave dimples are also fabricated on the silicon workpiece. As shown in Fig. 16, the independent dimples can be fabricated with a small value of nominal DOC that is less than half of the amplitude variation in the DOC direction. In the present fabrication process, the mean-to-peak amplitude in the nominal cutting direction is adopted as 2 $\mu m_{0\text{-}p}$, and the vibration amplitude in the nominal DOC direction is appropriately controlled by sinusoidal command wave and ramp command wave, which is changed from 0.5 $\mu m_{0\text{-}p}$ to 1 $\mu m_{0\text{-}p}$ at 100 Hz. The nominal cutting speed is set to be 200 mm/min. As the critical nominal DOC in ductile machining of silicon is 456 nm with vibration amplitudes A_c–A_d of 2–1 $\mu m_{0\text{-}p}$, the maximum DOC should be less than the above-mentioned value in order to realize the ductile sculpturing process. Figure 28 shows measured photographs of dimples sculptured on silicon [65]. The maximum nominal DOC in the microstructure fabrication is set to be about 0.45 μm, which is less than half of the amplitude variation of 0.5 μm. Consequently, the independent concave dimples with a maximum designed depth of about 0.45 μm are assumed to be machined in ductile cutting mode. The cross-section profile of the fabricated single pattern is measured by using a surface profiler (Taylor Hobson PGI 1250S). Although the measured structure height of the sinusoidal and ramp pattern is about 441 nm and 433 nm, respectively, the machining error along the nominal DOC direction is smaller than 20 nm for the current experimental setup. It demonstrates that the feasibility of high-quality and high-precision silicon microstructure fabrication by applying the amplitude control sculpturing method in EVC. The further development of the EVC technology is expected to promote the development of sophisticated silicon microstructures for advanced industrial applications.

6 Conclusion

Structured surfaces with sophisticated micro/nano-structures can provide advanced and useful functions for various industrial applications. For mass production of these functional devices, hardened steel and tungsten carbide are heavily demanded in molding industry. However, conventional diamond cutting is not applicable to steel materials and tungsten carbide machining due to extreme tool wear and brittle fracture generations in the workpiece. To overcome this problem, EVC has been applied to these difficult-to-cut materials machining successfully. Furthermore, utilizing the specific cutting process of EVC, a unique micro/nano-sculpturing method for difficult-to-cut materials was also introduced.

Section 2 clarified a criterion to indicate how to obtain ductile machining of tungsten carbide by applying EVC. The mechanism of ductile machining of tungsten carbide by applying EVC is clarified. In the EVC process, the tool cuts the surface that is finished in the previous vibration cycle. Thus, the actual uncut chip, i.e., instantaneous uncut chip thickness, becomes extremely thin. Based on this process, the actual cutting thickness becomes smaller than the critical value for ductile machining. In what follows, the influence of cutting conditions on the surface quality is investigated in an empirical manner. Based on these fundamental results, ultra-precision textured grooves and a dimple pattern are subsequently fabricated on tungsten carbide in ductile mode by applying the amplitude control sculpturing method in EVC.

Section 3 clarified the highly accurate micro/nano-structure fabrication on hardened steel by experimental investigations. It is confirmed that nano-structures can be successfully fabricated with a machining accuracy of about ±1 nm in the depth-of-cut direction. In what follows, accurate nano-sculpturing with a structure height of 20 nm and wavelength of 1 μm is attained by applying the amplitude control commands with sinusoidal, zigzag, and ramp waves. Finally, a sophisticated 3D micro/nano-structure, i.e., a sinusoidal grid surface, is subsequently carried out on a hardened steel workpiece. The experimental results verify the feasibility of efficient and accurate micro/nanomachining by the proposed sculpturing method.

In order to verify the feasibility of precision machining other difficult-to-cut materials, Sect. 4 investigates the ultra-precision machinability of single-crystal silicon by applying the EVC. With the nominal cutting speed of 200 mm/min and the mean-to-peak vibration amplitudes A_c–A_d of 2–1 $\mu m_{0\text{-}p}$, the critical DOC for ductile machining of silicon is increased into 475 nm, which is about 12.5 times higher than that of 38 nm in OC. Even though there are brittle cracks generated in the EVC at a nominal DOC that is higher than the critical value, the dominant defect size is significantly smaller than that in the OC. Based on the optimal machining conditions, high accuracy angle grid surface and dimple patterns are successfully sculptured on silicon. The experimental results demonstrate the feasibility of practical ultra-precision machining of silicon microstructures by applying EVC.

The research results presented in this chapter are also presented in the author's published papers [29, 30, 36, 37, 65, 70, 71].

References

1. Shieh HPD, Huang YP, Chien KW (2004) Micro-optics components for liquid crystal displays applications. In: Proceedings of the 24th international display research conference, Daegu, Korea, pp 56–58
2. Schumann M, Bückmann T, Gruhler N, Wegener M, Pernice W (2014) Hybrid 2D-3D optical devices for integrated optics by direct laser writing. Light: Sci Appl 3:e175. https://doi.org/10.1038/lsa.2014.56
3. Xie WT, Dai YJ, Wang RZ, Sumathy K (2011) Concentrated solar energy applications using Fresnel lenses: a review. Renew Sustain Energy Rev 15:2588–2606
4. Yang J, Luo FF, Kao TS, Li X, Ho GW, Teng JH, Luo XG, Hong MH (2014) Design and fabrication of broadband ultralow reflectivity black Si surfaces by laser micro/nano processing. Sci Appl 3:e185. https://doi.org/10.1038/lsa.2014.66
5. Miyoshi H, Adachi T, Ju J, Lee SM, Cho DJ, Ko JS, Uchida G, Yamagata Y (2012) Characteristics of motility-based filtering of adherent cells on microgrooved surfaces. Biomaterials 33:395–401
6. Lewis PR, McCutchen CW (1959) Experimental evidence for weeping lubrication in mammalian joints. Nature 184:1285
7. Bico J, Marzolin C, Quéré D (1999) Pearl drops. Europhys Lett 47:220–226
8. Callies M, Chen Y, Marty F, Pepin A, Quere D (2005) Microfabricated textured surfaces for super-hydrophobicity investigations. Microelectron Eng 78–79:100–105
9. Gao W, Kimura A (2007) A three-axis displacement sensor with nanometric resolution. CIRP Ann Manuf Technol 56:529–532
10. Kawasegi N, Sugimori H, Morimoto H, Morita N, Hori I (2009) Development of cutting tools with microscale and nanoscale textures to improve frictional behavior. Precis Eng 33:248–254
11. Evans CJ, Bryan JB (1999) "Structured", "textured" or "engineered" surfaces. CIRP Ann Manuf Technol 48:541–556
12. Bruzzone AAG, Coata HL, Lonardo PM (2008) Advances in engineered surfaces for functional performance. CIRP Ann Manuf Technol 57:750–769
13. Yan JW, Oowada T, Zhou TF, Kuriyagawa T (2009) Precision machining of microstructures on electroless-plated NiP surface for molding glass components. J Mater Process Technol 209:4802–4808
14. Paul E, Evans CJ, Mangamelli A, McGlauflin ML, Polvani RS (1996) Chemical aspects of tool wear in single point diamond turning. Precis Eng 18:4–19
15. Bulla B, Klocke F, Dambon O (2012) Analysis on ductile mode processing of binderless, nano crystalline tungsten carbide through ultra precision diamond turning. J Mater Process Technol 212:1022–1029
16. Brehl DE, Dow TA (2008) Review of vibration-assisted machining. Precis Eng 32:153–172
17. Shamoto E, Moriwaki T (1994) Study on elliptical vibration cutting. CIRP Ann Manuf Technol 43:35–38
18. Suzuki N, Yokoi H, Shamoto E (2011) Micro/nano sculpturing of hardened steel by controlling vibration amplitude in elliptical vibration cutting. Precis Eng 35:44–50
19. Shamoto E, Morimoto Y, Moriwaki T (1999) Elliptical vibration cutting (2nd report, study on effects of vibration conditions). J Jpn Soc Precis Eng 65:411–417 (In Japanese)
20. Shamoto E, Suzuki N, Hino R (2008) Analysis of 3D elliptical vibration cutting with thin shear plane model. CIRP Ann Manuf Technol 57:57–60
21. Shamoto E, Moriwaki T (1999) Ultaprecision diamond cutting of hardened steel by applying elliptical vibration cutting. CIRP Ann Manuf Technol 48:441–444
22. Suzuki N, Nakamura A, Shamoto E, Harada K, Matsuo M, Osada M (2003) Ultraprecision micromachining of hardened steel by applying ultrasonic elliptical vibration cutting. In: Proceedings of international symposium on micromechatronics and human science, Nagoya, Japan, pp 221–226
23. Brinksmeier E, Gläbe R (2001) Advances in precision machining of steel. CIRP Ann Manuf Technol 50:385–388

24. Zhang XQ, Kumar AS, Rahman M, Nath C, Liu K (2011) Experimental study on ultrasonic elliptical vibration cutting of hardened steel using PCD tools. J Mater Process Technol 211:1701–1709
25. Suzuki N, Haritani M, Yang J, Hino R, Shamoto E (2007) Elliptical vibration cutting of tungsten alloy molds for optical glass parts. CIRP Ann Manuf Technol 56:127–130
26. Suzuki N, Hino R, Masuda S, Shamoto E (2006) Ultraprecision cutting of sintered tungsten carbide by applying elliptical vibration cutting-study on ductile cutting mechanics. J Jpn Soc Precis Eng 72:539–545 (In Japanese)
27. Suzuki N, Masuda S, Shamoto E (2004) Ultraprecision machining of sintered tungsten carbide by applying ultrasonic elliptical vibration cutting. In: Proceedings of 4th European international conference, Glasgow, Scotland, pp 187–188
28. Nath C, Rahman M, Neo KS (2009) A study on ultrasonic elliptical vibration cutting of tungsten carbide. J Mater Process Technol 209:4459–4464
29. Zhang JG, Suzuki N, Wang YL, Shamoto E (2014) Fundamental investigation of ultra-precision ductile machining of tungsten carbide by applying elliptical vibration cutting with single crystal diamond. J Mater Process Technol 214:2644–2659
30. Zhang JG, Suzuki N, Shamoto E (2011) Micro machining of binderless tungsten carbide by applying elliptical vibration cutting technology. In: Proceedings of the Japan Society for Abrasive Technology Conference (ABTEC), Kasugai, Japan, pp 109–114
31. Xu WX, Zhang LC, Wu YB (2014) Elliptic vibration-assisted cutting of fibre-reinforced polymer composites: understanding the material removal mechanisms. Compos Sci Technol 92:103–111
32. Moriwaki T, Suzuki H, Mizugaki J, Maeyasu Y, Higashi Y, Shamoto E (2004) Ultraprecision cutting of Molybdenum by ultrasonic elliptical vibration cutting. In: Proceedings of the 19th annual meeting of the ASPE, Orlando, USA, vol 19, pp 621–624
33. Song YC, Park CH, Moriwaki T (2010) Mirror finishing of Co–Cr–Mo alloy using elliptical vibration cutting. Precis Eng 34:784–789
34. Ammouri AH, Hamade RF (2012) BUEVA: a bi-directional ultrasonic elliptical vibration actuator for micromachining. Int J Adv Manuf Technol 58:991–1001
35. Kim GD, Loh BG (2007) Characteristics of chip formation in micro V-grooving using elliptical vibration cutting. J Micromech Microeng 17:1458–1466
36. Zhang JG (2014) Micro/nano machining of steel and tungsten carbide utilizing elliptical vibration cutting technology, PhD dissertation, September, Nagoya University, Department of Mechanical Engineering
37. Zhang JG, Cui T, Ge C, Sui YX, Yang HJ (2016) Review of micro/nano machining by utilizing elliptical vibration cutting. Int J Mach Tools Manuf 106:109–126
38. Ahn JH, Lim HS, Son SM (1999) Improvement of micromachining accuracy by 2-dimensional vibration cutting. In: Proceedings of the 14th annual meeting of the ASPE, Monterey, USA, vol 20, pp150–153
39. Kim GD, Loh BG (2008) Characteristics of elliptical vibration cutting in micro-V grooving with variations in the elliptical cutting locus and excitation frequency. J Micromech Microeng 18:025002
40. Kim GD, Loh BG (2011) Direct machining of micro patterns on nickel alloy and mold steel by vibration assisted cutting. Int J Precis Eng Manuf 12:583–588
41. Kim GD, Loh BG (2010) Machining of micro-channels and pyramid patterns using elliptical vibration cutting. Int J Adv Manuf Technol 49:961–968
42. Brehl DE, Dow TA, Garrard K, Sohn A (2006) Microstructure fabrication using elliptical vibration-assisted machining. In: Proceedings of the 21th annual meeting of the ASPE, Monterey, USA, vol 39, pp 511–514
43. Brehl DE, Dow TA, Sohn A (2015) Micro-machining using EVAM, Precision Engineering Consortium in North Carolina State University, USA: https://www.pec.ncsu.edu/wp-content/uploads/sites/10/2015/03/1842.pdf (Cited in 2015)
44. Brehl DE, Dow TA (2006) Review of vibration-assisted machining methods for precision fabrication, North Carolina State University Raleigh. http://aspe.pointinspace.com/publications/Annual_2006/POSTERS/5PROCESS/2MACH/2013.PDF

45. Brehl DE, Dow TA (2007) 3-D microstructure creation using elliptical vibration-assisted machining. In: Proceedings of the 22th annual meeting of the ASPE, Dallas Texas, USA, pp 21–26
46. Brocato B, Dow TA, Sohn A (2004) Micro-machining using elliptical vibration assisted machining. In: Proceedings of the 19th annual meeting of the ASPE, Orlando, USA, vol 34, pp 80–83
47. Kim GD, Loh BG (2007) An ultrasonic elliptical vibration cutting device for micro V-groove machining: kinematical analysis and micro V-groove machining characteristics. J Mater Process Technol 190:181–188
48. Moriwaki T, Shamoto E (1995) Ultrasonic elliptical vibration cutting. CIRP Ann Manuf Technol 44:31–34
49. Shamoto E, Suzuki N, Naoi Y, Moriwaki T (2002) Development of ultrasonic elliptical vibration controller for elliptical vibration cutting. CIRP Ann Manuf Technol 51:327–330
50. Shamoto E, Suzuki N (2009) Development of elliptical vibration cutting technology and its application to ultraprecision/micro machining of hard/brittle materials. Adv Mater Res 69–70:133–137
51. Shamoto E, Suzuki N (2008) Elliptical vibration cutting of hard mold materials. In: Optical fabrication and testing, Optical Society of America, p OTuB1
52. Zhou M, Hu LH (2015) Development of an innovative device for ultrasonic elliptical vibration cutting. Ultrasonics 60:76–81
53. Li X, Zhang DY (2006) Ultrasonic elliptical vibration transducer driven by single actuator and its application in precision cutting. J Mater Process Technol 180:91–95
54. Brinksmeier E, Gläbe R (1999) Elliptical vibration cutting of steel with diamond tools. In: Proceedings of the 14th annual meeting of the ASPE, Monterey, USA, vol 31, pp 05.11
55. Guo P, Ehmann KF (2013) Development of a tertiary motion generator for elliptical vibration texturing. Precis Eng 37:364–371
56. Guo P, Ehmann KF (2013) An analysis of the surface generation mechanics of the elliptical vibration texturing process. Int J Mach Tools Manuf 64:85–95
57. Shamoto E, Suzuki N, Tsuchiya E, Hori Y, Inagaki H, Yoshino K (2005) Development of 3 DOF ultrasonic vibration tool for elliptical vibration cutting of sculptured surfaces. CIRP Ann Manuf Technol 54:321–324
58. Shamoto E, Suzuki N, Hino R, Tsuchiya E, Hori Y, Inagaki H, Yoshino K (2005) A new method to machine sculptured surfaces by applying ultrasonic elliptical vibration cutting. In: IEEE International Symposium on micro-nanomechatronics and human science. IEEE, Nagoya, Japan, pp 91–96
59. Suzuki N, Hino R, Shamoto E (2007) Development of 3 DOF ultrasonic elliptical vibration system for elliptical vibration cutting. In: Proceedings of ASPE spring topical meeting on vibration assisted machining technology, North Carolina, USA, pp 15–20
60. Yin L, Spowage AC, Ramesh K, Huang H, Pickering JP, Vancoille EYJ (2004) Influence of microstructure on ultraprecision grinding of cemented carbides. Int J Mach Tools Manuf 44:533–543
61. Yin L, Vancoille EY, Ramesh K, Huang H, Pickering JP, Spowage AC (2004) Ultraprecision grinding of tungsten carbide for spherical mirrors. Proc Inst Mech Eng, Part B: J Eng Manuf 218:419–429
62. Yamamoto Y, Suzuki H, Moriwaki T, Okino T, Hijikata Y, Roblee J, Miyashita T (2006) Development of cross and parallel mode grinding machine for high NA aspherical mold and die. In: Proceedings of the 21th annual meeting of the ASPE, Monterey, USA, vol 39, pp 499–502
63. Suzuki H, Moriwaki T, Okino T, Ando Y (2006) Development of ultrasonic vibration assisted polishing machine for micro aspheric die and mold. CIRP Ann Manuf Technol 55:385–388
64. Suzuki H, Hamada S, Okino T, Kondo M, Yamagata Y, Higuchi T (2010) Ultraprecision finishing of micro-aspheric surface by ultrasonic two-axis vibration assisted polishing. CIRP Ann Manuf Technol 59:347–350
65. Zhang JG, Zhang JJ, Cui T, Hao ZW, Zahrani A (2017) Sculpturing of single crystal silicon microstructures by elliptical vibration cutting. J Manuf Process 29:389–398

66. Casstevens JM (1983) Diamond turning of steel in carbon saturated atmospheres. Precis Eng 5:9–15
67. Evans C (1991) Cryogenic diamond turning of stainless steel. CIRP Ann Manuf Technol 40:571–575
68. Brinksmeier E, Gläbe R, Osmer J (2006) Ultra-precision diamond cutting of steel molds. CIRP Ann Manuf Technol 55:551–554
69. Song Y, Nezu K, Park C, Moriwaki T (2009) Tool wear control in single-crystal diamond cutting of steel by using the ultra-intermittent cutting method. Int J Mach Tools Manuf 49:339–343
70. Zhang JG, Suzuki N, Shamoto E (2013) Investigation on machining performance of amplitude control sculpturing method in elliptical vibration cutting. Procedia CIRP 8:328–333
71. Zhang JG, Suzuki N, Wang YL, Shamoto E (2015) Ultra-precision nano-structure fabrication by amplitude control sculpturing method in elliptical vibration cutting. Precis Eng 39:86–99
72. Chavoshi SZ, Goel S, Luo XC (2016) Influence of temperature on the anisotropic cutting behaviour of single crystal silicon: a molecular dynamics simulation investigation. J Manuf Process 23:201–210
73. Goel S, Luo XC, Agrawal A, Reuben RL (2015) Diamond machining of silicon: a review of advances in molecular dynamics simulation. Int J Mach Tools Manuf 88:131–164
74. Mohammadi H, Ravindra D, Kode SK, Patten JA (2015) Experimental work on micro laser-assisted diamond turning of silicon (111). J Manuf Process 19:125–128
75. Ayomoh M, Abou-El-Hossein K (2015) Surface finish in ultra-precision diamond turning of single-crystal silicon. In: Proceedings of SPIE/APOMA, vol 9633, p 96331
76. Chen YL, Cai YD, Shimizu Y, Ito S, Gao W, Ju BF (2016) Ductile cutting of silicon microstructures with surface inclination measurement and compensation by using a force sensor integrated single point diamond tool. J Micromech Microeng 26:025002
77. De CL, Kunzmann H, Peggs GN, Lucca DA (2003) Surfaces in precision engineering, microengineering and nanotechnology. CIRP Ann Manuf Technol 52:561–577
78. SoRef R (2010) Mid-infrared photonics in silicon and germanium. Nat Photonics 4:495–497
79. Mukaida M, Yan JW (2017) Ductile machining of single crystal silicon for microlens arrays by ultraprecision diamond turning using a slow tool servo. Int J Mach Tools Manuf 115:2–14
80. Komanduri R, Chandrasekaran N, Raft LM (1998) Effect of tool geometry in nanometric cutting: a molecular dynamics simulation approach. Wear 219:84–97
81. Yan JW, Yoshino M, Kuriagwa T, Shirakashi T, Syoji K, Komanduri R (2001) On the ductile machining of silicon for micro electro-mechanical systems (MEMS), opto-electronic and optical application. Mater Sci Eng A 297:230–234
82. Moriwaki T, Shamoto E, Inoue K (1992) Ultraprecision ductile cutting of glass by applying ultrasonic vibration. CIRP Ann Manuf Technol 41:141–144

Ultra-precision Machining of Hard and Brittle Materials with Coarse-Grained Grinding Wheels

Bing Guo and Qingliang Zhao

Abstract Many attempts have been made to achieve optical surfaces on hard and brittle materials by means of ductile diamond grinding using fine-grained diamond wheels. However, the large wear rate of the fine-grained grinding wheel, which is caused by dressing and by the grinding process, limited the achievable form accuracy and the maximum material removal volume, especially in case of ductile grinding of large optical surfaces. For solving this problem, a novel-type diamond wheel must be applied which features much higher grinding ratios than that of fine-grained diamond wheels to guarantee the machined surface form accuracy in ultra-precision grinding of hard and brittle materials. In this chapter, the precision conditioning methods of coarse-grained diamond wheels for ultra-precision grinding will be presented first. Subsequently, the ultra-precision grinding of hard and brittle materials with coarse-grained grinding wheels will be investigated in order to reveal the performance of these novel diamond wheels. To conclude the chapter, micro-structured surface will be machined on the diamond grain surfaces of coarse-grained wheels for improving ground subsurface damage.

1 Introduction

Ultra-precision grinding of hard and brittle materials such as optical glass, ceramics, cemented carbides, and crystal materials has been applied in a wide range of applications in order to obtain high-quality surfaces with nanoscale roughness, submicron form accuracy, and micron-level subsurface damage. Commonly, fine-grained grinding wheels with diamond grain diameter from several micrometers down to sub-micrometer are mainly used for ultra-precision grinding various hard and brittle materials [1]. According to many research results, the smaller the abrasive grain size, the better the final finished surface roughness and integrity [2–4]. However, the high

B. Guo (✉) · Q. Zhao
Center for Precision Engineering, School of Mechatronics Engineering,
Harbin Institute of Technology, Harbin 150001, China
e-mail: guobing@hit.edu.cn

J. Zhang et al. (eds.), *Simulation and Experiments of Material-Oriented Ultra-Precision Machining*, Springer Tracts in Mechanical Engineering,
https://doi.org/10.1007/978-981-13-3335-4_8

wear rate and self-sharpening effects of the wheel bonding system, wheel loading as well as periodic conditioning requirement of fine-grained grinding wheels deteriorate the form accuracy of the machined surface and simultaneously introduce a lower grinding efficiency [5–7].

For solving these problems, a novel diamond wheel must be applied which features much higher grinding ratios to guarantee the form accuracy in ultra-precision grinding of hard and brittle materials. The applied wheels should have the following characteristics:

1. Bigger grain size than fine-grained wheels for resisting abrasive wear;
2. Higher bonding strength to avoid the loss of abrasives;
3. Required constant wheel peripheral envelope surface for ensuring homogeneous undeformed chip thickness.

Monolayer coarse-grained diamond grinding wheel featuring grain sizes of more than 100 μm fulfills most of the above-mentioned requirements. Therefore, it could be a solution for the wheel wear problem incurred by fine-grained diamond wheels, provided that they are well conditioned in terms of minimized wheel run-out error, constant wheel peripheral envelope, as well as top-flattened diamond grains with required grain protrusion height of at least a few micrometers.

Generally, monolayer coarse-grained diamond wheel is not usually conditioned in the grinding process. On the one hand, they are mainly applied in rough grinding, which means the wheel conditioning is unnecessary. On the other hand, monolayer coarse-grained diamond wheels only have a single abrasive layer which easily led to damage of the abrasive layer before the conditioning of diamond grains themselves [8]. However, nowadays improved manufacturing technique can make the monolayer diamond wheel superior in terms of wheel's higher geometrical accuracy (base run-out and homogenous deposited bonding layer depth), the bigger bonding strength between base surface and bonding layer, as well as the holding strength of the grains inside bonding layer. Besides, if the on-site conditioning of monolayer coarse-grained diamond wheels could be precisely controlled, the coarse-grained diamond wheels will have a constant wheel peripheral envelope surface, then the ultra-precision grinding of hard and brittle materials could be realized.

2 On-site Precise Conditioning of Coarse-Grained Grinding Wheels

For conventional coarse-grained diamond wheels, the diamond grains with different protrusion heights and the wheel manufacturing deviation will result in a different depth of grinding. In grinding process, some diamond grains will penetrate beyond the critical depth of cut and induce brittle damage, as shown in Fig. 1. Hence, in order to realize the ultra-precision grinding of hard and brittle materials with coarse-grained diamond wheels, an efficient conditioning technique must be developed to guarantee diamond grains with a constant wheel peripheral envelope surface. To meet

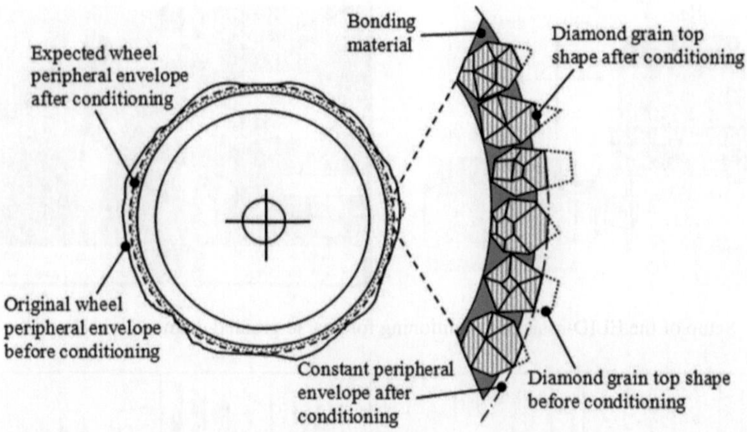

Fig. 1 Sketch map of coarse-grained diamond wheels for ultra-precision grinding of hard and brittle materials

this requirement, three kinds of on-site conditioning technique were adopted for conditioning the coarse-grained diamond wheels including electrolytic in-process dressing (ELID)-assisted conditioning with metal bond diamond truer, thermochemical conditioning with metal truer, and mechanical conditioning with vitrified diamond truer.

2.1 ELID-Assisted Conditioning of Coarse-Grained Grinding Wheels

Dressing of grinding wheels based on the oxidation of the metal bond was initially developed by Ohmori and Nakagawa [9]. Here, the ELID technique was used for the metal bond diamond truer in order to avoid the loading, leading to a continuous and stable conditioning process. Unlike the fine-grained diamond wheel, in this conditioning process, not the bond material but the outstanding tops of diamond grits were cut off in order to insure the diamond grains within a constant wheel peripheral envelope surface.

An experimental system illustrated in Fig. 2 was developed. The metal bond diamond truer with 91-μm grain size was mounted on the work spindle which was dressed in-process by ELID method. The interaction between the grinding wheel and the truer was a truncating mechanism along with the motion in X- and Z-directions, eventually resulting in flattened grains, until finally a constant grain peripheral envelope was obtained. The parameters of 60 V, 40% duty, and 20 A current were used for ELID system. The D151 monolayer nickel electroplated diamond wheels were adopted. The conditioning parameters were grinding wheel speed of 5000 rpm, infeed

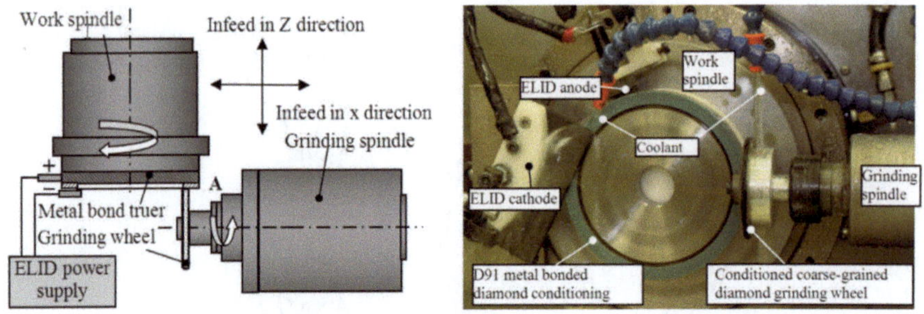

Fig. 2 Setup of the ELID-assisted conditioning for coarse-grained diamond wheels

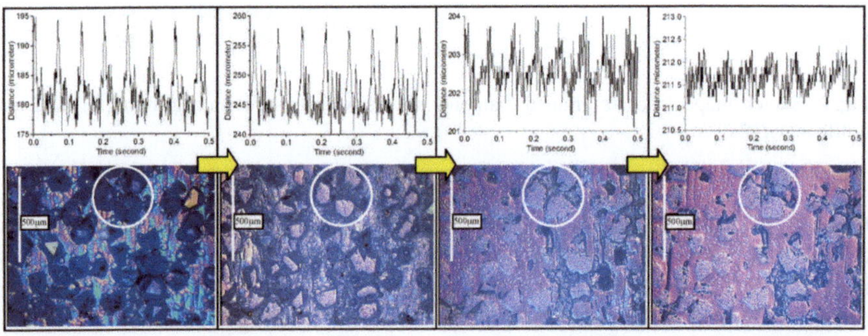

(a) Initial condition, 18μm run-out error.
(b) After 80μm infeed, 15μm run-out error.
(c) After 160μm infeed, 1.7μm run-out error.
(d) After 200μm infeed, 1μm run-out error.

Fig. 3 D151 coarse-grained diamond wheel run-out errors (top) and corresponding wheel topographies (bottom) [8]

depth per step of 1.5 μm, reciprocating speed of 6 mm/min, and truer speed of 1000 rpm.

Before conditioning, the run-out of coarse-grained diamond wheel exhibited a sine wave, as shown in Fig. 3a. The PV value of run-out error was about 20 μm. The run-out error decreases gradually following the conditioning process. Finally, the run-out error of 1 μm was realized after a total conditioning depth of 200 μm, which means a constant grain peripheral envelope was obtained. From the optical images, it can be found there were almost no flattened grains on the grinding wheel surface before the conditioning. As the conditioning process goes on, more and more diamond grains were truncated resulting in more flattened grain tops created. Hence, the average grain protrusion height was decreased, until the final stage shows that the bonding material was also contacted sharing the same level with the top-flattened diamond grains solidly embedded in the bonding material.

Figure 4a shows improved trend of run-out errors during the conditioning process. The changing trend of the conditioning forces is shown in Fig. 5b. Finally, the run-out errors decreased to 1.0–2.5 μm regarding different wheel peripheries from the original 16–18 μm. The change of conditioning forces, i.e., normal force Fn and

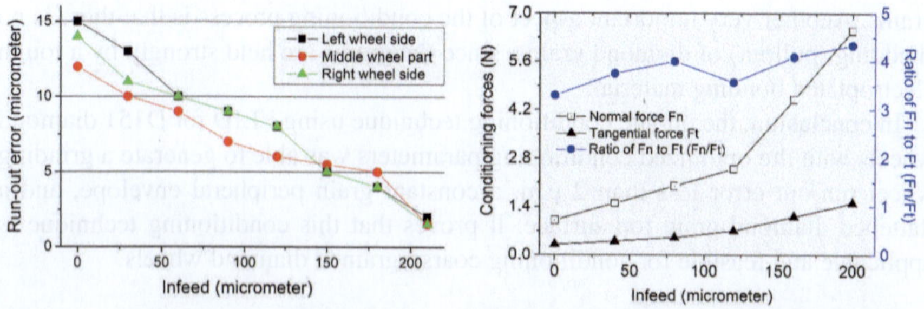

(a) Run-out error (b) Conditioning forces and their ratio.

Fig. 4 Improvement of the D151 diamond wheel run-out error during conditioning [8]

Fig. 5 SEM images of the D151 diamond grinding wheel surface topography after the conditioning process

tangential force Ft, as well as the force ratio Fn/Ft as a function of conditioning infeed depth is shown in Fig. 5b. As the infeed increases, both Fn and Ft increase with different speeds. Thus, the ratio Fn/Ft also increases from 3.3 to 4.35. This is caused by the increase of grain overlap rate, which means more flattened grain surfaces are generated on both the conditioning and grinding wheel, thereby creating a larger contact area between diamond grains, resulting in a bigger bearing ratio and hence larger normal forces and tangential forces. However, the tangential force does not increase as fast as the normal force; therefore, the measured increase of the ratio Fn/Ft indicates that a relatively slower truncating speed among diamond grains begins.

It can be seen from Fig. 5 that all diamond grains were conditioned with smoothly flattened top surfaces indicating that the mechanism was predominately attrition. However, there is still some tiny chipping, fracture, and cleavage present at the sides and edges of the diamond grains which are mainly attributed to the unique truncating process among super diamond abrasives, as well as the anisotropy of the diamond grains. Another very important aspect of the conditioning process is that there is no shedding (pullout) of diamond grains, since the grains are held strongly by a tough electroplated bonding material.

In conclusion, the on-site conditioning technique using ELID for D151 diamond wheels with the optimized conditioning parameters was able to generate a grinding wheel run-out error less than 2 μm, a constant grain peripheral envelope, and a flattened diamond grain top surface. It proves that this conditioning technique is applicable and feasible for conditioning coarse-grained diamond wheels.

2.2 *Thermochemical Conditioning of Coarse-Grained Grinding Wheels*

The above-mentioned ELID-assisted conditioning with metal bond diamond truer definitely requires the additional ELID equipment. Furthermore, the ELID electrolyte coolant is considered to erode the machining tool. Therefore, the thermochemical conditioning with metal truer was developed for coarse-grained grinding wheels further, in order to expand the application of coarse-grained grinding wheels.

Generally speaking, the thermochemical action is an important reason for diamond abrasive wear when machining steel materials. During the machining process, the chemical affinity interaction between iron and carbon atom could occur as follows [10]

$$C(\text{Diamond}) + xFe \rightarrow Fe_x - C \quad (1)$$

$$C(\text{Diamond}) \rightarrow C(\text{Graphite}) + xFe \rightarrow Fe_x - C \quad (2)$$

Diamond could succumb to graphitization at higher machining temperature due to its metastable feature. Therefore, the conversion from diamond to graphite becomes fast. Moreover, at the contact area between diamond and steel, the carbon atoms would penetrate to steel. The strong affinity between carbon and iron exacerbates the diffusion motion. In this process, the carbon atoms spread to and fill the vacancies in the metal lattice. In addition, the micro-collapse mechanical crushing also contributes to the diamond grit wear. Stress concentration results in that the specific cutting force may exceed the breaking strength of the diamond grits, and then the different forms of mechanical micro damage appear on the diamond grits. It suggests that the diamond wear mechanisms are mainly chemical reactions, diffusion motion, and mechanical factors when machining steel materials. This means steel would be the conditioning tool for coarse-grained diamond wheel.

Based on the above-mentioned analysis, the D151 electroplated coarse-grained diamond grinding wheel with a diameter of 200 mm was conditioned using the Cr12 steel by dry grinding process. Figure 6 shows the conditioning setup and the schematic diagram. SiC block was used for sharpening the clogging grinding wheel in conditioning process.

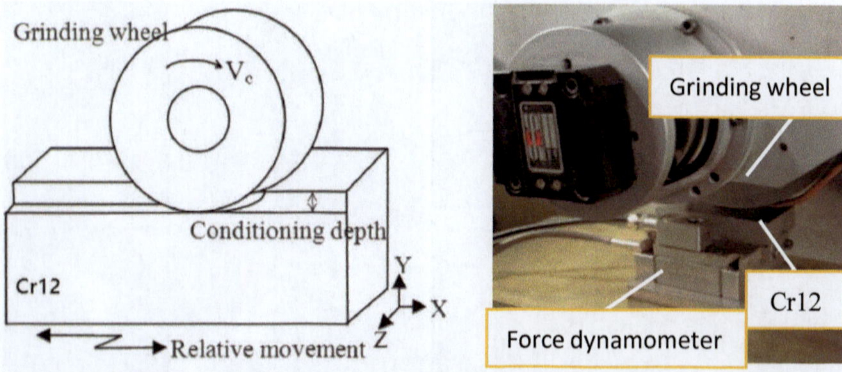

Fig. 6 Set-up of the thermochemical conditioning for coarse-grained diamond wheels

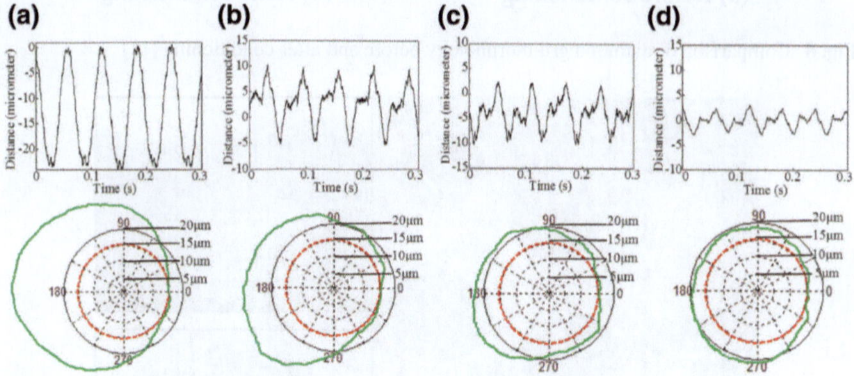

Fig. 7 Run-out error curves (top) and peripheral contour (bottom): **a** before conditioning, **b** conditioning for 2 h, **c** conditioning for 4 h, and **d** conditioning for 6 h [11]

During the conditioning, the run-out error of grinding wheel was measured, as shown in Fig. 7. The initial run-out error was mainly derived from the wheelbase manufacturing error and installation deviation. Following the conditioning, the measured circumferential contour became increasingly close to the ideal profile. The affinity interaction and chemical reactions between iron and carbon would occur on the contact area, which thus accelerate the conditioning speed.

Figure 8 shows the diamond grit morphology before and after conditioning. After conditioning, the diamond abrasive tips were flattened for reducing the run-out error. For the diamond grit A, the stress concentration that led to the abrasive internal stress was close to or even exceeded its fracture strength. Consequently, there generated the sectional micro-crushing, which would be beneficial to avoid the violent collision with the workpiece surface in the subsequent grinding. While for the grit B, the diamond tip was gradually blunt, and a widening abrasion plane perpendicular to

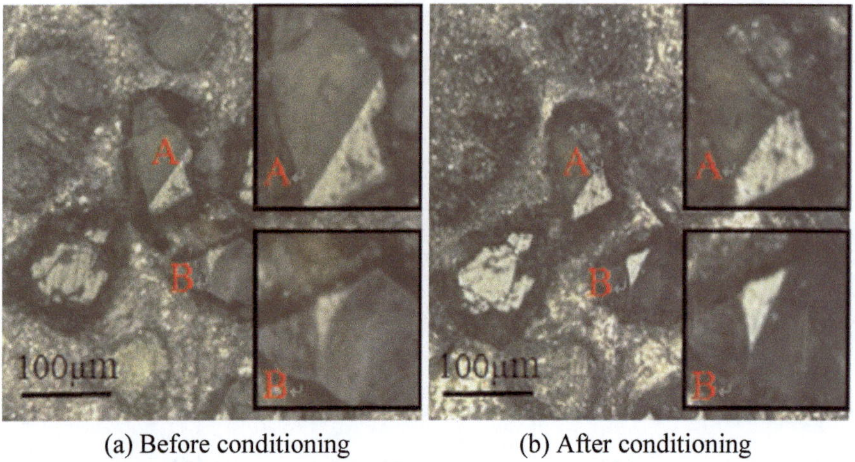

Fig. 8 Comparison of diamond grit morphology before and after conditioning [11]

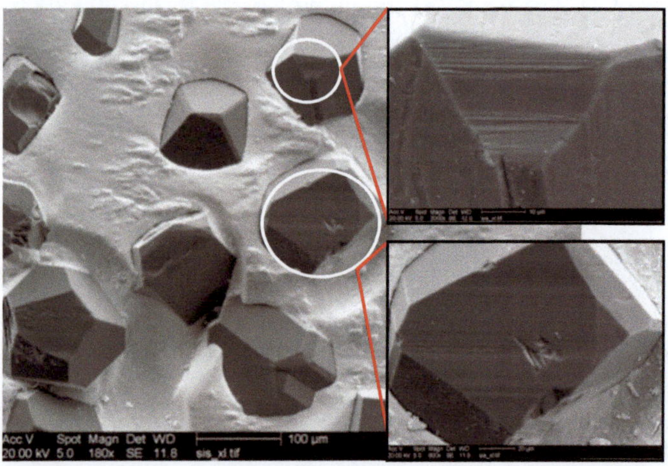

Fig. 9 Surface morphology of conditioned grinding wheel by SEM

the wheel radius direction was then formed. The detail of surface morphology of conditioned grinding wheel is shown in Fig. 9.

Raman spectra results have significant differences for various kinds of carbon materials. The Raman peaks at 1327 and 1582 cm^{-1} characterize sp3-hybridized diamond and sp2 hybrid graphite, respectively, and the spectral peak of Fullerene C60 is at 1469 cm^{-1} [12]. Figure 10a and b shows the laser Raman spectroscopy of the diamond abrasive surface before and after conditioning. After conditioning, graphite and C60 substances were generated on the abrasive surface. While the diamond and C60 substances were also found on Cr12 steel surface, as shown in Fig. 10c. That

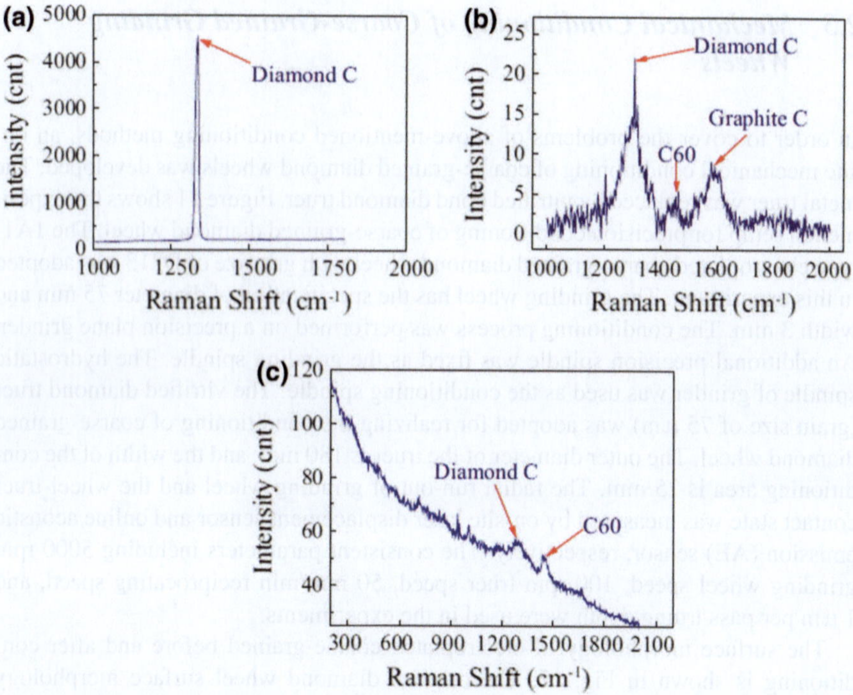

Fig. 10 Raman spectra results of abrasives and truer: **a** before conditioning (abrasive), **b** after conditioning (abrasive), and **c** after conditioning (truer) [11]

was caused by the affinity action between diamond abrasive and steel truer, as well as by the carbon atoms of abrasive gradually migrating and permeating to the truer.

It can be concluded that the electroplated coarse-grained diamond wheel could be effective and precision conditioned by the thermochemical conditioning with Cr12 metal truer. The run-out error quickly converged to 5.8 μm. Graphite and C60 were generated on the conditioned abrasive surface, showing that the conditioning mechanism is mainly involved in passivation, oxidation, and diffusion wear, as well as in a small amount of micro-crushing. It should be noted, however, the conditioning time was much longer than the ELID-assisted conditioning with metal bond diamond truer. Moreover, the thermal deformation of grinding wheel caused by conditioning heat would limit the further enhancement of conditioning accuracy.

2.3 *Mechanical Conditioning of Coarse-Grained Grinding Wheels*

In order to cover the problems of above-mentioned conditioning methods, an on-site mechanical conditioning of coarse-grained diamond wheels was developed. The metal truer was replaced by vitrified bond diamond truer. Figure 11 shows the experimental setup for precision conditioning of coarse-grained diamond wheel. The 1A1-type electroplated coarse-grained diamond wheel with grit size of D213 was adopted in this experiment. The grinding wheel has the specification of diameter 75 mm and width 3 mm. The conditioning process was performed on a precision plane grinder. An additional precision spindle was fixed as the grinding spindle. The hydrostatic spindle of grinder was used as the conditioning spindle. The vitrified diamond truer (grain size of 75 μm) was adopted for realizing the conditioning of coarse-grained diamond wheel. The outer diameter of the truer is 180 mm, and the width of the conditioning area is 25 mm. The radial run-out of grinding wheel and the wheel-truer contact state was measured by on-site laser displacement sensor and online acoustic emission (AE) sensor, respectively. The consistent parameters including 5000 rpm grinding wheel speed, 100 rpm truer speed, 50 mm/min reciprocating speed, and 1 μm per pass truing depth were used in the experiments.

The surface morphology of electroplated coarse-grained before and after conditioning is shown in Fig. 12. The original diamond wheel surface morphology exhibited embedded diamond grains with different geometries and different protrusion heights, which would lead to the brittle fracture of hard and brittle materials. Figure 12b indicates the conditioned wheel surface by vitrified diamond truer with 180 μm truing depth. The micro-cleavage, fracture, and flat top surfaces were generated on the diamond grains. Compared with the conditioning method with metal bond diamond truer and ELID, the conditioning process was operating in a more steady state. The conditioning was never interrupted due to the friability and self-

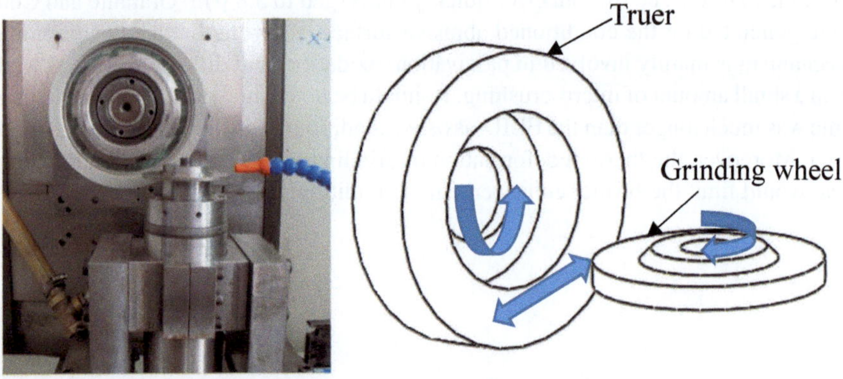

Fig. 11 Conditioning experimental setup of coarse-grained diamond wheel

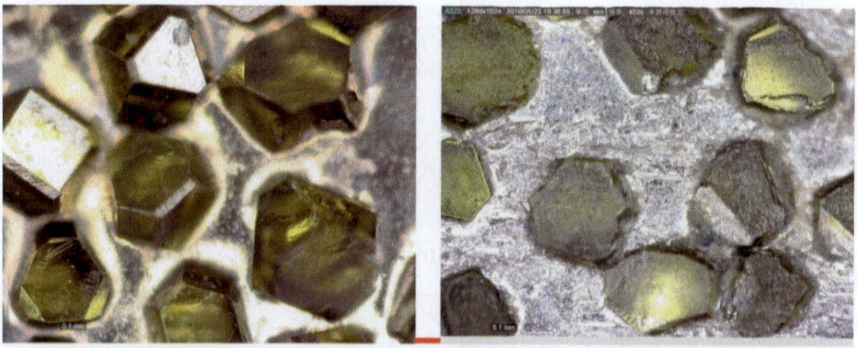

(a) Original wheel (b) conditioned wheel

Fig. 12 Surface morphology of D213 diamond wheels

sharpening of vitrified truer. The conditioning cost time was reduced to less than 3 h, and any additional equipment was not to be required.

To determine the conditioning threshold value of coarse-grained diamond wheel for realizing the ultra-precision grinding, the influences of the truing depth on ground surface quality, grinding wheel run-out was investigated. The BK7 glass workpiece was grounded with the grinding parameters (5000 r/min, 30 mm/min, 2 μm) after every 30-μm truing depth, as shown in Fig. 13. It reveals that the truing depth had an important influence on the ground surface topography. As described in Fig. 13a, there were obvious groove profiles and the brittle removed areas on the ground surface obtained by original coarse-grained diamond wheel, which means the material was removed mainly in brittle regime. As the truing depth increased from 30 to 120 μm, the surface groove depth and the brittle removed areas were decreased gradually, while ductile removed areas, on the contrary, were increased. When the truing depth increased to 150 μm, the workpiece surface was completely machined in ductile regime, as shown in Fig. 13f. There is no brittle crack, and grinding traces were observed on the ground surface. The surface roughness Ra values in the vertical and parallel to the grinding direction were less than 4 and 2 nm, respectively. The precision grinding of optical glass with nanometer-scaled surface roughness was realized.

In the previous conditioning experiments, the radial run-out of grinding wheel is a common judgment method for conditioning process, which is the same as that of fine-grained grinding wheels. However, grinding experimental results (as shown in Fig. 13) indicated the conditioned wheels with large run-out error could also often yield ductile ground surface with nanometer scale roughness. This proves for coarse-grained diamond wheel, the profile accuracy of contour circle which consists of well-conditioned diamond grit flat tops is the key for realization of precision grinding, but not the run-out error. For the run-out error of coarse-grained diamond wheel, not only the abrasive grains but also the bond surface between grains would contribute to the run-out measured results. When the diamond grain tops have been truncated

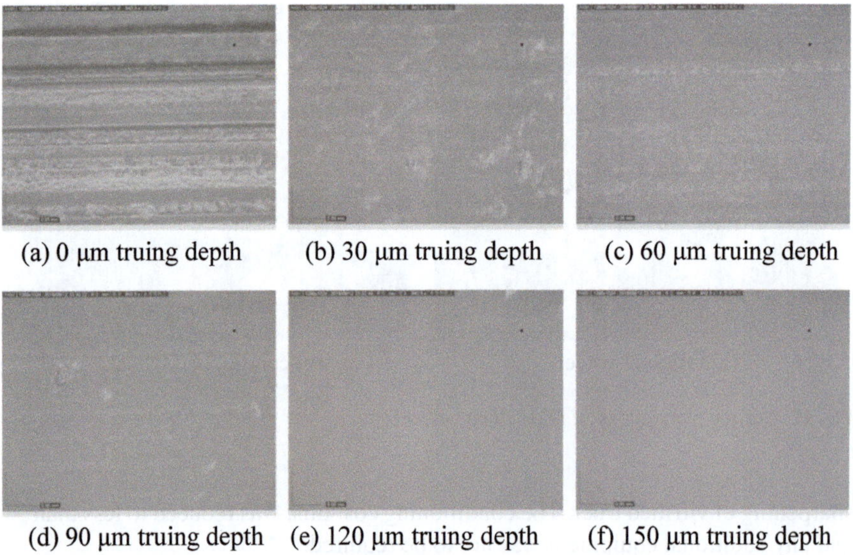

Fig. 13 Ground surface morphology by conditioned wheels with different truing depths (×900)

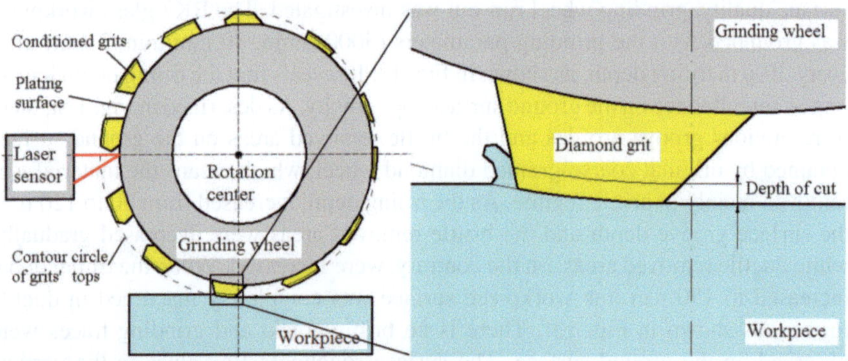

Fig. 14 Illustration of conditioned coarse-grained diamond wheel and the relation between grinding wheel run-out and depth of cut [13]

into a satisfactory contour circle, the radial run-out of grinding wheel was mainly determined by the variation of grit protrusion height from bond surface of wheel, as shown in Fig. 14 [13]. Therefore, the run-out error of coarse-grained diamond wheel would not be limited in microns like the conventional fine-grained wheel, unless the coarse-grained diamond grains are totally truncated off along the lowest plating surface resulting in a minimized run-out error within 2 μm, as presented by Zhao and Guo [8].

The effect of truing depth on the radial run-out of different surface positions of grinding wheel is shown in Fig. 15a. It reveals that the truing depth has a serious

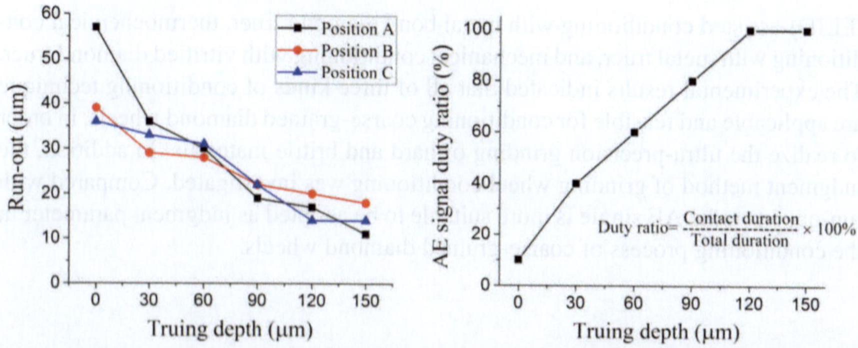

(a) grinding wheel run-out error (b) conditioning AE signal

Fig. 15 Effects of truing depth on grinding wheel run-out and conditioning AE signal

influence on grinding wheel run-out. The run-out values of different positions were reduced gradually with the increase of truing depth. The run-out of original wheel was 35–60 μm caused by both of wheel manufacture error and of wheel installation error. The diamond wheel run-out values after truing depth of 120 and 150 μm were 14–20 and 11–18 μm, respectively. That means that, unlike fine diamond wheel conditioning, the run-out values' deviation of different wheel surface positions was not converged at the end of conditioning process. In other words, the run-out variation caused by measure position was more than that by truing depth, which resulted from the irregular conditioned wheel surface. Therefore, the radial run-out of grinding wheel is hard to be selected as the conditioning evaluation parameter.

Compared the radial run-out, the AE signal judgment method of grinding wheel conditioning is more sensitive and can be measured online [14]. The relationship between truing depth and AE signal of conditioning process is shown in Fig. 15b. In the beginning of conditioning, the AE signal duty ratio was approximately equal to 10%, which implies only about 10% surface of diamond wheel was contacted with truer. As the conditioning process goes on, the AE signal duty ratio was increased, which means increasing surface of grinding wheel did contact with truer and more and more diamond grains were conditioned and truncated. It should be noted that the AE duty ratio reached 100%, when the truing depth was 120 μm. According to Fig. 13e, however, the conditioned coarse-grained diamond wheel with 120 μm truing depth did not realize entirely ductile precision grinding. This was because that the AE measurement method used in this experiment only indicated the peripheral conditioning status of diamond wheel, but the axial conditioning status could not be estimated by AE signal duty ratio. The axial deviation of grain protrusion height would also result in the cracks and brittle material removal in plane grinding. Therefore, an additional conditioning process with truing depth of 30 μm was carried out after the AE signal duty ratio was reached to 100%.

In conclusion, three kinds of conditioning technique were adopted for conditioning coarse-grained diamond wheels including electrolytic in-process dressing

(ELID)-assisted conditioning with metal bond diamond truer, thermochemical conditioning with metal truer, and mechanical conditioning with vitrified diamond truer. The experimental results indicated that all of three kinds of conditioning technique are applicable and feasible for conditioning coarse-grained diamond wheels, in order to realize the ultra-precision grinding of hard and brittle materials. In addition, the judgment method of grinding wheel conditioning was investigated. Compared with run-out error, the AE single is more suitable to be adopted as judgment parameter in the conditioning process of coarse-grained diamond wheels.

3 Ultra-precision Grinding of Hard and Brittle Materials

In order to solve the problems of wheel loading and wheel wear, avoid the periodic conditioning requirement of the fine-grained diamond wheels (grain size from several microns down to submicron scale), and achieve high-efficiency precision grinding of optical glasses, the deterministic grinding process with coarse-grained diamond wheels featuring grain sizes of approx. 90–300 μm have been paid more attention [15–17]. Prior to the grinding process, the protuberances of the diamond coarse grains have to be truncated with the specific conditioning method, aiming to guarantee diamond grains with a constant wheel peripheral envelope surface. Here, the ultra-precision grinding of hard and brittle materials with coarse-grained grinding wheels was investigated in order to reveal the performance of these coarse-grained grinding wheels.

3.1 Ductile Grinding Mechanism of Conditioned Coarse-Grained Wheels

In order to realize the ductile and precision grinding on hard and brittle materials by conditioned coarse-grained diamond grinding wheel, the maximum undeformed chip thickness must be less than the critical ductile–brittle transition depth of workpiece material. Unlike the conventional fine-grained wheel, in grinding process with a conditioned coarse-grained grinding wheel, the material is removed by the truncated abrasive grains. The characteristic chip shape for a truncated abrasive grain is shown in Fig. 16.

The chip cross-sectional profile is trapezoid. The chip width is determined by average grain cutting length b_s. The maximum undeformed chip thickness h_m for truncated grains could be obtained basing on the material balance between the material removal rate and the total volume of chips produced at each effective cutting grain. The product of the average volume per chip and the number of chips formed per unit time are equal to the total volumetric removal rate [19]

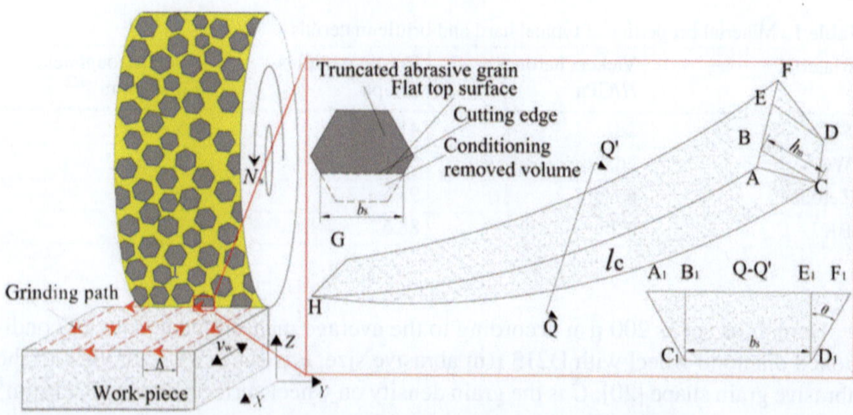

Fig. 16 Illustration of the chips generated by truncated abrasive grain [18]

$$(Cbv_s)V_c = a_p v_w b \tag{3}$$

where V_c is the average volume per chip, C is the grain density on the wheel surface, v_s is the grinding speed, v_w is the feed rate, and a_p is the grinding depth. Thus, (Cbv_s) is the number of chips per unit time for the whole wheel width b and $(a_p v_w b)$ is the theoretical volumetric removal rate.

As the maximum undeformed chip thickness h_m is much smaller than the abrasive cutting chip length l_c [19],

$$l_c = \sqrt{a_p \cdot d_s} \tag{4}$$

where d_s is equivalent wheel diameter.

As distinctly illustrated in Fig. 16, the chip could be divided into three parts: the pyramid BCDEHG and two tetrahedrons ABCH and DEFG.

$$V_c = V_{\mathrm{BCDEGH}} + V_{\mathrm{ABCH}} + V_{\mathrm{DEFG}} \tag{5}$$

Hence, the chip volume for a truncated abrasive grain can be expressed by

$$V_c = b_s \cdot h_m \cdot l_c / 2 + h_m^2 \cdot \tan\theta \cdot l_c / 3 \tag{6}$$

where θ is the inclination angle of the abrasive grain side edge ($\theta > 0$).

Substituting the chip volume and the cutting chip length into Eq. (3), we can obtain the theoretical maximum undeformed chip thickness h_m for grinding process with conditioned coarse-grained wheel, as:

$$h_m = -\frac{3b_s}{4\tan\theta} + \frac{1}{2}\left[\left(\frac{3b_s}{2\tan\theta}\right)^2 + \frac{12}{C\tan\theta}\left(\frac{v_w}{v_s}\right)\left(\frac{a_p}{d_s}\right)^{1/2}\right]^{1/2} \tag{7}$$

Table 1 Material properties of typical hard and brittle materials

Material	Vickers hardness H/GPa	Elastic modulus E/Gpa	Fracture toughness K_{IC}/Mpa·m$^{-1/2}$
SiC	28	410	4.5
WC	26.4	680	5.3
Zerodur	6	91	2.6
BK7	5.7	88.5	0.72

Here, b_s is set as 200 μm according to the average measurement value of conditioned diamond wheel with D213 μm abrasive size, θ is set as 35.3° in view of the abrasive grain shape [20], C is the grain density on wheel surface, and is 10.93/mm^2 for D213 μm wheel based on the experimental measurement.

The critical depth of cut for ductile machining hard and brittle materials could be determined by Bifano [21]

$$d_c = 0.15 \cdot \left(\frac{E}{H}\right) \cdot \left(\frac{K_{IC}}{H}\right)^2 \tag{8}$$

where E is the elastic modulus, H is the Vickers hardness, and K_{IC} is the fracture toughness. The material properties of typical hard and brittle materials are listed in Table 1. The calculated critical depths for SiC ceramic, WC cement, Zerodur glass ceramic, and BK7 optical glass are 60.2, 156, 91, and 37 nm, respectively.

As shown in Fig. 17, the chip thickness increases sharply with the feed rate. In contrast, the chip thickness exhibits an opposite tendency to the grinding wheel rotation. The ductile–brittle transition depth for BK7 is only 37 nm, so the grinding parameters for traverse surface grinding could be set at a rotation of 5000 r/min, feed rate of 50 mm/min, grinding depth of 2 μm. In the same way, the grinding parameters of 5000 r/min, 1.0 mm/min, and 5.0–25.0 μm could be used for ductile machining SiC. The corresponding h_m is less than 1.2 nm, which is adequate to generate the ground surface in ductile machining regime.

3.2 *Ground Surface Quality by Conditioned Coarse-Grained Diamond Wheels*

The ultra-precision grinding of BK7 optical glass with different grinding wheels including D151 conditioned coarse-grained grinding wheel, D7 fine-grained resin grinding wheel, and D7 fine-grained metal grinding wheel with ELID technology was investigated. The workpiece was 180 mm × 160 mm × 40 mm. The grinding parameters were 1600 r/min spindle speed, 1 μm depth of grinding, and 200 mm/min feed rate. The experimental setup is shown in Fig. 18.

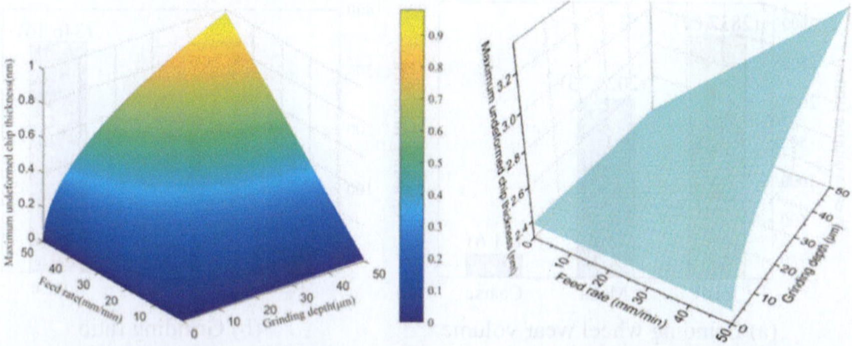

(a) For BK7 glass (b) For SiC ceramic

Fig. 17 Calculation of the maximum undeformed chip thickness

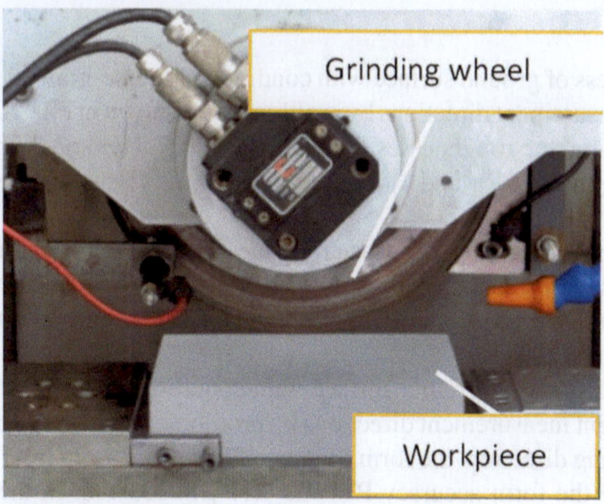

Fig. 18 BK7 glass grinding device

The grinding wheel's wear volume and grinding ratio with the same grinding parameters are shown in Fig. 19. Under the same glass removal volume of 14,400 mm^3, the wear volume of fine-grained resin grinding wheel, fine-grained metal grinding wheel, and conditioned coarse-grained grinding wheel was 2837.07, 2024.23, and 41.6 mm^3, respectively. Figure 19b indicates the grinding ratio which equals to the workpiece material removal volume divided by the grinding wheel's wear volume. The grinding ratio of fine-grained resin grinding wheel, fine-grained metal grinding wheel, and conditioned coarse-grained grinding wheel was 5.08, 7.11, and 346.16, respectively. The grinding ratio of conditioned coarse-grained grinding wheel was 50–70 times of that of fine-grained grinding wheels. This means that the conditioned coarse-grained grinding wheel has much greater wear resistant.

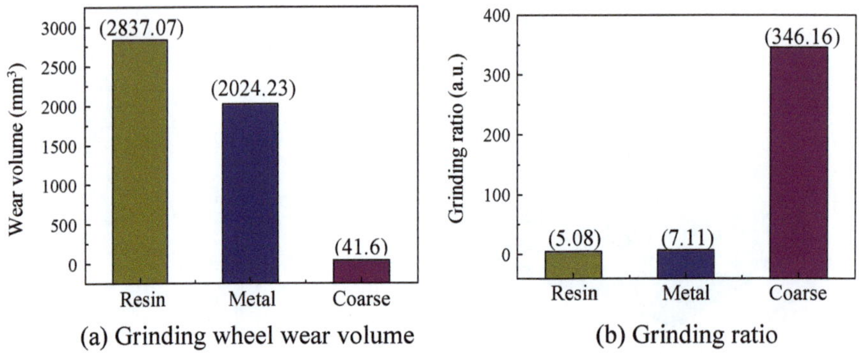

Fig. 19 Grinding wheel's wear volume and grinding ratio under the same glass removal volume. Resin: D7 fine-grained resin grinding wheel; metal: D7 fine-grained metal grinding wheel with ELID technology; and coarse: D151 conditioned coarse-grained grinding wheel

The roughness of ground surface with conditioned coarse-grained grinding wheel in different measurement directions by profilometer is shown in Fig. 20. Parallel to the grinding direction, the roughnesses Ra and Rq were 27.7 and 63.3 nm, respectively, while the roughnesses Ra and Rq were 54.1 and 110.5 nm in the perpendicular direction.

Figure 21 shows the ground surface roughness by different grinding wheels. The measure results indicated that, compared with the conventional fine-grained diamond wheels, the conditioned coarse-grained grinding wheel could obtain same ground surface roughness.

The form accuracy of ground surface with conditioned coarse-grained grinding wheel in different measurement directions by profilometer is shown in Fig. 22. Parallel to the grinding direction, the form accuracy PV was 2.28 μm in 175 mm measure distance, while the form accuracy PV was 4.17 μm/155 mm in the perpendicular direction.

Figure 23 shows the PV values compared results of ground surface with three different grinding wheels. It was found that the form accuracy with conditioned coarse-grained grinding wheel was significantly lower than the other fine abrasive grinding wheels. This is because, during the grinding process, the diamond abrasive grains of the fine abrasive grinding wheel were relatively easy to fall off, while the coarse-grained grinding wheel had a stronger holding force. Therefore, the preservation of form accuracy was better.

Furthermore, the grinding experiments were carried out on BK7 and Zerodur with three coarse-grained diamond wheels of three different grain sizes in order to further reveal the performance of these coarse-grained grinding wheels. The feed rate applied to BK7 was 4 mm/min, while for Zerodur was 10 mm/min. Prior to the grinding tests, the workpiece was grounded and polished in order to guarantee the good flatness, low surface roughness, and free subsurface damage.

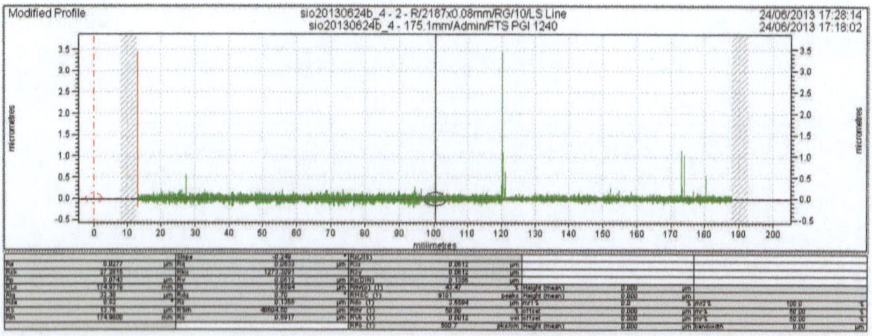

(a) Parallel to the grinding direction

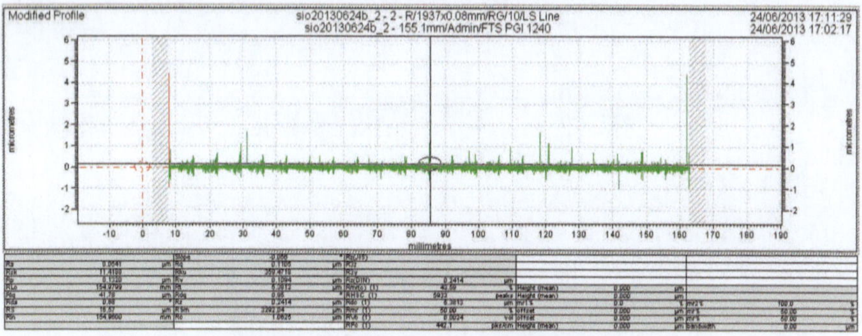

(b) Perpendicular to the grinding direction

Fig. 20 Surface roughness of BK7 ground by conditioned coarse-grained grinding wheel

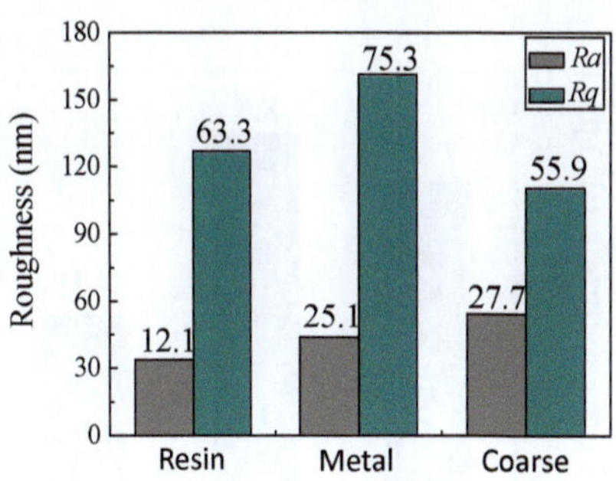

(a) Parallel to the grinding direction

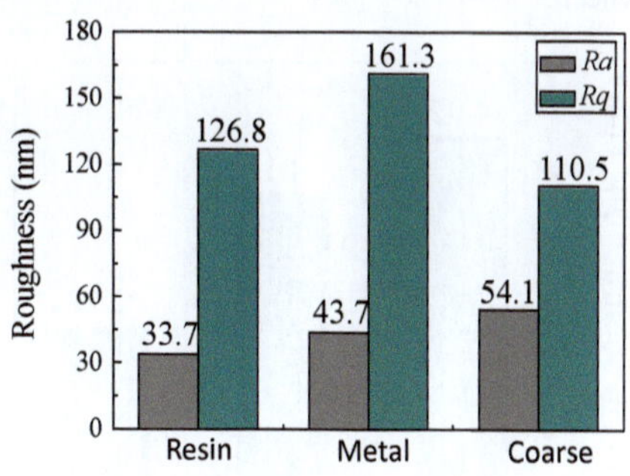

(b) Perpendicular to the grinding direction

Fig. 21 BK7 glasses' surface roughness results of each direction. Resin: D7 fine-grained resin grinding wheel; metal: D7 fine-grained metal grinding wheel with ELID technology; and coarse: D151 conditioned coarse-grained grinding wheel

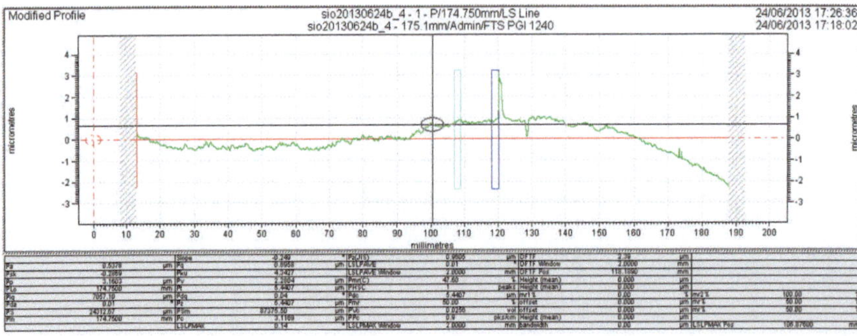

(a) Parallel to the grinding direction

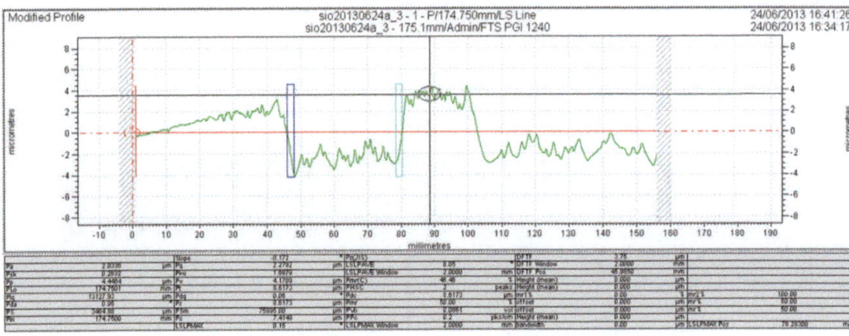

(b) Perpendicular to the grinding direction

Fig. 22 Surface form error results of BK7 glasses ground by conditioned coarse-grained grinding wheel

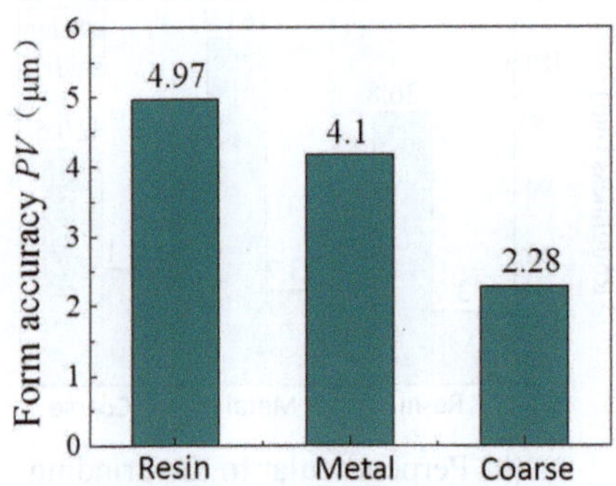

(a) Parallel to the grinding direction

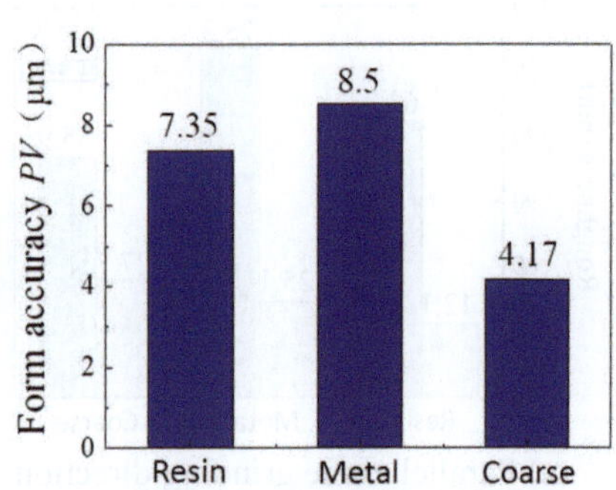

(b) Perpendicular to the grinding direction

Fig. 23 Ground surface form accuracy

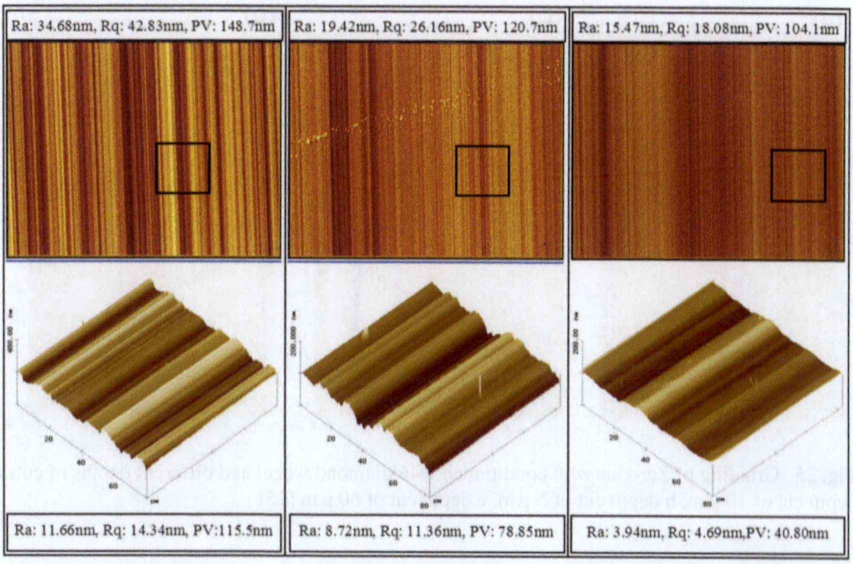

(a) Top: WLI image of the surface ground with a D46 grinding wheel. Bottom: AFM image of the area marked in the WLI image.

(b) Top: WLI image of the surface ground with a D91 grinding wheel. Bottom: AFM image of the area marked in the WLI image.

(c) Top: WLI image of the surface ground with a D151μm grinding wheel. Bottom: AFM image of the area marked in the WLI image.

Fig. 24 WLI (top) and AFM (bottom) images of ground surface with conditioned D46 (**a**), D91 (**b**), and D151 (**c**) diamond wheels with a feed rate of 10 mm/min and a depth of cut of 3 μm [15]

From Fig. 24(a, top), it can be seen that the ground BK7 surface had a surface roughness of 148.7 nm (PV) or 34.58 nm (Ra) as measured with a WLI, indicating a non-uniformity of protruding diamond grain heights of the grinding wheel over a length of 404.64 μm in axial direction. According to the AFM image shown in Fig. 24(a, bottom), over a scanned range of 80 μm × 80 μm, the corresponding surface roughness was 11.656 nm (Ra) or grain height non-uniformity was 115.5 nm (PV). Both WLI and AFM images proved that the precision-conditioned grinding wheel was capable of carrying out ductile grinding of BK7. From Fig. 24b, it can be seen that the ground BK7 with D91 wheel shared the similar surface topography with that ground by the 46 μm grain size diamond wheel, except the smaller surface roughness either in PV 120.7 nm by WLI and 78.85 nm by AFM, or in Ra 19.42 nm by WLI and 8.72 nm by AFM under the identical scanned ranges. This result indicated that the precision-conditioned 91 μm grain size diamond wheel was capable of realizing ductile grinding of BK7 with a smaller surface roughness when compared to 46 μm grain size diamond wheel. According to Fig. 24c, the ground BK7 surface by D151 wheel exhibits the smallest surface roughness (WLI, PV: 104.1 nm, Ra: 15.47 nm; AFM, PV: 40.8 nm, Ra: 3.94 nm) among that ground by three different grain size diamond wheels. There are also copied grinding marks from the wheel without cracks distributed on the ground surface, indicating a ductile grinding mode is mainly involved during the grinding process.

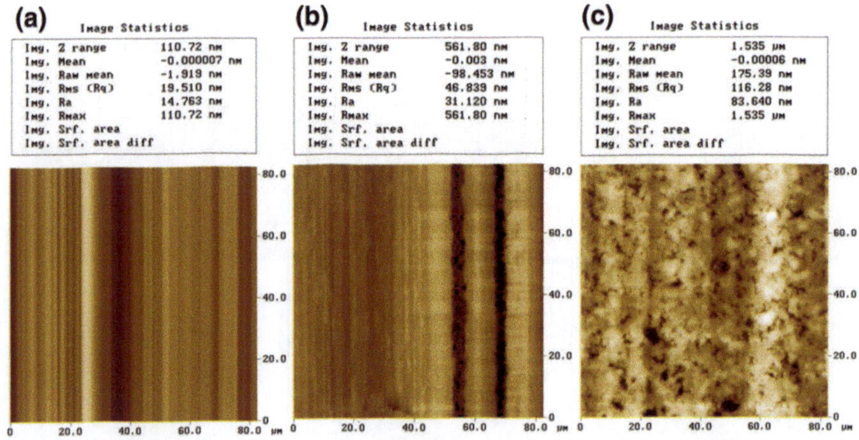

Fig. 25 Grinding of Zerodur with conditioned D46 diamond wheel and different depths of cut: **a** depth cut of 10 μm, **b** depth cut of 5 μm, **c** depth cut of 60 μm [15]

From the above results, it can be concluded that all three precision-conditioned grinding wheels were able to carry out ductile grinding of BK7. The D151 wheel generated surface features with the smallest roughness and lowest PV value, while the D91 and D46 wheels generated surfaces feature with relatively higher roughness and bigger PV value. This fact shall be resulted by the different overlap ratios exhibited by different coarse-grained diamond grinding wheels. D151 wheel has the biggest overlap ratio, while the D46 wheel has the smallest one.

From Fig. 25(a, left), it can be seen that under a depth of cut of 10 μm, the ground Zerodur surface with D46 diamond wheel exhibited a crack and fracture-free morphology resulted from a total ductile material removal mode, featuring a PV value of 110.72 nm and Ra of less than 15 nm. As the depth of cut increased to 15 μm, the ground Zerodur surface morphology became to present the cracks and fractures alongside with the ground groves. The ground morphology indicated a mixed ductile and brittle materials removal mode, resulting in a much big surface roughness values, as shown in Fig. 25b. While as shown in Fig. 25c, when the depth of cut finally reached to 60 μm, the ground Zerodur surface showed no similar morphology as indicated in Fig. 25a, b but fully distributed with propagated cracks and pits. The results indicated that the conditioned D46 wheel was capable of realizing ductile material mode grinding of Zerodur when the depth of cut was set in between 10 and 15 μm, at a feed rate of around 10 mm per minute.

When the same grinding parameters as of D46 wheel were applied to D91 wheel, the resulted surface quality (refer Fig. 26) was better than that resulted from D46. Under both depths of cuts of 10 and 15 μm, the ground Zerodur surfaces were totally of ductile material removal mode featuring grinding marks and quite small surface roughness values (PV of about 30 nm while Ra of about 4 nm). When the depth of cut increased to 15 μm, the surface morphology remains similar to that under 10 μm, except the surface roughness values correspond to a big PV of about 50 nm and Ra

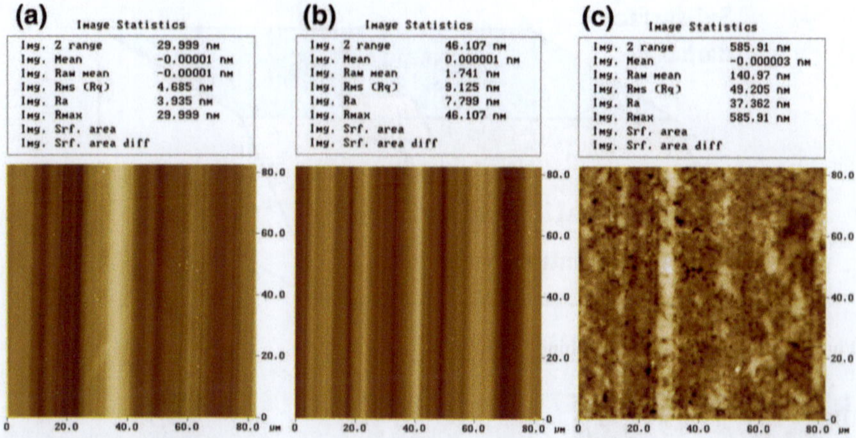

Fig. 26 Grinding of Zerodur with conditioned D91 diamond wheel and different depths of cut: **a** depth cut of 10 μm, **b** depth cut of 15 μm, **c** depth cut of 60 μm [15]

of about 8 nm. This fact indicated that the conditioned D91 had a superior grinding performance for Zerodur than D46 in terms of smaller surface roughness values, as well as the total ductile material removed surface under identical depth of cuts and feed rates with D46.

In the grinding of BK7 and Zerodur using three conditioned coarse-grained diamond wheels with different grain sizes (D151, D91, and D46), the bigger the diamond grain size was, the smaller the ground surface roughness would be obtained, due to the different overlap ratios. The surface roughness achievable on BK7 and Zerodur differed due to a different machinability of these materials. However, for both BK7 and Zerodur, the ground surface roughness mainly depended on the depth of cut other than the feed rate.

3.3 Subsurface Damage by Coarse-Grained Diamond Wheels

Cross-sectional microscopy is the often used method for the measurement of subsurface damage. Optical glasses' sample preparation generally consists of four steps: cutting, lapping, polishing, and etching [15]. (1) Cutting is done in 45° to the machined surface by a diamond saw; (2) the surface of the cut sample is then polished with diamond abrasive papers to remove enough material from the inspected cross section to ensure that any damage incurred during cutting is removed; (3) the polished surface is refined by polishing with loose diamond abrasive of 1μm grain size; (4) the test surface is placed into HF acid solution (HF 5%) for 20 s at room temperature. This will make the subsurface cracks more discernible for microscopic inspection. When the sample is prepared, it needs goad coating for inspection with a scanning electron

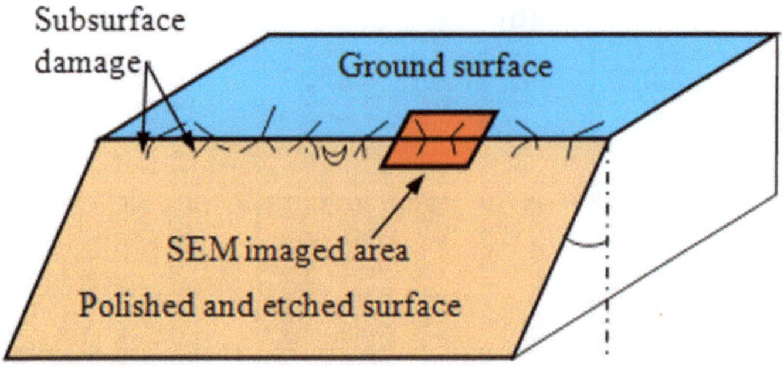

Fig. 27 Diagram of angle polishing method [15]

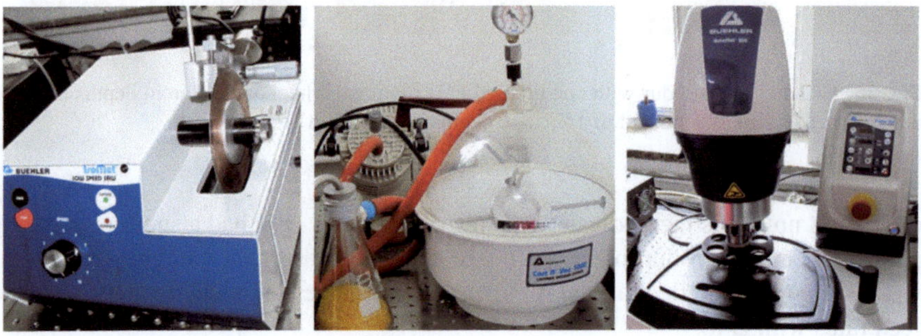

(a) Low-speed sawing slicer (b)Vacuum soaked device (c) Polishing machine

Fig. 28 Equipment used for sample preparation

microscope. A prepared specimen ready for SEM imaging is shown in Fig. 27. The equipment used for sample preparation is shown in Fig. 28.

The subsurface damage of ground surface by various diamond wheels are shown in Fig. 29. From Fig. 29a, it can be seen that the depth of subsurface cracks by D7 resin bond diamond grinding wheel was relatively low due to the elasticity of resin bond. The elasticity of resin bond reduced the rigid impact of the grinding wheel on workpiece. The depth of subsurface damage crack was less than 2 μm, while the depth of subsurface cracks by the D7 metal bond diamond grinding wheel was slightly larger than that by resin bond wheel because of the higher rigid of metal bond system. The depth of subsurface damage crack was about 5 μm.

Figure 29c shows the subsurface damage topography of ground surface by conditioned coarse-grained diamond wheel. It can be seen that the subsurface damage caused by conditioned coarse-grained diamond wheel was more serious than that of the fine abrasive grinding wheel, especially for BK7 glass and fused silica. The depth of subsurface damage crack was about 8 μm. In the preconditioning process, the abrasive grain protrusion height of coarse-grained diamond wheel was uniform in order to realize ductile grinding. However, the flat tops of diamond grits created in preconditioning process lead to greater specific grinding normal force. In addi-

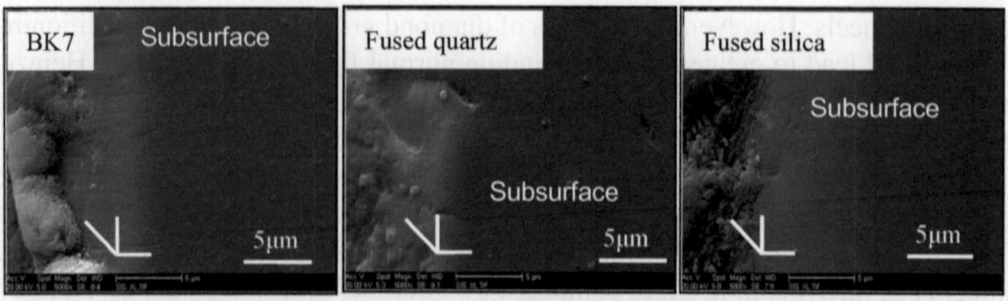

(a) Subsurface damage ground by D7 resin bonded diamond wheel

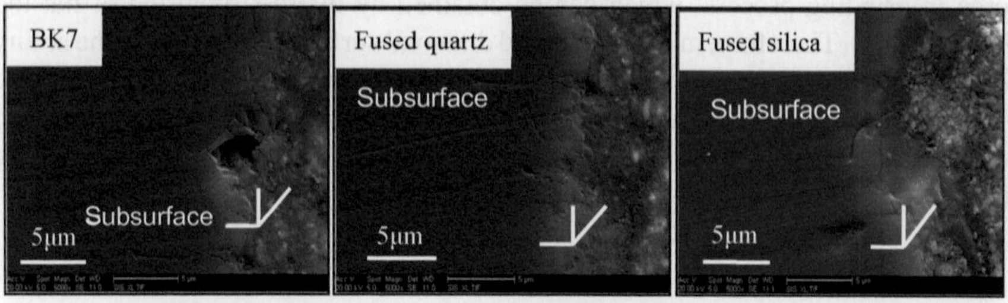

(b) Subsurface damage ground by D7 metal bonded diamond wheel

(c) Subsurface damage ground by D151 conditioned coarse-grained diamond wheel

Fig. 29 Optical glasses' subsurface damage ground by three kinds of wheels

tion, the flat tops of diamond grits did increase the friction between the workpiece and grinding wheel. Therefore, more serious subsurface damage was introduced by conditioned coarse-grained diamond wheel in grinding process.

4 Micro-structured Coarse-Grained Grinding Wheels for Improving Ground Subsurface Damage

The conditioned coarse-grained diamond grinding wheels are able to achieve the identical surface roughness, higher form accuracy, and larger material removal rate on hard and brittle materials due to their better wear resistance ability, longer grinding wheel life, and bigger chip space when compared to the traditional fine-grained diamond wheels. However, the flat tops of diamond grits created in preconditioning process will lead to greater specific grinding normal force as presented by Heinzel and Rickens [22, 23] indicating that a serious subsurface damage will be introduced by coarse-grained diamond wheel in optical glass grinding process.

It has been shown that the micro-structured surface of diamond would reduce the normal grinding force as compared with traditional electroplate grinding block [24–26]. This is associated with the stability of active cutting elements and the integrity of micro-cutting edges composed by a number of microstructures. Besides, the increased number of cutting edges would also lead to a decrease of specific cutting force in grinding process, which can be obtained by micro-structuring of flat top diamond grits. Therefore, micro-structured diamond surface seems to have the ability to improve grinding performance of coarse-grained diamond wheels and reduce the specific grinding normal force. Based on the above, the micro-structured surface was adopted for coarse-grained diamond wheels aiming to improve the grinding performance, especially subsurface damage.

4.1 Laser Micro-structuring on Conditioned Coarse-Grained Diamond Wheel

A sub-nanosecond pulsed laser of wavelength 532 nm, pulse width 0.8 ns, pulse repetition frequency 1–5 kHz, initial laser beam diameter 3 mm, divergence angle 7 mrad, and Gaussian energy distribution was taken as the laser source for micro-structuring the diamond surface and grinding wheels. As shown in Fig. 30, the laser micro-structuring setup was composed of laser source, mirrors, focus lens, CCD cameras, laser distance sensor, and a four-axis precision translation stage. The focus lens was a 10 × microscope objective and could focus the laser beam to less than 20 μm on the focus plane.

The influence of laser power on the micro-machining of diamond material was researched with machining diamond surface, as shown in Fig. 31. Both of the groove

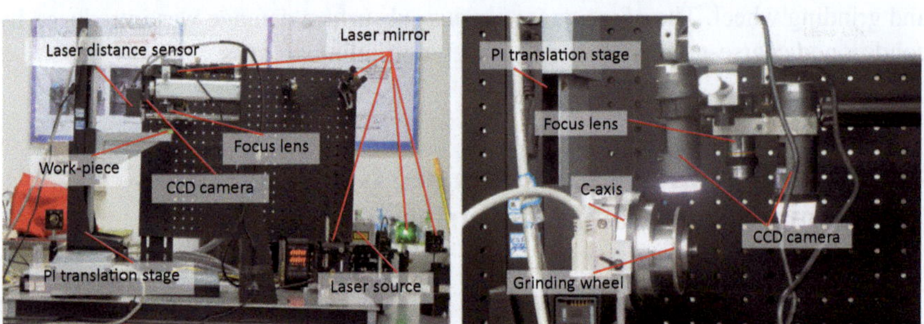

Fig. 30 Laser micro-machining setup

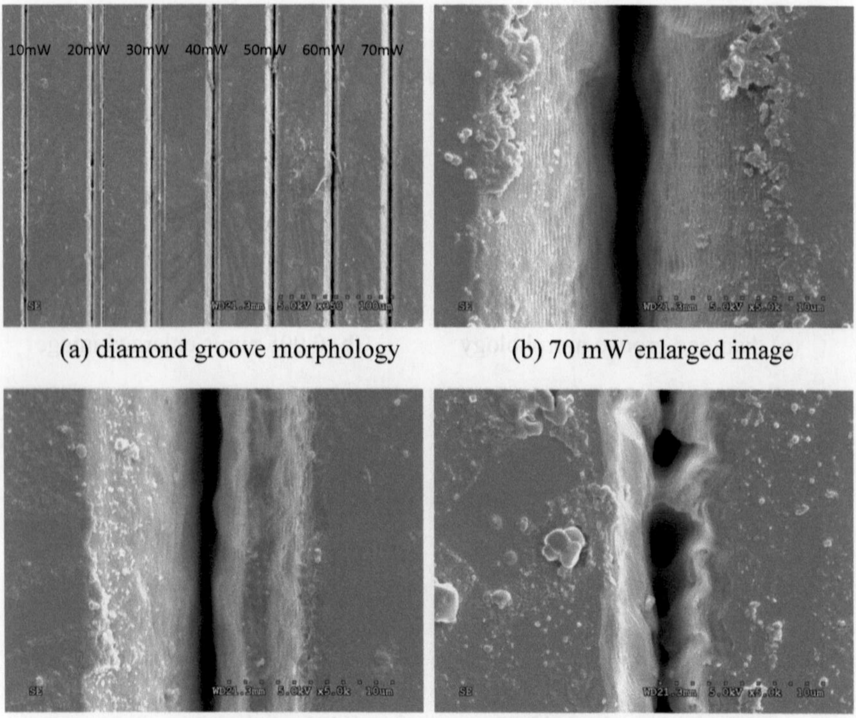

(a) diamond groove morphology (b) 70 mW enlarged image

(c) 40 mW enlarged image (d) 10 mW enlarged image

Fig. 31 Effects of the laser power on diamond groove morphology

width and depth increased with a gradually decreasing rate as the laser power rising from 10 to 70 mW. Owing to Gaussian profile of the laser pulse energy, the laser fluence gradually reduces with increasing the distance to the laser spot center. The laser spot center owns the peak of laser fluence in the whole spot. Consequently, the ablation depth in the groove profile nonlinearly decreases as the increase of the distance from the groove centerline and achieves the maximum depth in the groove centerline. The change of the groove depth can be attributed to the well-known logarithmic dependence of the ablation depth and the laser fluence. In laser beam machining, the ablation depth would increase logarithmically depending on the laser fluence. Hence, the ablation depth fast increased with the increase of laser power. However, the ablation depth seems to reach a limit when the average laser power is larger than 40 mW. This is caused by the shielding effect of plasma in the depth groove.

Moreover, the groove width also shows a similar tendency with the groove depth. According to the laser ablation regime, material removal occurs in any position where the laser fluence exceeds the material damage threshold. Therefore, a smaller groove width would appear when the laser pulse energy is insufficient to assure the laser

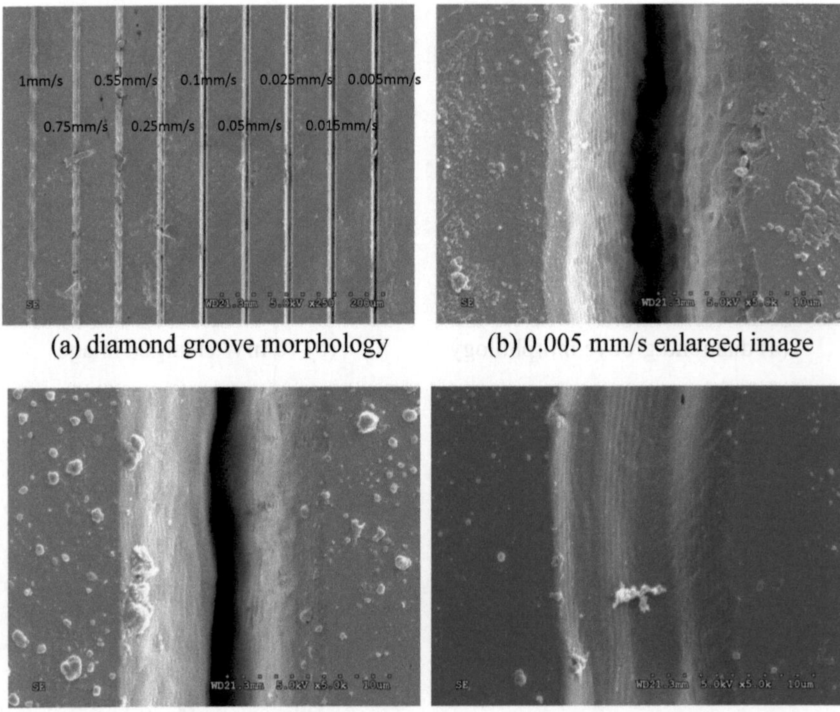

(a) diamond groove morphology (b) 0.005 mm/s enlarged image

(c) 0.1 mm/s enlarged image (d) 1 mm/s enlarged image

Fig. 32 Effects of the scanning speed on diamond groove morphology

fluences in the whole spot exceed the damage threshold. The reduced ablation area with lower laser power such as 10 mW induces not only a decrease of the groove width, but also a certain extent of the discontinuity at the groove bottom. As the laser power increases, the ablation width gradually tends to the spot diameter when the laser power is bigger than 40 mW and this is clearly reflected in Fig. 31. In some extent, adjusting the laser power provides us an effective method to control the groove width and depth. To obtain a larger and deeper microgroove, higher average laser power is necessary and inevitable. However, the research results also demonstrate that an excessive increase of the average laser power would result in a reduction of the energy efficiency and a further increase of the average laser power after the laser power exceeds a certain value just yields a tiny addition of the groove width or depth.

Figure 32 clearly demonstrates the influence of the scanning speed on diamond groove morphology. The groove width and depth exist a significant correlation with the scanning speed. The groove depth drops firstly with a large rate until the scanning speed reaches 0.1 mm/s and further decreases with a much lower rate as the scanning speed increased to 1.0 mm/s. In contrast, the groove width looks like keeping a stable value, while the scanning speed increases from 0.005 to 1.0 mm/s.

A further examination of the equivalent accumulated laser pulses indicates us that the increase of the scanning speed would result in a continuous reduction of the equivalent accumulated laser pulses which directly influences the total ablation depth. Moreover, the variation tendency of the accumulated laser pulses about the scanning speed is consistent with the groove depth and an obvious transition also occurs at the speed 0.1 mm/s. Therefore, the decrease of the groove depth can be attributed to the reduction of the accumulated laser pulses as the scanning speed increases. The accumulated laser pulse number just mainly influences the ablation depth, while the groove width is mainly determined by the distribution of the laser fluence. The radius where the laser fluence equals to the material damage threshold determines the ablation area and the groove width, and this radius is decided by the average laser power and independent with the accumulated laser pulses.

Based on the above experimental results, three kinds of micro-structured conditioned coarse-grained diamond grinding wheels with different abrasive grain sizes were prepared as shown in Fig. 33. The groove width of these patterns was 30 μm. The interval between two adjacent grooves was 120 μm for the parallel-line pattern. The groove direction had an angle 45° with the wheel axis direction.

The conditioned coarse-grained diamond grinding wheels were micro-structured by sub-nanosecond pulsed laser, successfully. Continuous microgroove arrays with 30 μm width were obtained on the peripheral surface of grinding wheel. The protruding parts of most diamond grits were cut-through by microgrooves. The broken diamond grit and falling off of grit were not found during laser machining.

4.2 The Effects of Micro-structured Surface on Grinding Performance

The influences of micro-structured surface on grinding performance of conditioned D213 coarse-grained diamond grinding wheels were investigated. Several optical glass BK7 workpieces were ground with the grinding wheels of abrasive grain size D91 before and after micro-structuring. The ground workpiece surface roughness Ra was measured by AFM. And the grinding force was recorded by dynamometer.

The Influence of Microstructures on Roughness

For comparatively analyzing the influence of microstructures on surface roughness, the surface roughness of single factor experiments of spindle rotation, feed rate, depth of grinding with and without microstructures are presented in Fig. 34. The surface roughness ground with non-micro-structured wheel was more insensitive to the change of the grinding parameters. Most of the roughness values were less than Ra 10 nm, while the surface roughness obtained by micro-structured wheel showed much bigger undulation. And it existed a several times increase compared to surface ground with non-micro-structured wheel. Nevertheless, grinding BK7 glass by micro-structured wheel with surface roughness less than Ra 40 nm was still feasible.

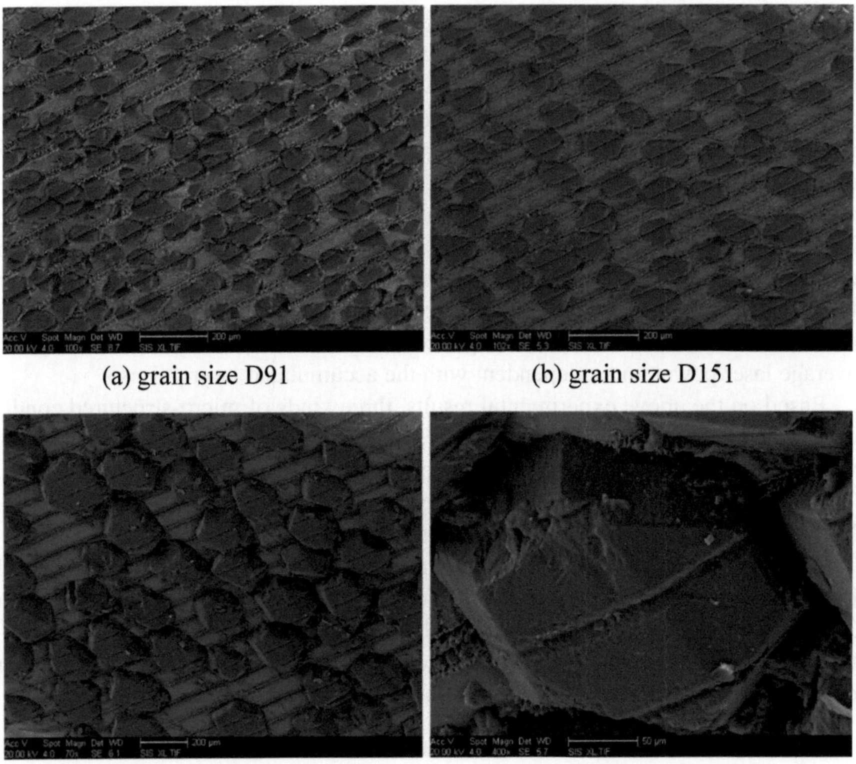

(a) grain size D91 (b) grain size D151

(c) grain size D213 (d) grain size D213 enlarged image

Fig. 33 Laser micro-structured conditioned coarse-grained diamond grinding wheels

The Influence of Microstructures on Grinding Force

As described in Fig. 35, the normal grinding force was sensitive to the change of the grinding parameters in both wheel conditions. Increasing the wheel rotation speed resulted in a gentle reduction of the normal grinding force, while the feed rate showed the opposite tendency. The normal grinding force almost linearly increased with the increase of the grinding depth in both wheel conditions. The increase of grinding depth resulted in a corresponding rise of the geometric contact length between the wheel surface and the workpiece surface. Therefore, more grain flat tops would get in contact with workpiece surface and the normal force increased.

The microstructures effectively reduced the contact area of the coarse-grained diamond tops and the workpiece surface by 33.3% thereof resulted in a decrease of the normal grinding force. Moreover, more grinding fluid could be transported into the grinding zone by the grooves, which would also be helpful for reducing the normal grinding force. The normal grinding force obtained with micro-structured wheel had a considerable reduction compared to not micro-structured wheel. Therefore, the

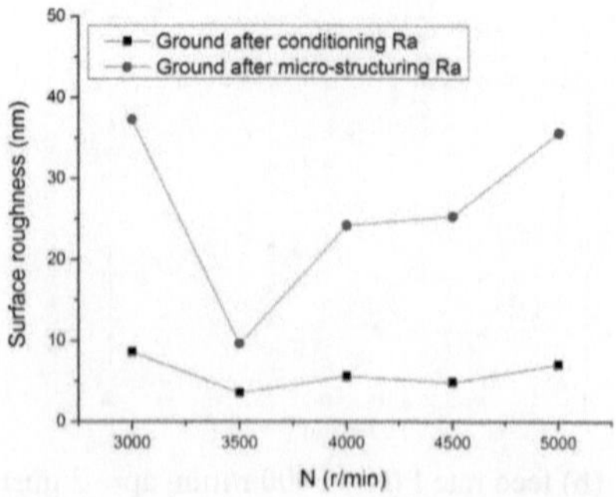

(a) spindle rotation (f = 10 mm/min, ap= 2 μm)

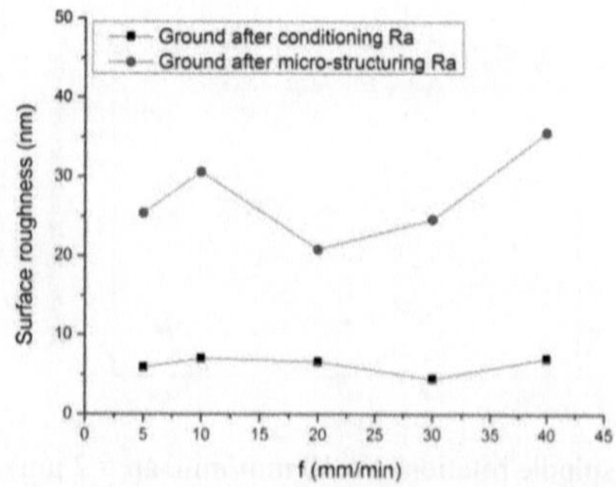

(b) feed rate f (N = 5000 r/min, ap = 2 μm)

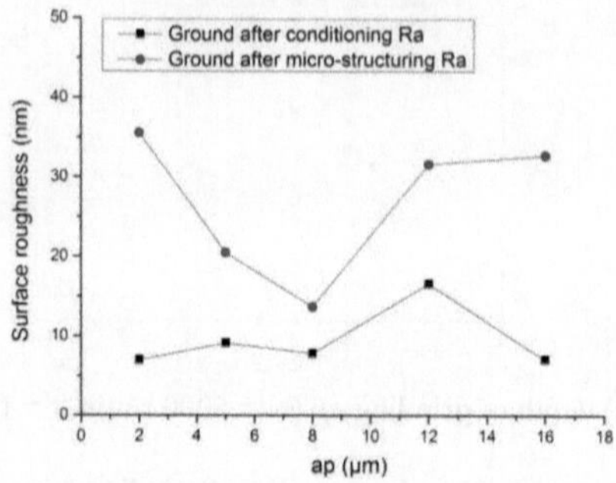

(c) depth of grinding *ap* (N = 5000 r/min, f = 10 mm/min)

Fig. 34 Influence of micro-structured surface on surface roughness

normal grinding forces of grinding with surface micro-structured wheel decreased by 35–74% of grinding with surface not micro-structured wheels in the experiments.

The influence of the microstructures on tangential grinding force also shows a similar tendency with the normal grinding force, as shown in Fig. 36. The results that the tangential force of grinding operation with surface micro-structured wheel also existed a markedly decrease by 30–60% of that with non-micro-structured wheel, which proves the feasibility of using surface micro-structured wheel to reduce the grinding force once again. Furthermore, the fluctuation of the tangential grinding force of grinding operation with micro-structured wheel also decreased.

The normal-to-tangential force ratio is an important factor for analyzing the grinding process. Figure 37 shows the force ratios of grinding with wheel just conditioned and micro-structured in the order of wheel rotation, feed rate, and grinding depth. It is interesting that the force ratio almost kept consistency except in the conditions with large grinding depth. This phenomenon indicates that the grinding force of the two-wheel condition is mainly composed of the normal contact force and the friction force between the abrasive grain and workpiece surface and the force could be attributed to material removal just a small part of the total grinding force. There-

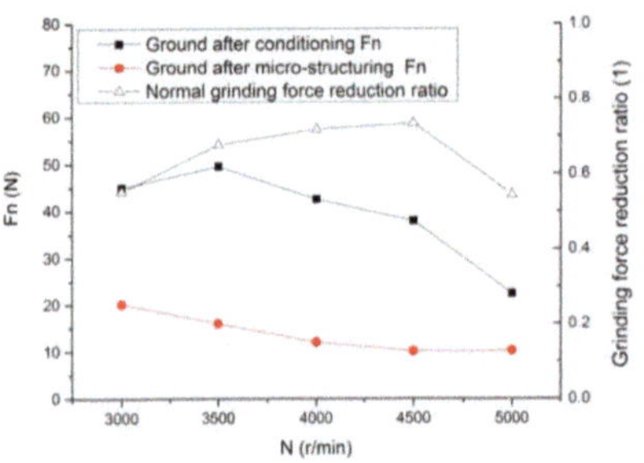

(a) spindle rotation (f = 10 mm/min, ap = 2 μm)

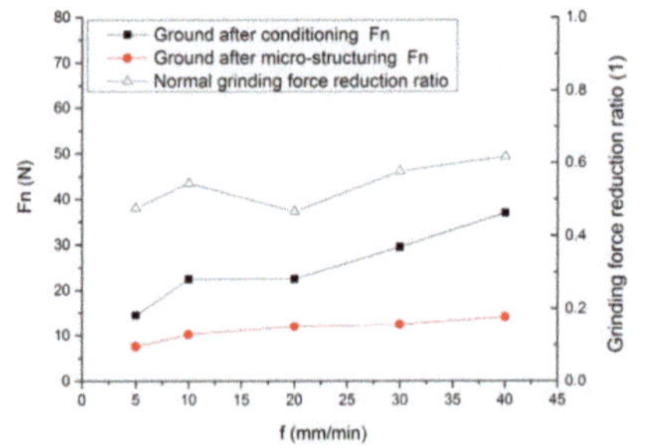

(b) feed rate f (N = 5000 r/min, ap = 2 μm)

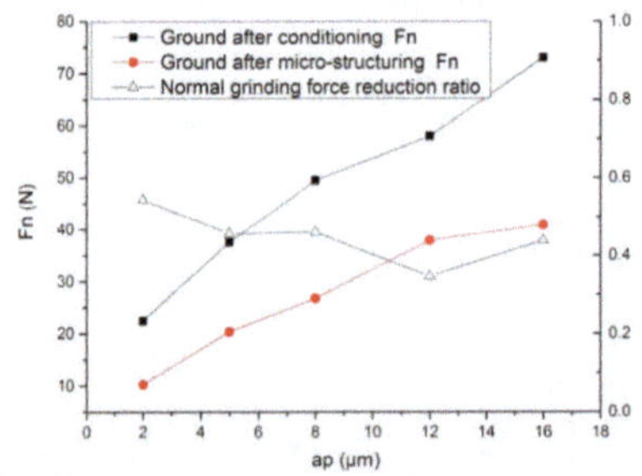

(c) depth of grinding *ap* (N = 5000 r/min, f = 10 mm/min)

Fig. 35 Influence of micro-structured surface on normal grinding force

fore, micro-structuring grinding wheel surface with larger structuring ratio maybe is acceptable.

The Influence of Microstructures of Wheel Surface on Subsurface Damage
The influence of micro-structured surface on subsurface damage was investigated on BK7 glass. Four micro-structured conditioned D151 coarse-grained diamond grinding wheels with different interval microgroove arrays were machined by laser. The intervals of these grinding wheels were 30, 79, 90, and 150 μm, respectively. The grinding parameters were 3000 r/min spindle speed, 2 μm depth of grinding, and 2 mm/min feed rate. The subsurface damage was also measured by angle polishing methods. The measured results and their averaged values of subsurface crack depth are provided in Table 2.

And the SEM images of Sample 1 are shown in Fig. 38 for revealing the detail of subsurface damage. The subsurface damage depth was improved from about 10 μm (by original coarse-grained wheel, as shown in Fig. 38a) to 5 μm because of the uniform grit protrusion height due to preconditioning. The subsurface damage depth by micro-structured wheel with 150 μm interval groove arrays was further decreased to about 3 μm. Compared with the conventional coarse-grained diamond wheel, the subsurface damage depth was reduced effectually when using the micro-structured coarse-grained diamond wheel. The effect of microgroove interval on subsurface damage could be observed from Fig. 38c–f. The subsurface damage depth was

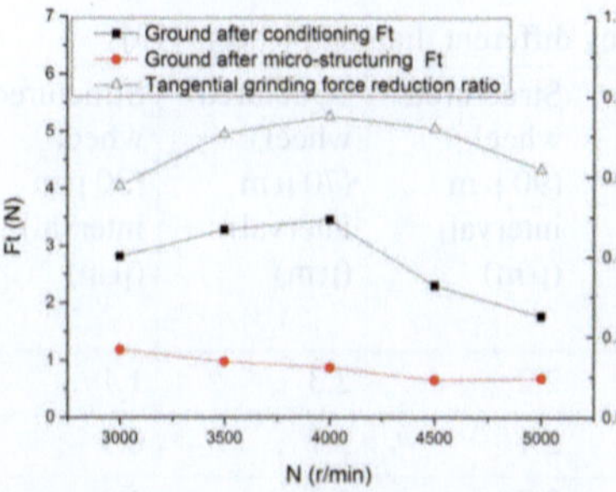

(a) spindle rotation (f = 10 mm/min, ap = 2μm)

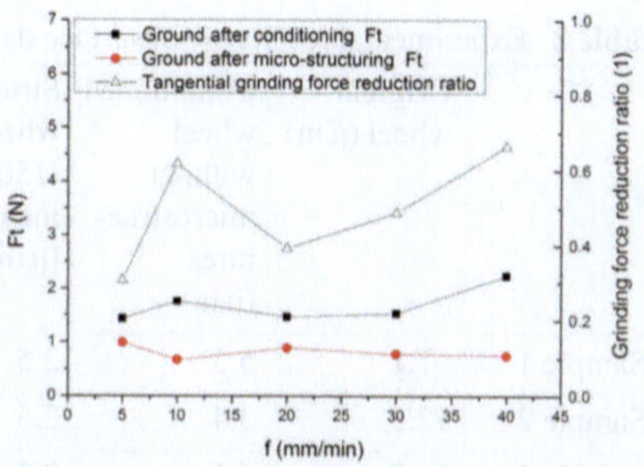

(b) feed rate f (N = 5000 r/min, ap = 2 μm)

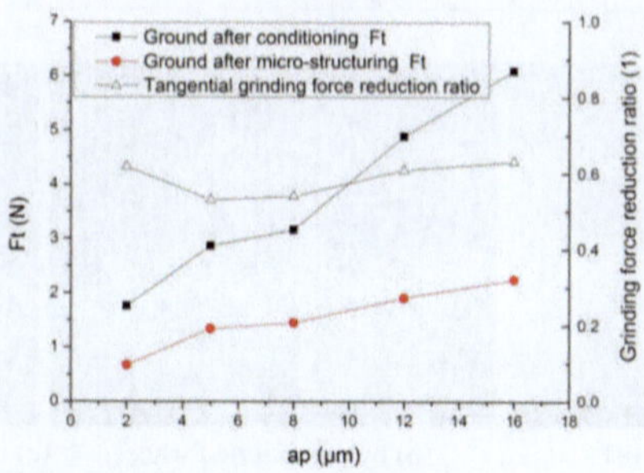

(c) depth of grinding *ap* (N = 5000 r/min, f = 10 mm/min)

Fig. 36 Influence of micro-structured surface on tangential grinding force

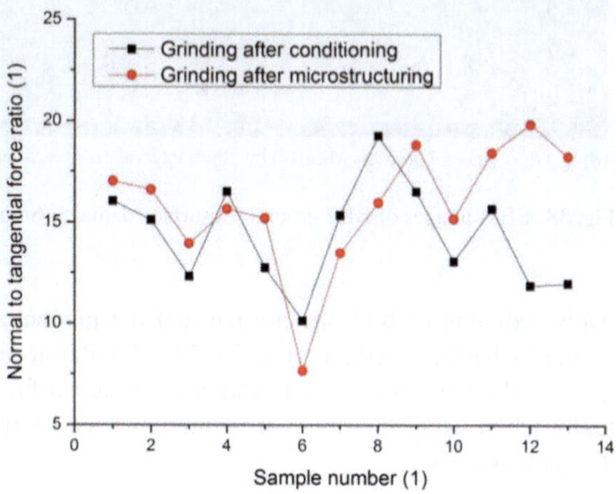

Fig. 37 Influence of micro-structured surface on normal-to-tangential force ratio

reduced with the decreasing interval. At the interval of 30 μm, the subsurface damage depth of 1.5 μm was obtained. This is mainly caused by the increase of active cutting edges and hence decreased individual chip load when using smaller interval.

It could be concluded that although grinding with micro-structured wheel induced an increase of the surface roughness, micro-structured surface was capable for effec-

Table 2 Experimental results of subsurface damage by different diamond wheels [13]

	Original wheel (μm)	Conditioned wheel without microstructures (μm)	Structured wheel (150 μm interval) (μm)	Structured wheel (90 μm interval) (μm)	Structured wheel (70 μm interval) (μm)	Structured wheel (30 μm interval) (μm)
Sample 1	7.1	5.3	2.5	2.2	2.3	1.1
Sample 2	12.5	3.4	2.5	2.1	1.7	0.9
Sample 3	11.7	6.1	3.5	2.7	2.3	1.3
Average	10.4	4.9	2.8	2.3	2.1	1.1

(a) by original wheel (b) by conditioned wheel (c) by 150μm interval grooves wheel

(d) by 90μm interval grooves wheel (e) by 70μm interval grooves wheel (f) by 30μm interval grooves wheel

Fig. 38 SEM images of BK7 ground subsurface damage (Sample 1) [13]

tively reducing of both the normal and the tangential grinding forces for coarse-grained wheels. A reduction of 35–74% for the normal force and 30–60% for the tangential force was achieved. And thus, the subsurface damage depth reduced effectually when using the micro-structured surface on the conditioned coarse-grained diamond wheels.

5 Conclusions

In this chapter, a novel coarse-grained diamond wheel was applied which features much higher grinding ratios than that of fine-grained diamond wheels to guarantee the machined surface form accuracy in ultra-precision grinding of hard and brittle materials.

Firstly, three kinds of conditioning technique were adopted for conditioning coarse-grained diamond wheels including ELID-assisted conditioning with metal bond diamond truer, thermochemical conditioning with metal truer, and mechanical conditioning with vitrified diamond truer. The experimental results indicated that all of three kinds of conditioning technique are applicable and feasible for conditioning coarse-grained diamond wheels, in order to realize the ultra-precision grinding of hard and brittle materials.

Secondly, the ultra-precision grinding of hard and brittle materials with coarse-grained grinding wheels was investigated. The experiments indicated the conditioned coarse-grained diamond wheels could achieve the identical surface roughness, higher form accuracy, and better wear resistance ability compared with the traditional fine-grained diamond wheels. However, deeper subsurface damage was introduced by coarse-grained diamond wheel.

Finally, for improving ground subsurface damage, the conditioned coarse-grained diamond grinding wheels were micro-structured by sub-nanosecond pulsed laser, successfully. Although the grinding process with micro-structured wheel induced an increase of the surface roughness, micro-structured surface was capable for effectively reducing of both the normal and the tangential grinding forces for coarse-grained wheels. And thus, the subsurface damage depth reduced effectually when using the micro-structured surface on the conditioned coarse-grained diamond wheels.

Acknowledgements The authors sincerely acknowledge the Natural Science Foundation of China (51875135, 51405108), Natural Science Foundation of Heilongjiang Province (E2018037) and the Alexander von Humboldt (AvH) Foundation of Germany for their financial support. Sincere thanks also go to Dr. Lingling Zhao, Dr. Mingtao Wu, Mr. Wei Zhang, Mr. Qianyu Jin from Harbin Institute of Technology, and Prof. Brinksmeier, Dr. Preuss, Dr. Riemer, Dr. Rikens, Mr. Horst Kosenski, Mr. Frank Karstens from Bremen Univerisity for their technical support.

References

1. Brinksmeier E, Mutlugünes Y, Klocke F et al (2010) Ultra-precision grinding. CIRP Ann—Manufact. Technol. 59(2):652–671
2. Bifano TG (1991) Ductile-regime grinding: a new technology for machining brittle materials. J Trs Asme Ser 113(2):184–189
3. Ngoi BKA, Sreejith PS (2000) Ductile regime finish machining—a review. Int J Adv Manuf Technol 16(8):547–550
4. Guo B, Zhao Q (2015) Wheel normal grinding of hard and brittle materials. Int J Adv Manuf Technol 79(5–8):873–880
5. Chen X, Rowe W B, Mills B et al (1998) Analysis and simulation of the grinding process. Part IV: Effects of wheel wear. Int J Mach Tools and Manuf 43(38):41–49
6. Zhao Q, Chen J, Yao J et al (2009) Investigation of surface and subsurface damage in diamond grinding of optical glass using hybrid copper-resin-bonded diamond wheel. J Vac Sci Technol, B 27(3):1489–1495

7. Guo B, Zhao Q (2015) Mechanical truing of V-shape diamond wheels for micro-structured surface grinding. Int J Adv Manuf Technol 78(5–8):1067–1073
8. Zhao Q, Guo B (2015) Ultra-precision grinding of optical glasses using mono-layer nickel electroplated coarse-grained diamond wheels. Part 1: ELID assisted precision conditioning of grinding wheels. Precis Eng 39:56–66
9. Ohmori H, Nakagawa T (1990) Mirror surface grinding of silicon wafers with electrolytic in-process dressing. CIRP Ann—Manuf Technol 39(1):329–332
10. Ke-Yong BI (2005) Technology of cast iron processing with diamond tool. Jewellery Sci Technol
11. Ling LZ, Zhao QL, Guo WJ et al (2013) Precision grinding of BK7 glasses using conditioned coarse-grained diamond wheel. Proc Instit Mech Eng Part B J Eng Manuf 227(10):1571–1577
12. Fateley B (1991) The handbook of IR and raman characteristic frequencies of organic molecules. Academic Press
13. Guo B, Zhao Q, Fang X (2014) Precision grinding of optical glass with laser micro-structured coarse-grained diamond wheels. J Mater Proc Tech 214(5):1045–1051
14. Guo B, Yu X, Liu X, Zhao Q et al (2018) Ultra-precision cutting of linear micro-groove array for distributed feedback laser devices. Int J Nanomanuf 14(1):9
15. Zhao Q, Guo B (2015) Ultra-precision grinding of optical glasses using mono-layer nickel electroplated coarse-grained diamond wheels. Part 2: investigation of profile and surface grinding. Precis Eng 39:67–78
16. Guo B, Wu M, Zhao Q (2017) The FEM simulation of ultra-precision grinding of optical glass with micro-structuring coarse-grained diamond wheels. Freeform Optics. JTu5A.26
17. Ling LZ, Zhao QL, Guo WJ et al (2013) Precision grinding of BK7 glasses using conditioned coarse-grained diamond wheel. Proc Inst Mech Eng Part B J Eng Manuf 227(10):1571–1577
18. Wu M, Guo B, Zhao Q et al (2018) Precision grinding of a microstructured surface on hard and brittle materials by a microstructured coarse-grained diamond grinding wheel. Ceramics Int 44(7)
19. Malkin S (1989) Grinding technology: theory and applications of machining with abrasives. SME
20. Vargas GE, Wegener K, Kuster F et al (2014) Simulation of the hone broaching process with diamond tools. J Braz Soc Mech Sci Eng 36(2):325–333
21. Bifano TG (1991) Ductile-regime grinding: a new technology for machining brittle materials. J Trs Asme Ser B 113(2):184–189
22. Heinzel C, Rickens K (2009) Engineered wheels for grinding of optical glass. CIRP Ann—-Manuf Technol 58(1):315–318
23. Butler-Smith PW, Axinte DA, Pacella M et al (2013) Micro/nanometric investigations of the effects of laser ablation in the generation of micro-tools from solid CVD diamond structures. J Mater Process Technol 213(2):194–200
24. Guo B, Zhao Q, Yu X (2014) Surface micro-structuring of coarse-grained diamond wheels by nanosecond pulsed laser for improving grinding performance. Int J Precis Eng Manuf 15(10):2025–2030
25. Guo B, Wu M, Zhao Q et al (2018) Improvement of precision grinding performance of CVD diamond wheels by micro-structured surfaces. Ceram Int
26. Guo B, Zhao Q (2017) Ultrasonic vibration assisted grinding of hard and brittle linear micro-structured surfaces. Precision Eng 48:98–106

Technology Precision Machining of Ceramic Balls in the V-Grooves of Variable Curvature

Oleg Pasichnyi

Abstract Many industries employ a great variety of ball bearings, pumps, and other mechanisms, whose performance, service life, and reliability essentially depend on the quality of manufacture of precision ceramic balls. This chapter describes the traditional methods of processing precision balls, their advantages, and disadvantages in relation to the precision processing of ceramic balls. Several modern methods of precision machining of ceramic balls with the controlled position of the balls' rotation axis during processing are considered and present a promising method of processing balls in guide V-grooves with variable curvature, which allows you to control both the quality and performance of processing. A mathematical model of processing balls with the variable position of the balls' rotation axis is suggested. The mathematical model takes into account the forces, pressures, and wears of the guide V-grooves and allows determining the performance of the tool and assessing the quality of the balls surface coating by traces of processing. Examples of tools with guide V-grooves with variable curvature are shown, and the results of the proposed method of precision processing of boron carbide balls are given.

O. Pasichnyi (✉)
V.N. Bakul Institute for Superhard Materials, National Academy of Science of Ukraine, Kiev, Ukraine
e-mail: olegpasichny@gmail.com

J. Zhang et al. (eds.), *Simulation and Experiments of Material-Oriented Ultra-Precision Machining*, Springer Tracts in Mechanical Engineering,
https://doi.org/10.1007/978-981-13-3335-4_9

1 Introduction

Demand for high-precision, high-performance, and highly automated equipment is increasing along with the rapid development of industrial technologies. All the industries have a high number of ball bearings, valves, pumps, chemical apparatus, and other mechanisms, the resource and reliability of which are largely determined by the efficiency of parts such as precision balls. Currently, precision balls are mainly made of steel. But steel balls quickly fail at high temperatures and loads, as a result of abrasive, corrosive, chemical and other types of wear.

One of the promising ways to solve this problem is the use of precision balls made of modern ceramic materials in the construction of machines and devices. Currently, the widespread use of ceramic materials in mechanical engineering is constrained by insufficient development of technologies to ensure their efficient processing. First of all, it concerns the prospects of the application material based on boron carbide and silicon carbide for the manufacture of precision balls. Those are ones of the hardest ceramics possessing high hardness, strength and unique wear resistance in the conditions of the influence of aggressive chemicals and abrasives.

However, today even the leading manufacturers of chemical, oil and gas equipment, and ball bearings do not use boron carbide ceramics for manufacturing precision ceramic balls and silicon carbide ceramics are very rarely used. This is largely due to the high hardness of materials based on boron carbide and, as a consequence, the high complexity and cost of precision machining when using traditional methods of grinding and finishing. Due to the high hardness of ceramics, its processing is possible only with the use of diamonds and diamond tools, and diamond consumption is high enough.

Despite a large number of patent materials, research, and development in the field of precision balls processing, usually in industry precision processing is produced in the way proposed by Friedrich Fischer in 1883 [1].

Processing is carried out by rolling of balls between the two plates, wherein one plate rotates. The balls are thus located in annular V-grooves formed on the surface of one of the plates (Fig. 1a). When processing large batches of balls used elevator processing, with one of the plates has a window for loading and unloading balls in the elevator or hopper in which the balls are mixed.

This way of processing is simple to implement and allows achieving a stable, fairly high quality. Nevertheless, it had some disadvantages, which are particularly prevalent in the processing of ceramic materials:

- Low productivity;
- High consumption of diamonds;
- High wear of the lapping plates;
- The lack of a uniform mesh of tool marks on the ball surface;
- High diameter deviation processed balls.

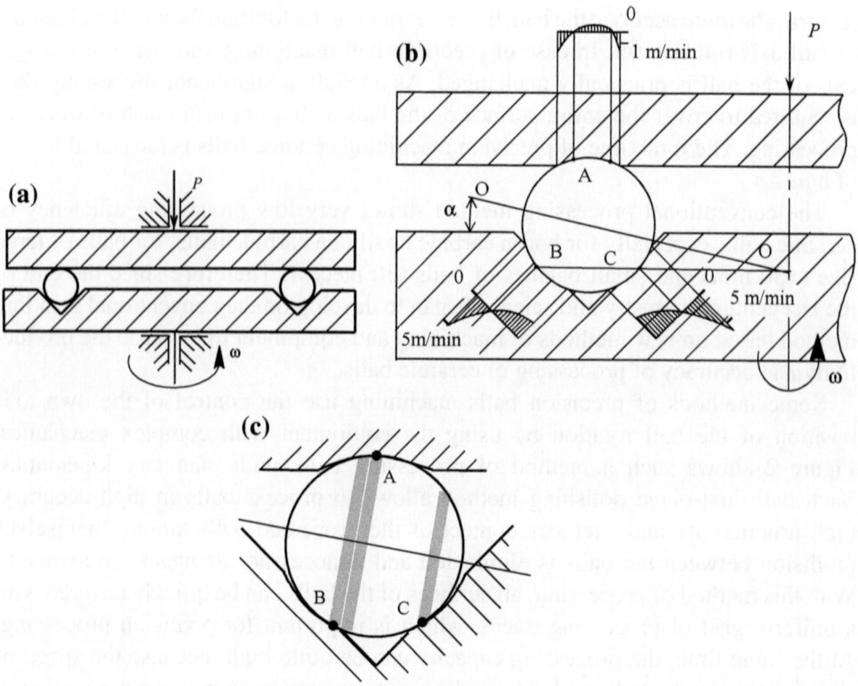

Fig. 1 Processing in the annular V-grooves. **a** Schema of the processing, **b** the rotation axis of the ball and sliding speed, and **c** traces of processing

The kinematic feature of the traditional processing method—low speed in the contact areas—can be increased only by increasing the drive plate speed rotation. But the range of change variation of the speed of rotation is limited by the capabilities equipment in the range of 30–120 rpm. With a greater increase in the speed of the plate rotation, vibrations occur and the quality of processing decreases.

Real average slip speed between the surfaces of balls and plates in the contact area usually does not exceed 1 m/min. Figure 1b shows the calculated velocity of the sliding speed between plates and ball in the contact areas with a ball diameter of 17 mm that is located in the V-groove with an angle of 90° and a diameter of 600 mm, the lower plate with groove rotates at a speed of 80 rpm.

The accuracy of balls processing is determined by the uniformity of the balls surface coating by traces of processing. The main factor that has a significant impact on the process of coating the ball surface traces of processing is the changing of the own axis position of the ball rotation during processing. During a single run, the ball between the plates machined only a small part of the surface of the ball, and each of the three zones of contact leaves the surface of the ball track in the form of a circle (Fig. 1c) [2, 3]. The width of these circles is small.

When processing in circular guide V-grooves, the change in the position of the ball rotation axis is very small, occurs uncontrollably, only under the action of random factors. The more accurate the ball, the fewer random factors and the smaller changing of ball axis rotation tilt. In case of precision ball machining, the angle of the spin axis of the ball is practically unchanged. As a result, a significant processing time is required to cover the entire surface of the ball with a uniform mesh of traces of processing. The total time of precision machining ceramic balls is measured in tens of hours.

The conventional processing method shows very low processing efficiency of ceramic balls, especially for boron carbide or silicon carbide balls, for balls of large size (>30 mm) and small batches of balls (30 pieces). Therefore since the end of the last century, industry and science began to develop on the conventional grinding method based on new methods of machining and equipment to increase the productivity and accuracy of processing of ceramic balls.

Some methods of precision balls machining use the control of the own axis position of the ball rotation by using the equipment with complex mechanics. Figure 2 shows such a method of processing balls with planetary kinematics. Such ball dual-plane polishing method allows to process balls in high accuracy, high productivity and excludes contact of the processed balls among themselves (collision between the balls is eliminated and unnecessary damages are avoided). With this method of processing, all surfaces of the balls can be quickly covered with a uniform grid of processing traces, which is important for precision processing. At the same time, the processing capacity can be quite high, because the speed of sliding between the balls and the grinding plates is greater than when processing in the conventional V-groove grinding method. This method has shown a sufficiently high efficiency in the finishing of ceramic balls (polishing) [4].

Also a known method of processing with an independent drive of each of the three machining tools, Fig. 3 shows an example of such processing. This processing method allows to simply and effectively control the position of the ball rotation axis; as a consequence of this, it is possible to implement a very fast coating of all surface of the ball with a uniform grid of processing traces [5]. This method of processing also allows some control of the sliding speeds in the processing areas. The increase in these speeds leads to an increase in processing performance.

However, such methods of processing with the complex kinematics of the equipment, due to the increased number of independent moving components and equipment parts, increase the instability of the results of precision processing. Therefore, in these methods of processing, high requirements for the rigidity of the processing equipment are imposed. This is especially true for the workpieces of balls with large deviations from the round shape, which is typical for the workpieces of ceramic balls.

Other methods of ceramic balls effective processing are precision processing methods based on the use of magnetic fluid [6]. These processing methods are based on the magnetohydrodynamic behavior of the magnetic fluid, which can levitate all non-magnetic materials suspended in it and the damping properties of the magnetic fluid.

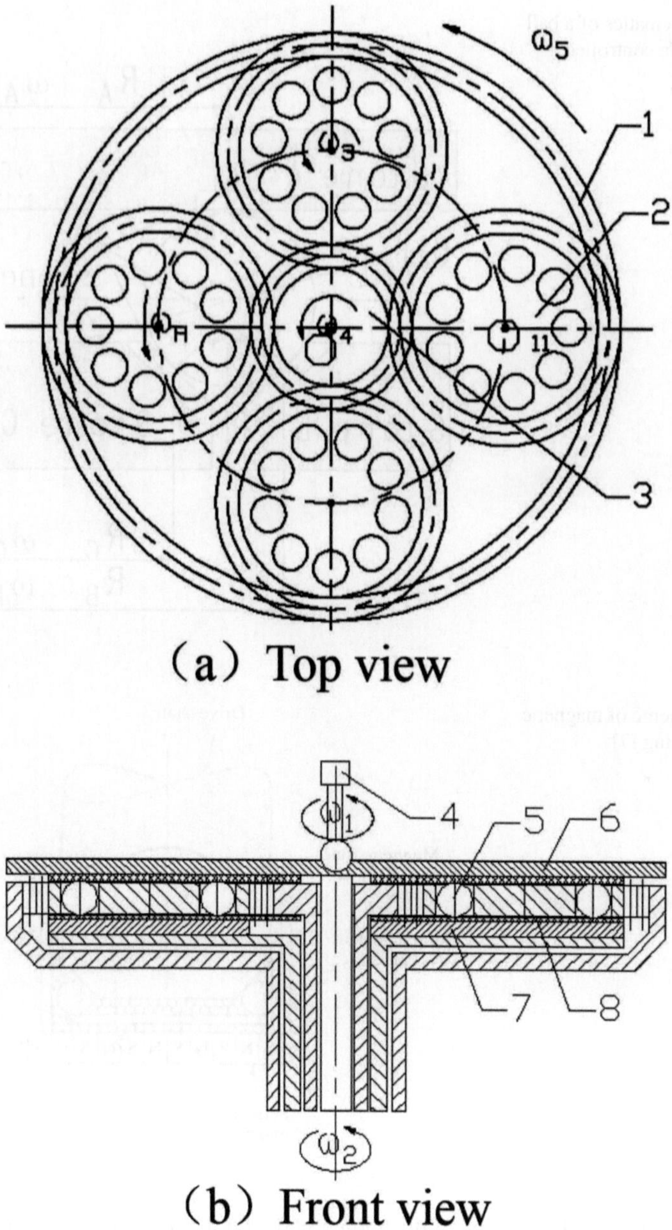

Fig. 2 Schematic of dual-plane polishing apparatus [4]

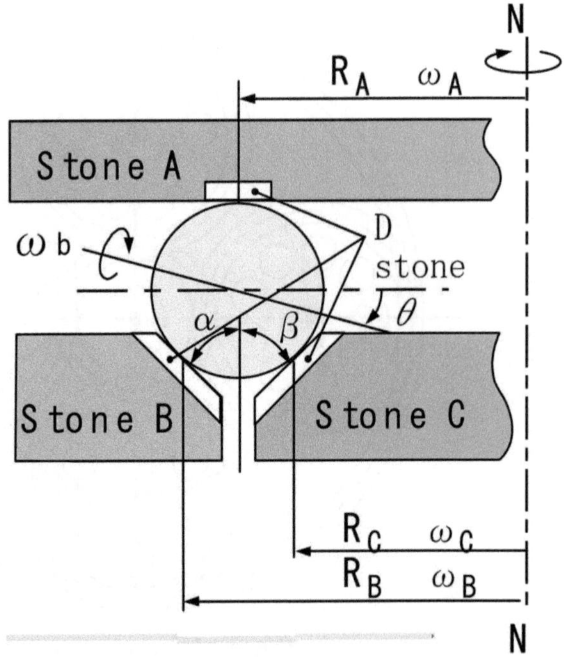

Fig. 3 Kinematics of a ball in spin angle controlled method [5]

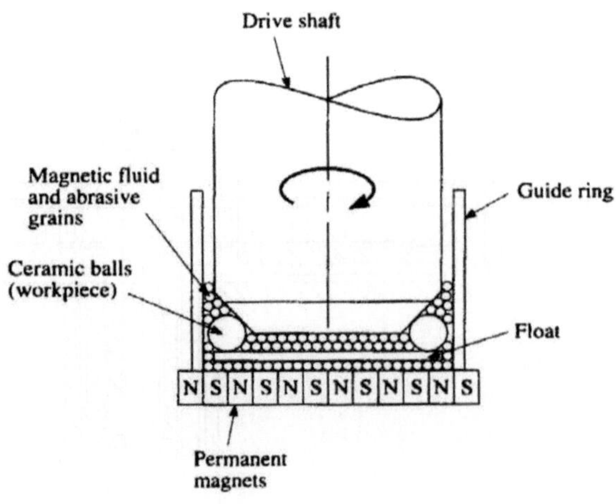

Fig. 4 Scheme of magnetic float polishing [7]

Figure 4 shows the scheme of the magnetic float polishing apparatus based on the magnetohydrodynamic behavior [7]. Magnetic fluid consists of a colloidal dispersion in water or kerosene of small ferromagnetic particles, usually Fe_3O_4 magnetite, and abrasive particles.

When the magnetic field is applied, the magnetic particles are attracted down toward the area of the higher magnetic field and to each other; the result increases the viscosity of the magnetic fluids containing abrasive. When the drive shaft rotates,

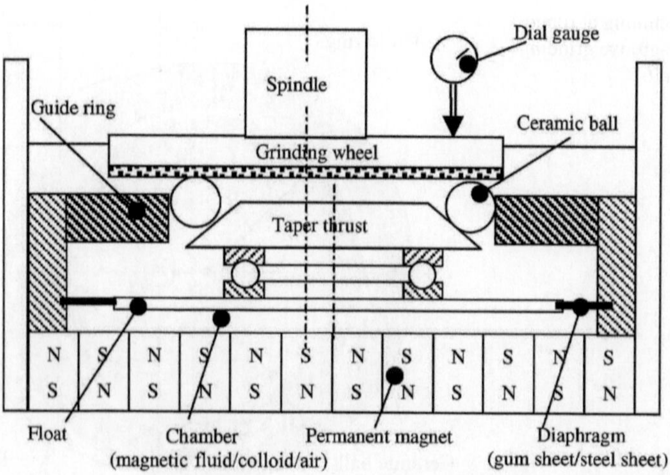

Fig. 5 Schematic view of magnetic fluid support grinding machine of ceramic balls [10]

abrasive particles treat the ball from all sides; the stronger the magnetic field, the more intense the processing [8, 9].

Zhang and his research group studied the relationship between the vibration of the support system and the error in the surface generation and discussed the effect of the stiffness and damping [10]. They concluded that high-precision grinding required high damping to increase the magnification factor and minimize the phase difference.

The magnetic fluid under the influence of the magnetic field becomes viscous and increases the damping capacity of the system, which can lead to improved processing quality. Zhang and colleagues suggested separating the magnetic fluid from the grinding process and using it only as extinguishing vibration damping system. The magnetic fluid has been sealed in a chamber under the float with a rubber membrane, to ensure soft support and damping. Balls, thus, are grounded by quickly rotating the abrasive grinding wheel. A scheme illustrating the use of magnetic fluid as a damping system is shown in Fig. 5.

Precision machining with magnetic fluid allows, in some cases, to significantly increase the quality of ball machining and machining performance. However, it requires the use of special equipment and does not always provide a rapid improvement of the spherical shape of the balls and no difference in the size of the balls in the party.

Most of all known methods of precision processing of ceramic balls are based on the conventional processing scheme in the annular grooves, but with innovations that can improve the efficiency of processing.

In recent years, precision machining of balls in eccentric V-grooves has attracted much attention of researchers, Fig. 6 [11–14]. This method of machining is often used for precision processing of ceramic and cemented carbide balls and remains relevant today. Machining in this way can improve both the performance and quality

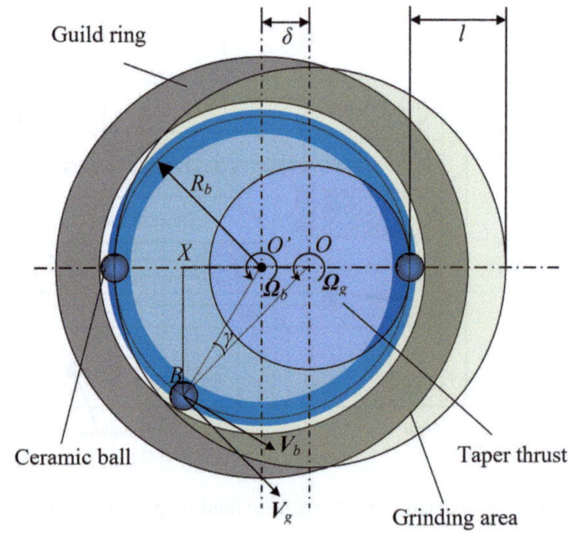

Fig. 6 Machining at round eccentric V-groove grinding mechanism [6]

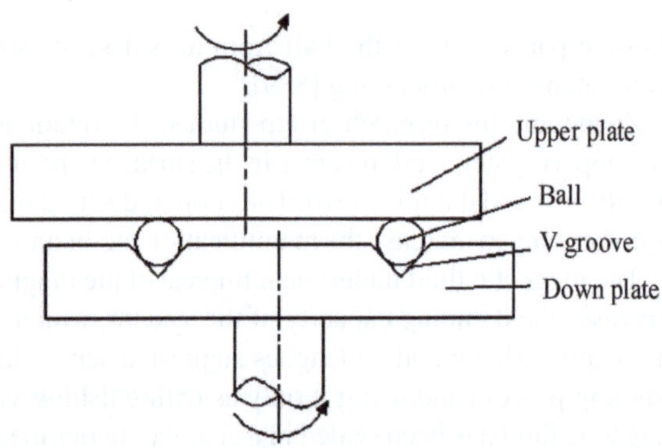

Fig. 7 Schematic illusion of eccentric dual-rotating V-groove lapping apparatus [12]

of processing. With the eccentric position of the guide V-groove, the direction of the ball movement V_b does not coincide with the direction of the velocity vector of the upper flat plate V_g at the point of contact, at the processing in the annular V-grooves (Fig. 2b), so the speed of sliding between the treated ball and the plate increases. As a result, processing performance is improved. The eccentricity of the guide groove position increases the grinding area, in consequence of this reduces the wear of the upper plate, and increases the stability of the process of precise grinding of the balls.

One of the processing variants with eccentricity is eccentric dual-rotating V-groove lapping mode (Fig. 7). This method differs from traditional processing by independent drives of rotation of the upper and lower plates. According to the research, the ball would be lapped more uniformly under eccentric dual-rotating V-groove lapping mode by adjusting the velocities of the lower plate and upper plate [12].

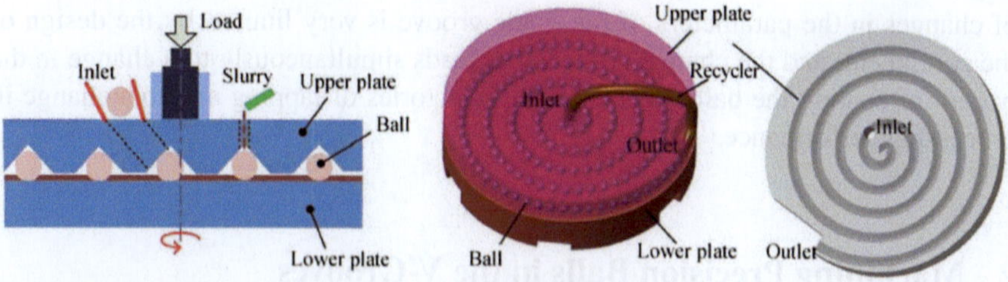

Fig. 8 Ball processing with a spiral V-groove plate [16]

When processed in round eccentric V-grooves, each of the contact points leaves a lapping track on the surface of the ball in the form of a circle, same as Fig. 1c. However, the mismatch of the direction of the ball movement with the direction of the upper flat plate velocity vector increases the random factors that lead to the ball turning during machining. As a result, the entire surface of the ball is covered with a tool grid faster than when machining in concentric grooves and improves the quality of processing.

The disadvantages of this method include a limitation on the choice of the value of the permissible eccentricity of the guide groove position. With increasing eccentricity of the guide V-groove position, the volume size of the guide groove decreases, accordingly, and the number of balls processed simultaneously decreases. In addition, the offset and asymmetric position of the guide V-groove, especially with a large eccentricity, places increased demands on the accuracy and stiffness of the equipment. Despite the known analytical calculations for processing with an eccentric guide groove, the pattern of change in the position of the own axis of rotation of the balls remains poorly controlled and difficult to predict.

More opportunities for improving the efficiency of precision processing of ceramic balls have a processing technology based on spiral V-shaped grooves [15, 16].

The principle of this precision method of processing balls in series is based on that the radius of curvature of the guide groove continuously changes along with it. When changes the radius of the V-groove, the position of ball axis rotation changes accordingly. As a result, the ball surfaces can be enough quickly and evenly covered with processed traces. At the same time, the processing productivity is also increased because the direction of the ball movement does not coincide with the direction of the velocity vector of the upper flat plate at the point of contact and the speed of sliding between the treated ball and the plate is increased. Figure 8 shows the schematic diagram of the system for the spiral V-groove plate method for processing high-precision ceramic balls [16].

The upper plate may be installed with eccentricity. With the increase of eccentricity, the variation range of spin angle increases during a single lapping period, and the lapping trajectory can be more uniformly distributed on the surface of a ball [15].

This method of processing in spiral grooves involves the use of a loading mechanism and a recycler basket for collecting balls in it, between their runs along the groove, so this method is not suitable for processing small batches of balls. The range of changes in the parameters of the guide groove is very limited by the design of the equipment, and the change in the spiral leads simultaneously to a change in the pattern of coating the ball's surface with trajectories of lapping and to a change in processing performance.

2 Machining Precision Balls in the V-Grooves with Variable Curvature

Analysis of the above methods shows that the most promising, especially in terms of precision processing of ceramic balls, is the processing of balls in the guide V-groove with variable curvature. To improve the efficiency of precision processing of ceramic balls, for more complete use of this scheme potential, a new balls processing method in the guide V-groove with variable curvature is proposed.

This method of processing allows you to more widely control both the position of the axis of rotation of the balls and the sliding speeds at the points of contact of the balls with the upper plate. That is, this method of processing allows you to control the process of forming a grid of tool marks on the surface of the balls and the processing capacity, independent of each other [17].

Figure 9 shows the proposed processing scheme. The ceramic balls 2 are located in the guide V-groove 4, made on the lower plate 3, with a variable curvature along it. The upper plate 1 is pressed against the balls with a force P, and the lower plate is rotated at a speed ω.

This scheme can be used for both precise grinding and finishing of balls. For grinding, plate 1 can be used a superabrasive diamond grinding wheel, and for finishing (polishing)—iron lapping. Plate 3 in both cases can be cast iron. When using this scheme for precision processing of ceramic balls removal allowance is provided by the use of polishing paste containing diamonds as an abrasive. In the particular case shown in Fig. 9, the shape of the guide V-groove in the plan consists of interfaced sections of circles arc with radius R1 and R2.

The position of the own axis of the ball rotation OO at any point depends on various factors, but the greatest influence on it has the value of the curvature radius of the guide V-groove at this point. Since the curvature radius of the guide V-groove changes along with it according to a certain law, the rolling of the balls on it changes accordingly and the position of their own axis rotation, characterized by the value of the angle α. In the process of processing according to this scheme, the balls make a complex rotation, and the spherical surface is formed by intersecting traces of processing. That is, the processing according to the proposed scheme allows implementing such kinematics of the process, in which all spherical surface of the balls will be quickly covered with a uniform grid of tool traces.

The discrepancy of the velocity vectors of the balls V_b and the plate V_p, which is provided by the trajectory of the balls, leads to a significant increase in the velocity of sliding V_c between them at the point of contact A. The rule of change in the

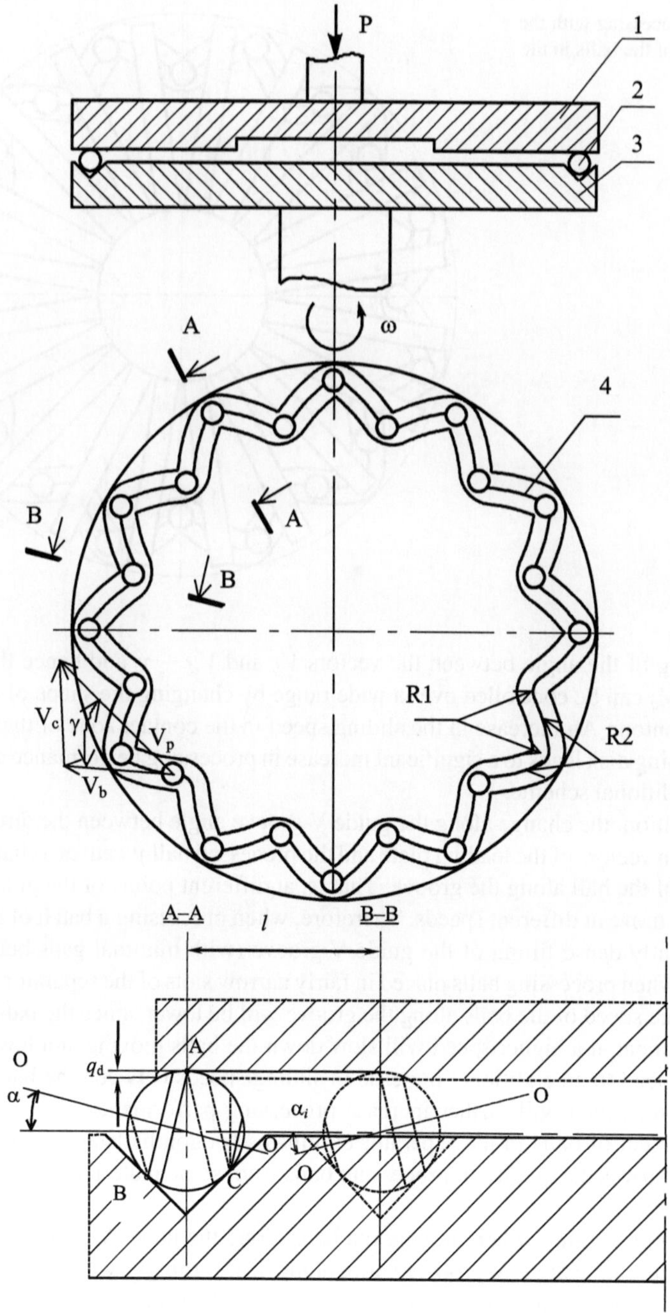

Fig. 9 Processing balls in a V-groove with a variable radius of curvature along it

Fig. 10 Processing with the placement of the balls in the separator

processing of the angle between the vectors V_b and $V_d - \gamma$ and hence the sliding velocity V_c can be controlled over a wide range by changing the shape of the guide groove contour. An increase in the sliding speed in the contact zone of the ball with the clamping disk leads to a significant increase in processing performance compared to the traditional scheme.

In addition, the change along the guide V-groove angle between the directions of the motion vectors of the leading plate and the processed ball γ causes a change in the velocity of the ball along the groove. That is, at different points of the guide groove balls will move at different speeds. Therefore, when processing a batch of balls with a sufficiently dense filling of the guide V-groove (with minimal gaps between the balls) or when processing balls placed in fairly narrow slots of the separator (Fig. 10), the average speed of the balls along the groove will be lower, since the balls moving at the moment at a higher speed will slow down the balls moving at a lower speed. This will lead to an additional increase in sliding speeds between the balls and the upper plate, which will further increase processing performance. Processing with the placement of balls in the separator eliminates friction balls against each other, which improves the quality of processing balls from some materials, such as silicon carbide.

During processing, when moving the balls along the guide V-grooves, since the radius of curvature of the V-grooves varies along it, accordingly, will change the angle of the balls rotation axes tilt a. Thus, by changing the trajectory of the V-grooves can be in a wide range to control the process of forming a grid of processing traces on the surface of the ball, its uniformity, density, and removal rate of the allowance for processing. With optimal V-groove parameters, the surface of the ball is quickly covered with a uniform grid of processing traces, thus achieving high quality and productivity.

Continuous change of position of the processed balls in the radial direction with an amplitude l (Fig. 9) allows a more rational use of the surface of the upper plate, which creates conditions for reducing its linear wear q_d, as well as provides automatic alignment of the diameters of the balls during processing. This allows for machining of small batches with high requirements for the accuracy of the diameters of all balls in the party, which is important, as noted earlier, for precision ceramic balls.

Thus, changing the characteristics of the guide groove, namely the rule of change of the radius of curvature of its contour in the circumferential direction, it is possible in a wide range to control both the process of forming a spherical surface and the processing capacity. The developed scheme of processing of balls in the closed guide V-grooves with variable curvature favorably differs from the known methods of processing. Its application allows creating conditions for the implementation of highly effective precision processing of ceramic balls.

3 Mathematical Model of Kinematics of the Process of Precision Processing of Balls

Recently, many works are devoted to the analysis of movements of the ball in the process of precision machining [4, 6, 8, 13, 16]. Nevertheless, the literature describes the kinematic dependence of precision ball machining based on large assumptions, without taking into account important factors. In particular, they do not take into account the influence of the guide V-groove wear value, the distribution of forces along the arcs of the contact of the balls with the side forming grooves, friction between the balls and plates, and other factors affecting the movement of the balls. Therefore, in this work, the kinematics analysis of the ball's motion in the process of precision machining is carried out in general form, taking into account the influence on the balls processing kinematics of the tool geometric parameters, the characteristics of the finishing paste, the material of the balls and the tool.

In particular, the dependences obtained from the analysis are applicable in the case of precision diamond machining of ceramic balls in V-grooves with a variable radius of curvature.

(a) The balls are absolutely not a deformable body.
(b) The normal reaction force that occurs at the elementary site of contact of the ball with the tool is proportional to its displacement in the tool body.
(c) The area of the contact of the ball with the V-grooves is determined by the amount of wear.
(d) Forming a wear of the guide groove is described by circular arcs with the radius equal to the radius of the balls being processed.

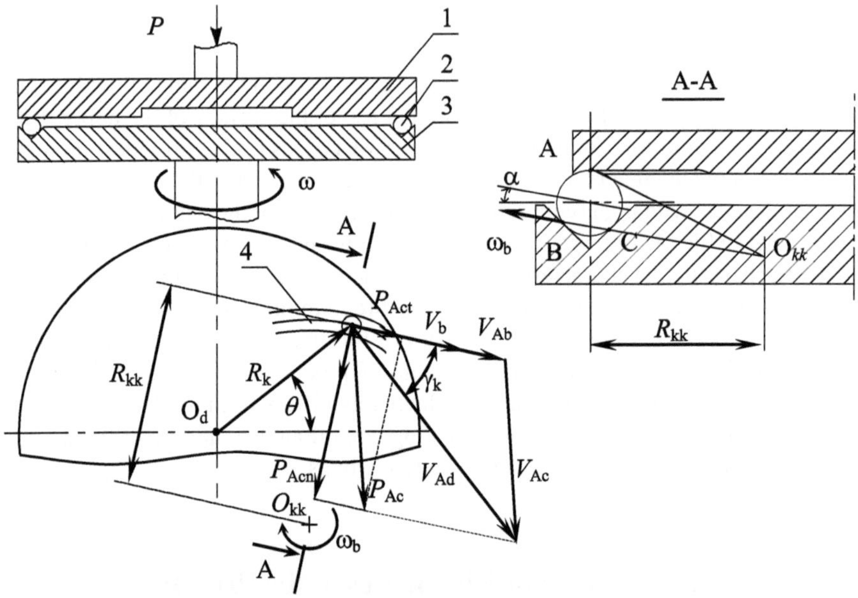

Fig. 11 Scheme of precision balls processing

(e) The balls during the machining do not touch each other. The following assumptions were made in the development of the mathematical model.

Figure 11 shows the investigated scheme of precision balls processing. The ceramic balls 2 are located in the guide V-groove 4, made on the surface of the lower plate 3. Processing is carried out in the environment of polishing paste, wherein the upper plate 1 is pressed to the balls with force P, and lower plate 3 is rotated at a speed ω. To simplify the calculations, we assume that the lower plate 3 with the guide V-groove 4 is stationary, and the processing of the balls occurs as a result of the rotation of the upper plate 1.

The form of guide V-grooves (guide grooves contour) in general is convenient to define polar coordinates with the center in the point of O_d as $R_k = F(\theta)$, where R_k is the polar radius of an arbitrary point of the vertex of the groove and θ is polar angle defining the position of this point.

At any time, the movement of the processed ball can be considered as a rotation at a speed of ω_b around an inclined angle α instantaneous axis of rotation passing through the points B and C, located within the corresponding contact arcs. The position of these points depends on many factors, such as the nature of the pressure distribution along the contact arcs and their magnitude, the ratio of the friction coefficients of the ball on the surface of the upper and lower plate.

When analyzing the movement of the ball during processing, it is convenient to decompose the rotation of the ball around the instantaneous axis of rotation BC into two interrelated movements: rotation around its own axis of rotation tilted at an angle

α, at a speed ω_b, and movement along the guide groove. In this case, the portable velocity of the ball along the guide groove can be expressed either through the linear velocity of the ball along the V-groove V_b or through its angular velocity ω_b rotation around the center of curvature of the contour of the guide groove at this point O_{kk}. These speeds are connected as $V_b = R_{kk} \cdot \omega_b$, where R_{kk} is the radius of curvature of the contour of the guide grooves at a given point.

The radius of the guide V-groove curvature can be determined by the well-known formula [2]:

$$R_{kk} = \frac{((F(\theta) \cdot \cos\theta)'^2 + (F(\theta) \cdot \sin\theta)'^2)^{3/2}}{|(F(\theta) \cdot \cos\theta)' \cdot (F(\theta) \cdot \sin\theta)'' - (F(\theta) \cdot \sin\theta)' \cdot (F(\theta) \cdot \cos\theta)''|}$$

At the point of contact of the ball with the upper plate A, the surface area of the ball is moving along the direction of the grooves at a speed of V_{Ab}, and the surface area of the top plate at a speed of V_{Ad}: $V_{Ad} = R_k \cdot \omega_d$. The angle between the direction of the velocities V_{Ab} and V_{Ad} is equal to the angle of attack of the V-groove at this point $\gamma_k = \mathrm{arctg}(F'(\theta))$.

As a result of the mismatch of the velocity vectors V_{Ab} and V_{Ad}, the surface of the ball in contact with the upper plate at point A slips with respect to it with the speed of sliding of V_{Ac}. In the direction of the velocity vector between the contacting surfaces, there is a friction force P_{Ac}. The tangential component of this friction force P_{Act} is directed toward the movement of the ball, which provides rolling of the ball along the guide groove.

As a result of pressing the processed balls to the lower plate with a force P, in the contact zones B and C, there are reaction forces P_B and P_C, Fig. 12. These forces are the result of the arc-distributed contact of normal pressures on the side of the lateral forming guide V-grooves on the balls dP_B and dP_C.

As a result of the force P_A, ball is moved in the direction of the lower plate on the value of d_O (d_O is small value) and along the arcs of contact occur distributed pressure whose intensity $\mathrm{d}P_B(\phi_{Bb})$ and $\mathrm{d}P_C(\phi_{Cc})$ is proportional to the movement of these points normal to the contact surface direction $d_B(\phi_{Bb})$ и $d_C(\phi_{Cc})$:

$$\mathrm{d}P_B(\phi_{Bb}) \equiv \mathrm{d}_B(\phi_{Bb}), \mathrm{d}P(f_{Cc}) \equiv \mathrm{d}_C(\phi_{Cc}).$$

In the upper contact zone A of the ball with the upper plate, their touch also occurs on some of the surface area. But the value of the contact area of the ball with the upper plate is very small, and the pressure distribution $\mathrm{d}P_A$ on the contact area has no significant influence on the kinematics of the movements of the ball. Therefore, the pressure force of the upper plate to the balls will take into account as the concentrated resultant reaction force of the surface of the disk P_A, applied to point A.

From the triangle AGH (Fig. 13), we make the equations relating the velocity of the ball V_{Ab} and the sliding velocity between the surfaces of the ball and the upper plate V_{Ac}, with the rotation speed of the upper plate ω_d:

$$V_{Ac} = V_{Ad} \cdot \frac{\sin\gamma}{\sin\gamma_c} = \omega_d \cdot R_d \cdot \frac{\sin\gamma}{\sin\gamma_c}, V_{Ab} = V_{Ab} \cdot \frac{\sin(\gamma_c - \gamma)}{\sin\gamma_c} = \omega_d \cdot R_d \frac{\sin(\gamma_c - \gamma)}{\sin\gamma_c},$$

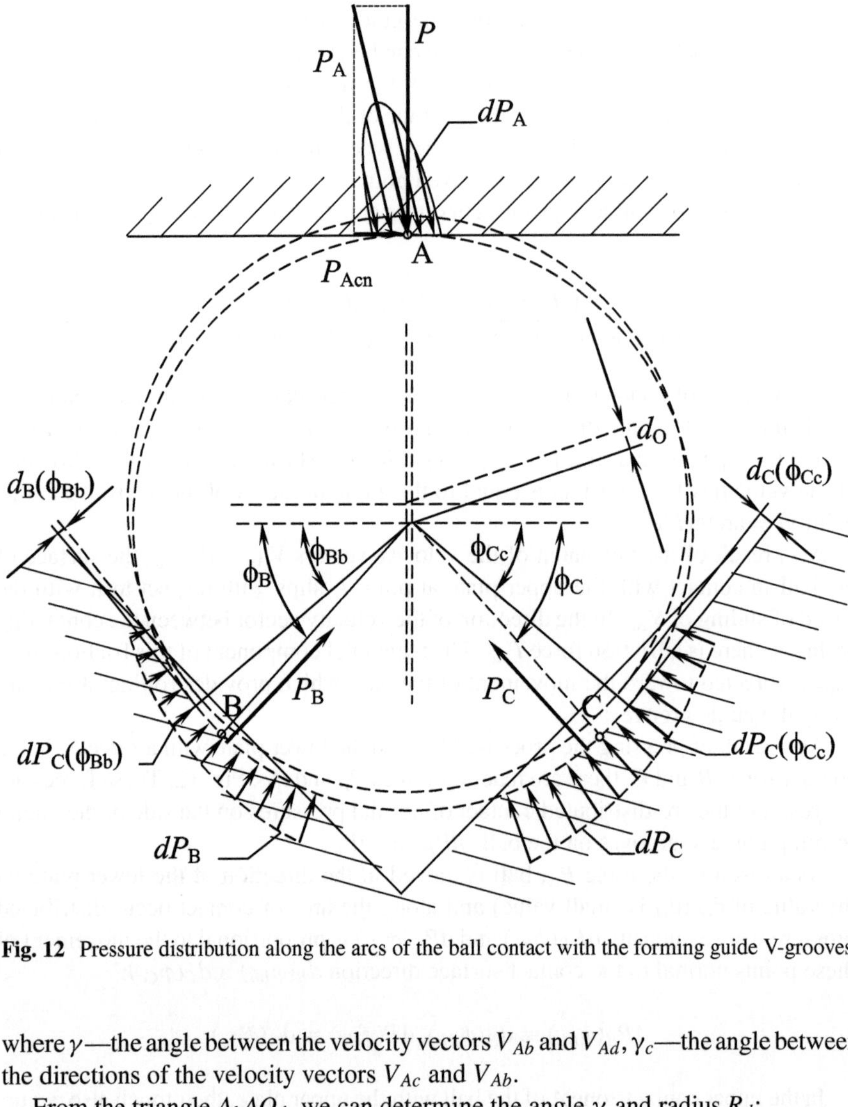

Fig. 12 Pressure distribution along the arcs of the ball contact with the forming guide V-grooves

where γ—the angle between the velocity vectors V_{Ab} and V_{Ad}, γ_c—the angle between the directions of the velocity vectors V_{Ac} and V_{Ab}.

From the triangle A_kAO_d, we can determine the angle γ and radius R_d:

$$\gamma = \gamma_k + \text{arctg}\left(\frac{\Delta R \cdot \sin\gamma_k}{R_k - \Delta R \cdot \cos\gamma_k}\right),\ R_d = \sqrt{R_k^2 - 2 \cdot R_k \cdot \Delta R \cdot \cos\gamma_k + \Delta R^2},$$

where ΔR—displacement of the ball position in the V-groove.

The displacement of the balls ΔR according to Fig. 13 is:

$$\Delta R = \frac{\sqrt{r^2 - \left(\frac{d_{pB}}{2}\right)^2} - \sqrt{r^2 - \left(\frac{d_{pC}}{2}\right)^2}}{2 \cdot \cos\beta},$$

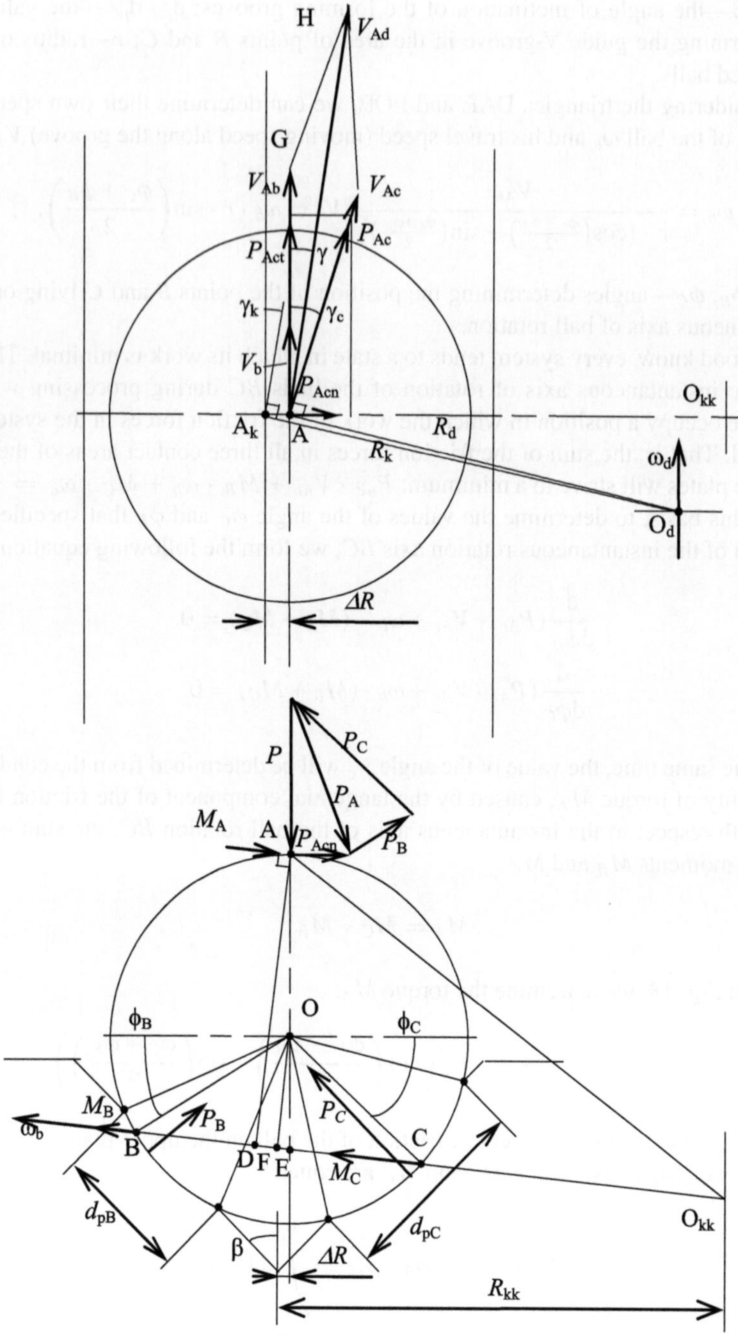

Fig. 13 Forces and speeds in the contact area

where β—the angle of inclination of the forming grooves; d_{pB} d_{pC}—the value of wear forming the guide V-groove in the area of points B and C; r—radius of the processed balls.

Considering the triangles DAE and FOE, we can determine their own speed of rotation of the ball ω_b and his travel speed (moving speed along the groove) V_b:

$$\omega_b = \frac{V_{Ab}}{r \cdot \left(\cos\left(\frac{\phi_C - \phi_B}{2}\right) + \sin\left(\frac{\phi_C + \phi_B}{2}\right)\right)}, \; V_b = \omega_b \cdot r \cdot \sin\left(\frac{\phi_C + \phi_B}{2}\right),$$

where ϕ_B, ϕ_C—angles determining the position of the points B and C lying on the instantaneous axis of ball rotation.

As good know, every system tends to a state in which its work is minimal. Therefore, the instantaneous axis of rotation of the balls BC during processing will at any time occupy a position in which the work of the friction forces in the system is minimal. That is, the sum of the friction forces in all three contact areas of the ball with the plates will strive to a minimum: $P_{Ac} \cdot V_{Ac} + M_B \cdot \omega_b + M_C \cdot \omega_b \Rightarrow \min$.

On this basis, to determine the values of the angle ϕ_C and ϕ_B that specifies the position of the instantaneous rotation axis BC, we form the following equation:

$$\frac{\mathrm{d}}{\mathrm{d}\phi_B}(P_{Ac} \cdot V_{Ac} + \omega_b \cdot (M_B + M_C) = 0, \tag{1}$$

$$\frac{\mathrm{d}}{\mathrm{d}\phi_C}(P_{Ac} \cdot V_{Ac} + \omega_b \cdot (M_B + M_C) = 0, \tag{2}$$

At the same time, the value of the angle γ_c will be determined from the condition of equality of torque M_A, caused by the tangential component of the friction force P_Act with respect to the instantaneous axis of the ball rotation BC, the sum of the friction moments M_B and M_C:

$$M_A = M_C + M_B \tag{3}$$

From Fig. 13, we determine the torque M_A:

$$M_A = \mu_d \cdot P \cdot \cos\gamma_C \cdot r \cdot \left(\cos\left(\frac{\phi_C - \phi_B}{2}\right) + \sin\left(\frac{\phi_C + \phi_B}{2}\right)\right) \tag{4}$$

where $\mu_д$—the given coefficient of friction of the ball on the upper plate.

Points of friction spinning M_B and M_C are equal:

$$M_B = \int\limits_{\phi_{Bn}}^{\phi_{Bk}} \mathrm{d}M_B(\phi_{Bb}) \cdot \mathrm{d}\phi_{Bb}, \; M_C = \int\limits_{\phi_{Cn}}^{\phi_{Ck}} \mathrm{d}M_C(\phi_{Cc}) \cdot \mathrm{d}\phi_{Cc},$$

$\phi_{Bn}, \phi_{Bk}, \phi_{Cn}, \phi_{Ck}$—the angles which determine the start and end positions of the contact arcs in areas *B* and *C*, respectively; $\mathrm{d}M_B(\phi_{Bb})$ and $\mathrm{d}M_C(\phi_{Cc})$—the intensity distribution in the contact arcs of the friction torques.

Angles $\phi_{Bn}, \phi_{Bk}, \phi_{Cn}, \phi_{Ck}$ are defined as:

$$\phi_{Bn} = \arcsin\left(\beta - \frac{d_{pB}}{2 \cdot r}\right), \phi_{Bk} = \arcsin\left(\beta + \frac{d_{pB}}{2 \cdot r}\right),$$
$$\phi_{Cn} = \arcsin\left(\beta - \frac{d_{pB}}{2 \cdot r}\right), \phi_{Ck} = \arcsin\left(\beta + \frac{d_{pB}}{2 \cdot r}\right).$$

The intensity of the friction moments $\mathrm{d}M_B(\phi_{Bb})$ and $\mathrm{d}M_C(\phi_{Cc})$ distributed along the contact arcs are equal to:

$$\mathrm{d}M_B(\phi_{Bb}) = \mu_\kappa \cdot \mathrm{d}P_B(\phi_{Bb}) \cdot r \cdot \left|\sin\left(\frac{\phi_C + \phi_B}{2}\right) - \sin\left(\phi(\phi_{Bb}) + \frac{\phi_C - \phi_B}{2}\right)\right| \tag{5}$$

$$\mathrm{d}M_C(\phi_{Cc}) = \mu_\kappa \cdot \mathrm{d}P_C(\phi_{Cc}) \cdot r \cdot \left|\sin\left(\frac{\phi_C + \phi_B}{2}\right) - \sin\left(\phi(\phi_{Cc}) - \frac{\phi_C - \phi_B}{2}\right)\right| \tag{6}$$

where μ_κ—the relative coefficient of friction of the ball on the forming guide grooves [2].

According to the accepted assumptions, the intensity of the distributed pressure on the arcs of contact $\mathrm{d}P_B(\phi_{Bb})$ и $\mathrm{d}P_C(\phi_{Cc})$ is given as: $\mathrm{d}P_B(\phi_{Bb}) = K{\cdot}d_B(\phi_{Bb})$, $\mathrm{d}P_C(\phi_{Cc}) = K{\cdot}d_C(\phi_{Cc})$,

where *K* is the coefficient of proportionality, $d_B(\phi_{Bb})$ and $d_C(\phi_{Cc})$ are the values of displacements of elementary points as a result of the force P.

The values of displacements of elementary points $d_B(\phi_{Bb})$, $d_C(\phi_{Cc})$ are equal to:

$$d_B(\phi_{Bb}) = r \cdot \left(\frac{\cos\left(\arcsin\left(\frac{d_O}{r}\right) \cdot \cos(\phi_{Bb} - \phi_O)\right) + \phi_O - \phi_{Bb}}{\cos(\phi_{Bb} - \phi_O)} - 1\right),$$

$$d_C(\phi_{Cc}) = r \cdot \left(\frac{\cos\left(\arcsin\left(\frac{d_O}{r}\right) \cdot \cos(\phi_{Cc} + \phi_O)\right) - \phi_O - \phi_{Cc}}{\cos(\phi_{Cc} + \phi_O)} - 1\right),$$

where d_O is the value displacement of the ball in the direction of the lower plate, ϕ_O is the direction of this displacement.

Given that d_O is small enough:

$$d_B(\phi_{Bb}) = \lim_{d_O \to 0} d_B(\phi_{Bb}) = d_O \cdot \sin(\phi_{Bb} - \phi_O),$$
$$d_C(\phi_{Cc}) = \lim_{d_O \to 0} d_C(\phi_{Cc}) = d_O \cdot \sin(\phi_{Cc} + \phi_O)$$

Thus, the intensity of the distributed pressure on the arcs of contact $\mathrm{d}P_B(\phi_{Bb})$ и $\mathrm{d}P_C(\phi_{Cc})$ is equal to:

$$\mathrm{d}P_B(\phi_{Bb}) = Kd_O \sin(\phi_{Bb} - \phi_0) \tag{7}$$

$$\mathrm{d}P_C(\phi_{Cc}) = Kd_O \sin(\phi_{Cc} - \phi_0) \tag{8}$$

The system of Eq. (9), composed of Eqs. (1)–(8), is a mathematical model of the kinematics of precision processing of balls in a groove with a variable radius of curvature along it:

$$\begin{cases} \frac{d}{d\phi_B}(P_{Ac} \cdot V_{Ac} + \omega_b \cdot (M_B + M_C)) = 0 \\ \frac{\mathrm{d}}{\mathrm{d}\phi_C}(P_{Ac} \cdot V_{Ac} + \omega_b \cdot (M_B + M_C)) = 0 \\ M_A = M_C + M_B \\ M_A = \mu_d \cdot P \cdot \cos(\gamma_C) \cdot r \cdot \left(\cos\left(\frac{\phi_C - \phi_B}{2}\right) + \sin\left(\frac{\phi_C + \phi_B}{2}\right)\right) \\ M_B = \int\limits_{\phi_{Bn}}^{\phi_{Bk}} \mu_\kappa \cdot \mathrm{d}P_B(\phi_{Bb}) \cdot r \cdot \left|\sin\left(\frac{\phi_C + \phi_B}{2}\right) - \sin\left(\phi(\phi_{Bb}) + \frac{\phi_C - \phi_B}{2}\right)\right| \cdot \mathrm{d}\phi_{Bb} \\ M_C = \int\limits_{\phi_{Cn}}^{\phi_{Ck}} \mu_\kappa \cdot \mathrm{d}P_C(\phi_{Cc}) \cdot r \cdot \left|\sin\left(\frac{\phi_C + \phi_B}{2}\right) - \sin\left(\phi(\phi_{Cc}) - \frac{\phi_C - \phi_B}{2}\right)\right| \cdot \mathrm{d}\phi_{Cc} \\ \mathrm{d}P_B(\phi_{Bb}) = Kd_O \sin(\phi_{Bb} - \phi_0) \\ \mathrm{d}P_C(\phi_{Cc}) = Kd_O \sin(\phi_{Cc} - \phi_0) \end{cases} \tag{9}$$

Solving the resulting system of equations, it is possible to define the main kinematic parameters of the process of finishing balls, such as the position of the instantaneous axis of the ball rotation, speed of balls along the groove, the distribution of sliding velocities and pressures in the arcs of contact. By changing the form of the guide V-groove, it is possible in a wide range to control the process of forming a grid of processing traces on the surface of the ball, its uniformity and density by controlling the position of its own axis of rotation of the balls, as well as the rate of removal allowance by changing the value of the slip rate between the surfaces of the treated ball and the tool.

Thus, using the given dependencies, it is possible to optimize these parameters and thus to improve performance and quality of the process of finishing balls.

4 Determination of the Geometric Parameters of the V-Grooves

To select the optimal parameters of the guide V-groove, first of all, it is necessary to determine the permissible range of changes in these parameters, at which stable processing of balls is possible.

The considered scheme of processing balls in V-grooves with a variable radius of curvature has kinematic features that, under certain circumstances, contribute to the creation of conditions that do not allow the processing [3]. There are two types of processing termination; under certain parameters and processing conditions, it is possible that:

- Balls are stopping in the V-grooves.
- Balls are rolled out from the guide V-grooves.

These phenomena do not occur in the processing of the conventional method, so they are not described in the literature and the conditions of their occurrence have not previously been studied.

Balls during processing will cease to move along the groove, if the torque M_A will be less than the sum of the friction moments M_B and M_C of the ball on the guide V-groove (Fig. 13).

The emergence of such a situation contributes to three factors:

- A reduction in the tangential component of the friction force P_{Act} (Fig. 14a), which determines the torque M_A. This occurs when increasing the angle between the V_{Ab} and V_{Ad}, i.e., by increasing the angle of attack of the grooves γ_k.

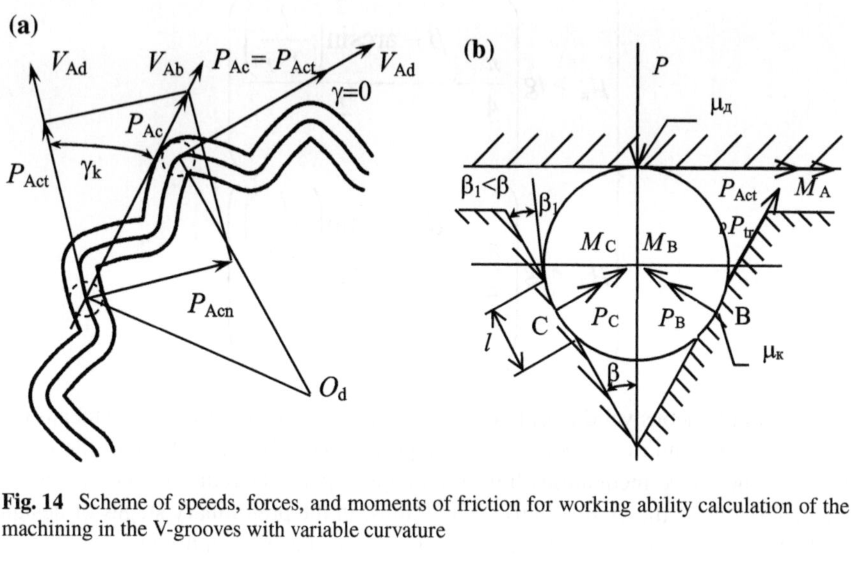

Fig. 14 Scheme of speeds, forces, and moments of friction for working ability calculation of the machining in the V-grooves with variable curvature

- Increase the value of the friction torque of the ball on the surface of the guide V-groove M_C and M_B at the expense of increasing the length of arc of contact l as a result of wear of the V-groove (Fig. 14b).
- Increase the relationship between coefficients of friction of the surface of the balloon on the guide groove and the upper plate μ_k/μ_d.

Another type of violation of the normal course of the processing process is the rolling out of the guide groove balls. Balls during processing will roll out of the guide groove, in case of increase above the critical value of the angle of inclination β or by increasing above a certain value, the value of the friction coefficients μ_k and μ_d (Fig. 14b).

To determine the conditions under which there will be an interruption of the processing of balls, the following inequalities were compiled.

For the case of stopping balls in the guide V-groove, if the inequality is correct, it means the occurrence of such conditions under which the ball ceases to move along the guide groove leads to disruption of the processing process.

$$\mu_d \cdot P \cdot \cos\gamma \cdot r \cdot \left(\cos\left(\frac{\phi_C - \phi_B}{2}\right) + \sin\left(\frac{\phi_C + \phi_B}{2}\right)\right) > \int\limits_{\phi_{Bn}}^{\phi_{Bk}} \mathrm{d}M_B(\phi_{Bb}) \cdot \mathrm{d}\phi_{Bb} + \int\limits_{\phi_{Cn}}^{\phi_{Ck}} \mathrm{d}M_C(\phi_{Cc}) \cdot \mathrm{d}\phi$$

To determine the conditions under which the balls are rolled out of the guide V-groove, the corresponding system of inequalities has been compiled:

$$\begin{cases} \mu_{\text{к}} > tg\left(\dfrac{\pi}{4} - \dfrac{\beta - \arcsin\left(\dfrac{l}{2 \cdot r}\right)}{2}\right) \\ \mu_{\text{д}} > tg\left(\dfrac{\pi}{4} - \dfrac{\beta - \arcsin\left(\dfrac{l}{2 \cdot r}\right)}{2}\right) \end{cases}$$

The simultaneous validity of both inequalities in any part of the groove means the emergence of conditions under which the processed balls will roll out of the guide groove. The above inequalities limit the range of possible values of the geometric characteristics of the guide V-groove and, thus, in the preliminary stages of design of the tool can act as criteria for its performance in specific conditions.

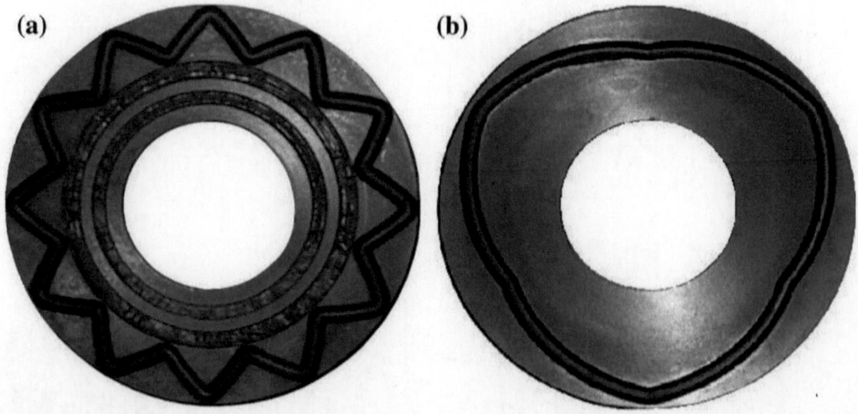

Fig. 15 Plates with a guide V-groove of the conjugate sections that made: **a** close to the spiral Archimedes; **b** close to the exponent

The proposed method of processing, by choosing the shape of the guide groove, allows regard to control both the formation of the grid traces of processing and processing performance.

Therefore, for the experimental verification of the proposed method, we made two types of grooves having the same process of forming a grid of processing traces on the surface balls, but different sliding speeds and speeds of moving the balls along the groove. Using the developed mathematical model of balls processing, the design of plates with guide V-grooves of the following forms is calculated:

– In the form of conjugate areas close to the Archimedes spiral (Fig. 15a);
– In the form of conjugate areas close to the exponential law (Fig. 15b).

The plate with a guide groove in the form of conjugated sections of the Archimedes spiral, based on the properties of the Archimedes spiral, is characterized by a constant angle of attack (except for the conjugation sections). The greater the attack angles of the V-groove, the higher the sliding speed in the contact area and, accordingly, the processing performance.

The plate with a guide groove in the form of conjugate sections made by the exponent, according to the mathematical model, is characterized by an almost constant rolling speed of the balls along the guide V-groove (except for the interface sections). This makes it possible to eliminate or at least significantly reduce the friction of the processed balls against each other.

Despite the fact that these grooves look very different, they implement the same algorithm for covering the entire surface of the ball with a uniform grid of processing traces. This algorithm consists in the fact that on the mating sections of the groove the balls make half a turn around their axis, and on the main sections—an integer number of turns: one turn for the case of the Archimedes spiral, three turns for the case of exponential sections.

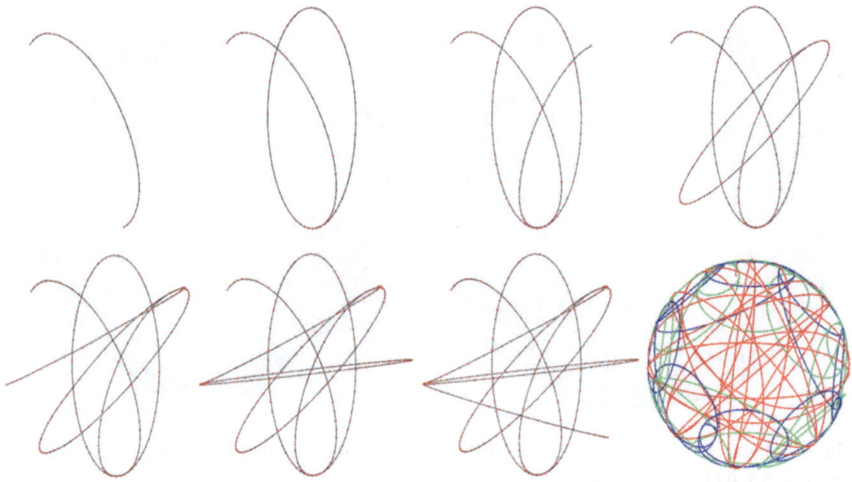

Fig. 16 Forming a grid of processing traces on the surface of the ball

The sequence of the trace grid formation is shown in Fig. 16. This illustration shows a grid of tool marks left by the contact point of the ball with the top plate after passing through the next of the mating V-groove elements. Since at three full rotations of the ball it becomes in the same position as after one full rotation, for both types of grooves, the scheme of formation of the grid traces of the tool is the same.

The last picture in Fig. 16 shows a grid of tool marks left by all three points of contact, after passing through three repeating elements, which corresponds to the run across entire V-groove for the exponential case and ¼ entire V-groove for the Archimedes spiral case. Red represents the trace of contact with the top plate, and blue and green represent traces of contact ball with the V-groove.

Figure 17a shows the calculated values of the sliding velocities between the balls and the upper plate for the case of a guide V-groove composed of conjugated sections made in the Archimedes 1 spiral and exponentially 2. It shows that in the first case, the sliding speed is much higher. When processing balls in the guide V-groove of the conjugate sections of the spiral Archimedes average sliding speed is 40 m/min and of the conjugate sections of the exponents is 14 m/min. Since the value of the sliding speed between the surfaces of the workpiece and the tool is associated with the processing performance, such a significant increase in these speeds allows us to argue that the productivity of processing balls in guide grooves with variable curvature will also be much higher than in the conventional method.

As can be seen from Fig. 17b, the velocity of the balls along the guide V-groove, made up of conjugate sections of the exponents, is almost constant (except for the conjugation sections), while the grooves, made up of conjugate sections of the Archimedes spiral, varies in a fairly wide range of 20–37 m/min. Respectively, in the first case, the friction of the processed balls of each other during processing will be significantly less.

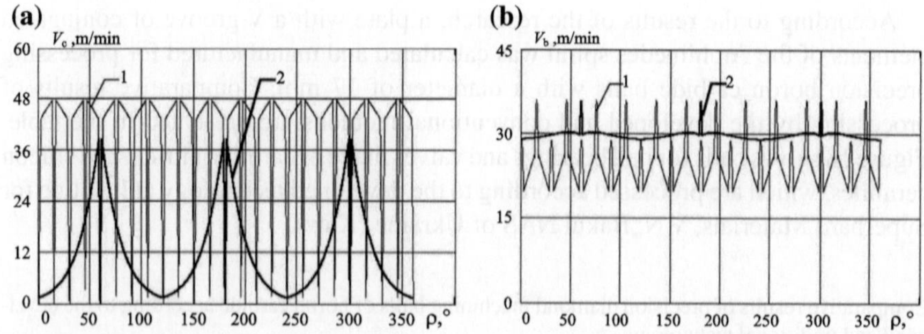

Fig. 17 Calculated speed: **a** ball slides with top plate; **b** movement of the balls along the guide V-grooves. V-Groove of elements made: 1—Archimedes spiral; 2—exponentially

For the experiment with the manufactured tool were used ceramic balls of boron carbide and silicon carbide with a diameter of 22 mm and a roundness of 20 μm.

The experiment showed that the processing of silicon carbide balls in the guide V-grooves made by Archimedes spiral, there is such intense friction between of the balls that the processing performance is 45 μm/h, and the deviation from the roundness of the balls increasing to 40 μm. At the same time, when they were processed in the guide V-groove, made exponentially, the roundness of the balls after processing was about 1.0 μm, and the processing performance was lower by about 3.5 times—13 μm/h.

When processing boron carbide balls using these tools, it was found that the productivity and quality of processing are higher when using a plate with guide V-grooves made in a spiral Archimedes, in comparison with the exponent. The non-roundness of the treated balls was 0.5 μm, with a productivity of 24 μm/h, while the non-roundness in the V-grooves made exponentially was 0.8 μm, and the productivity was about 3.5 times lower—7 μm/h.

Such results of using a tool with a guide V-groove from the conjugate elements of the Archimedes spiral are explained by the fact that the wear products of silicon carbide are more abrasive than the wear products of boron carbide, so the intense friction between the balls of silicon carbide leads to a decline in the quality of processing. When using a plate with guide V-grooves from the conjugate elements of the exponent, due to lower sliding speeds, the processing performance of the balls and boron carbide and silicon carbide, less about 3.5 times.

The use of a separator (Fig. 10) allows excluding the friction of the balls each other. As shown by the results of the experiment, it can improve both the performance and quality of processing in both types of V-grooves.

The productivity of processing balls in the application of the separator increased by 15% in the processing using the V-groove of the Archimedes spiral elements and 8% using the exponent. The greater increase in productivity in the first case is explained by the more uneven speed of the balls along the V-groove, which, as described earlier, when using the separator contributes to an additional increase in sliding speeds in the processing areas and, accordingly, processing performance.

According to the results of the research, a plate with a V-groove of conjugated elements of the Archimedes spiral was calculated and manufactured for processing precision boron carbide balls with a diameter of 17 mm. Comparative results of processing by the developed and conventional methods are presented in the table. Figure 18 shows balls for ball bearings and valves made of carbide, nitride, and silicon ceramics, which are processed according to the developed technology at Institute for Superhard Materials, V.N. Bakul NAS of Ukraine (Kiev).

Comparative results of precision diamond machining balls of boron carbide according to the developed and traditional technology

Name of indicator	Unit	Basic process	Developed process
Processing performance	μm/h	3.2	38
Processing time	h	33	1.2
Minimum allowance for processing	μm	120	30
Deviation from spherical form	μm	1.8	0.2
The difference of the size balls in the party	μm	5.3	0.2
The surface roughness, Ra	μm	0.06	0.03
Consumption of diamonds, qk	carat/ball	1.12	0.72
The relative wear of the lapping wheel	mm^3/mm^3	0.75	0.3
The durability of the disk tool, before needs dressing	PCs	<1000	>2000

Fig. 18 Examples of finished precision ceramic balls

5 Conclusion

The known methods of precision processing of ceramic balls are analyzed. For processing of high-precision ceramic balls, it is offered to use the developed method of processing in the closed V-grooves with variable curvature.

After the mathematical analysis of processing in this way, the mathematical model of processing in the guide V-grooves with variable curvature is presented. The mathematical model allows to define the main kinematic parameters of the process of finishing balls, such as the position of the instantaneous axis of the ball rotation, speed of balls along the V-groove, the distribution of sliding velocities and pressures in the arcs of contact. By changing the form of the guide V-groove, it is possible in a wide range to control the process of forming a grid of processing traces on the surface of the ball, its uniformity and density by controlling the position of its own axis of rotation of the balls, as well as the rate of removal allowance by changing the value of the slip rate between the surfaces of the treated ball and the tool.

Inequalities limiting the range of possible values of the geometric characteristics of the guide V-groove are proposed. These inequalities can be a criterion for the operability of the tool at the stage of its design.

The results of the experiments confirmed the ability to independently control the quality and performance of the precision processing process of ceramic balls and the effectiveness of the proposed processing method.

References

1. https://www.pressebox.com/pressrelease/schaeffler-technologies-ag-co-kg/The-ball-grinding-machine-from-Friedrich-Fischer-was-patented-125-years-ago/boxid/760392
2. Pasichnyi OO (2001) Theoretical analysis of spheres finishing process kinematics. Up-to-date processes of machining processing by instruments of SHM and quality of machines details surface. Collector of Scientifics works (series G). V. Bakul Institute for Superhard Materials of the NAS of Ukraine, Kiev, vol 179, pp 87–97
3. Pasichnyi OO (2000) Kinematic analysis of finishing process balls. Progressive technologies and systems of mechanical engineering. International digest of scientific works, DonNTU, Donetsk, vol 13, 272, pp 72–76
4. Feng KP, Zhou ZZ, Lv BH, Yuan JL (2013) Study on dual-plane ball polishing method for finishing ceramics ball. In: Advanced materials research. trans tech publications, vol 797, pp 444–449
5. Kurobe T, Morita T, Tsuchihashi N 回転スピン制御法による ④ 化ケイ素球 α 超精密研磨.
6. 馬聞倬.(2013) High efficiency ultra-precision grinding of ceramic balls. Doctoral dissertation, 佐賀大学
7. Kato K, Zhang B, Umehara N, Childs THC, Jones D (1994) Kinematics of balls in magnetic fluid grinding(predictions of the onset of skidding motion between balls and driving shaft). Nippon Kikai Gakkai Ronbunshu, C Hen/Transactions of the Japan Society of Mechanical Engineers, Part C, 60(572):1433–1439

8. Kirtane TS (2004) Finishing of silicon nitride (Si3N4) balls for advanced bearing applications by large batch magnetic float polishing (MFP) apparatus. Doctoral dissertation, Oklahoma State University
9. Zhang B, Uematsu T, Nakajima A (1998) High efficiency and precision grinding of Si3N4 ceramic balls aided by magnetic fluid support using diamond wheels. JSME Int J Ser C 41(3):499–505
10. Zhang B, Nakajima A (2003) Dynamics of magnetic fluid support grinding of Si3N4 ceramic balls for ultraprecision bearings and its importance in spherical surface generation. Precision Eng 27(1):1–8
11. Zhang B, Umehara N, Kato K (1995) Effect of the eccentricity between the driving shaft and the guide ring on the behavior of magnetic fluid grinding of ceramic balls. J-Jpn Soc Precision Eng 61:586–586
12. Yuan JL, Tang KF, Wang ZW, Lv BH, He XH (2009) Lapping of WC-Co cemented carbide ball by eccentric dual-rotating V-groove lapping mode. In: Advanced Materials Research. Trans Tech Publications, vol 69, pp 287–290
13. Lv CC, Sun YL, Zuo DW (2015) A novel eccentric lapping method with two rotatable lapping plates for finishing cemented carbide balls. A A 2:2
14. Kang J, Hadfield M (2005) The polishing process of advanced ceramic balls using a novel eccentric lapping machine. Proce Inst Mech Eng Part B: J Eng Manuf 219(7):493–503
15. Zhou F, Yuan J, Lyu B, Yao W, Zhao P (2016) Kinematics and trajectory in processing precision balls with eccentric plate and variable-radius V-groove. Int J Adv Manuf Technol 9(84):2167–2178
16. Feng M, Wu Y, Yuan J, Ping Z (2017) Processing of high-precision ceramic balls with a spiral V-groove plate. Frontiers Mech Eng 12(1):132–142
17. Shepelev AA, Pasichnyi OO (2000) A process of diamond machining of ball-shaped workpieces of structural ceramics. J Superhard Mater C/C Sverkhtverdye Mater 22(2):76–79

The Contribution of Ion Plasma Sprayed Coating to Performance of Precision Diamond Dressing Tools

V. Lavrinenko and M. Sheiko

Abstract This chapter describes the effect of ion plasma sprayed coating on performance of diamond precision dressing tools under diamond-abrasive machining conditions. The use of thick wear-resistant titanium nitride and carbide coating with a compensating interplayer of plastic metal (cobalt) is demonstrated to be a promising method for improving efficiency of dressing tools manufactured by electroplating and electroforming.

1 Introduction

Coated tools are known to be widely used in metal-working industries, provide a higher machining efficiency and better accuracy parameters of machined workpieces, extend ranges of technological tool application, and improve tool wear resistance and reliability Kharkov [1]. However, almost all of these achievements have been made for cutting tools (single-point tools, drills, milling cutters, and indexable cutting inserts). On the other hand, some R&D efforts have come up with coating to be applied to abrasive tools, which improve the grinding ratio 1.5–2-fold [2]. It was shown in [3] that the presence of special films on the surface of a diamond wheel can essentially change its performance. Therefore, a study of coating deposition methods for diamond tools has opened up new possibilities for improving machining efficiency.

The present-day industries rely extensively on various tool coating methods depending on the tool functionality and special features of production to involve this tool. The coating has a grain size ranging from 40 to 80 μm and microhardness from 13 to 14 GPa, which makes them suitable as wear-resistant coatings for cutting tools [4–6]. The detonation-type VN20 carbide coating is quite efficient in the hardening of die parts. Deforming tools hardened with such coating have a good

V. Lavrinenko (✉) · M. Sheiko
V. Bakul Institute for Superhard Materials National Academy of Sciences of Ukraine, Kiev, Ukraine
e-mail: lavrinenko@ism.kiev.ua

J. Zhang et al. (eds.), *Simulation and Experiments of Material-Oriented Ultra-Precision Machining*, Springer Tracts in Mechanical Engineering,
https://doi.org/10.1007/978-981-13-3335-4_10

performance and durability [5]. However, despite the benefits offered by the coating applied by detonation, implementation of this technique is rather slow, not the least reason being the high noise level during the coating application (from 120 to 145 dB) [5].

These drawbacks can be essentially avoided by employing a method of material condensation in vacuum through ion bombardment [7]; it can produce single- and multilayer composite coating based on titanium nitrides and carbonitrides. In this case, the lower layer of the coating, which is immediately adjacent to the diamond-containing surface, is deposited at high condensation temperature, thus providing a good adhesion with the tool surface, while the upper layer at the lowest possible temperature ensures a high hardness and wear resistance of the coating. This method produces coated tools that offer a 2–3 times better performance.

More attention has been given recently to the ion plasma coating method. It was shown in [8] that currently important are the research efforts aimed at surface modification by either coating deposition on or treatment of working surfaces of tools, engineering components and electronic parts in plasma of metals, inert and active gases. For example, plasma sprayed coating has a good experience of tool hardening by vacuum ion, submitted in work [9]. The quality of coatings being sprayed can be greatly improved by using the plasma source with separation of micron droplet component of the flow; in that way, it is only the ionized particles which make up the coatings. The layers produced by condensation of a separated plasma flow contain no defects typical of the direct-flow deposits; the resulting coatings are more uniform and droplet-free, and therefore, the coating material has superior physical–mechanical properties. An investigation of droplet-free TiN-based coatings has revealed that they have a 1.5 times higher microhardness in comparison to the coating deposited without any separation of the plasma flow. Moreover, in addition to the high hardness, the coatings possess good plasticity as well. According to [9], the coating produced from a separated flow features a fine dense structure with columnar crystallites measuring 80–200 nm.

Although there have been numerous publications on the ion plasma spraying of coatings onto cutting tools, the use of such coatings on abrasive tools has not been adequately covered [2, 10]. Specifically, Gabler et al. [10] studied 0.3–0.8-μm-thick CVD diamond coatings on 80-mm-diam. 1A1 and 125-mm-diam. 6A2 wheels and tested them in the finish grinding of glass, where a surface roughness Ra = 5 nm was achieved.

In addition, take note of the work [11] where with the aim of improving the grinding ability of diamond tools, partially Ni-coated diamond abrasives were developed. Comparing with non-coated and Ni-coated diamond abrasives, the grinding ability and the electroplating characteristics of partially Ni-coated diamond abrasives were investigated. It was confirmed that the grinding force of the tool used the partially Ni-coated abrasives is smaller than Ni-coated abrasive and the tool life is longer than that of tools fabricated with non-coated or Ni-coated diamond abrasives. The partially Ni-coated diamond abrasives are most suitable to fabricate the diamond tools [11].

Fig. 1 Slavutich-type dresser prior to coating deposition

The objective of the present work has been to clarify the influence of ion plasma sprayed (IPS) coatings on the diamond tool's life under diamond-abrasive machining conditions.

2 Methodology of Research

There are two possible options of improving an abrasive tool's performance through coating application: (i) Reduce the friction coefficient in operation by using "soft" coatings [2] or (ii) harden the tool surface by means of a "hard" coating [10]. We checked the efficiency of various coatings for dressing tools—Slavutich-type diamond dressers (Fig. 1).

The coatings' application on various test tools (specimens) was carried out using a Mod. VU-2MBS-m apparatus (Fig. 2) modified for this particular purpose.

The coated tools were tested on a bench based on a Mod. 3E171VF cylindrical grinder. They were used to dress a straight 1 (PP) 600 × 35 × 305 25A25NSM27K6 abrasive wheel. The dressing conditions were as follows: wheel speed $v_w = 35$ m/s, traverse rate $S_{\mathrm{tr}} = 1$ m/min, and infeed $S_{\mathrm{infeed}} = 0.02$ mm.

3 The Application of «Soft» Ion Plasma Sprayed Coatings

The results summarized in Table 1 show that copper coatings of various modifications (treated with ions Cu^+ and N^+, Cu^+ and Ti^+, N^+) fail to provide any improvement in the dresser performance. It was only in the case—copper and titanium nitride + cobalt coating—that the dresser's wear resistance increased by a factor of 1.6. This

Fig. 2 Upgraded installation of microarc vacuum ion plasma coating VU-2MBS-m

means that the benefits of the tool coating should be sought for among "hard" coating rather than "soft" ones.

The tests have shown that the coating of refractory compounds of transition metals, in particular titanium carbide and nitride, turns out to be more efficient (see Table 1).

4 The Application of «Hard» Ion Plasma Sprayed Coatings

The tool's performance is known to improve owing to multilayer coatings rather than single-layer ones [1, 12].

The multilayer coatings were sprayed onto various tools by means of the above-mentioned Mod. VU-2MBS-m apparatus (see Fig. 2) equipped with two electric arc heat sources for successive or simultaneous microarc vacuum ion plasma evaporation of one or two materials. They enable treatment to proceed at a higher arc current and thus intensify the spraying process.

It has been found out that the substrate temperatures should be at least 300 and 400 °C for the deposition of TiN and TiC coatings, respectively. At these temperatures, the coatings contain no residual titanium. Therefore, prior to deposition of multilayer coatings based on titanium carbide or nitride, the substrate was heated

Table 1 Performance test results for Slavutich-type diamond dressers

Coating	Abrasive consumption (kg)	Specimen weight change (mg)	Relative working consumption (mg/g)	Relative diamond consumption (mg/g)
Uncoated	0.703	27.5	39.12	2.93
"Soft" coating				
Cu (1.5 h)	0.327	17.0	51.99	3.90
Cu and surface treatment with ions				
Cu (1.5 h) + Cu^+	0.350	14.8	45.32	3.40
Cu, N, and surface treatment with ions				
Cu + N (1.5 h)	0.341	15.3	44.87	3.37
Cu + N (1.5 h) + Cu^+ and N^+ (1.5 h)	0.333	27.0	81.08	6.08
Cu, Ti, and surface treatment with ions				
Cu + Ti (1.5 h)	0.282	17.0	60.28	4.52
Cu + Ti (1.5 h) + Cu^+ and Ti^+ (1.5 h)	0.246	76.0	308.94	23.17
"Hard" coating				
Cu (0.5 h) and TiN/Co (10:1) (1.5 h)	0.346	8.5	24.57	1.84
TiN/Co	0.651	15.7	24.12	1.81
TiC/Co	0.705	10.7	15.18	1.13

to 400 ± 10 °C and kept at this temperature throughout the formation of the entire multilayer coating. It is well known [13] that the level of internal stresses in the layer being deposited increases as the layer thickness grows; it is somewhat higher in titanium nitride coating than that in the titanium carbide ones, and thus, the critical thickness value is lower in a single-layer titanium nitride coating. We kept this fact in mind during the preparation of multilayer coatings of these materials with a soft interlayer—the calculated critical layer thickness was 3.5 μm for a TiC-based multilayer coating and 1.5 μm for a TiN-based multilayer coating.

A thin cobalt layer served as an interlayer. With a 0.1-μm-thick interlayer, the critical thickness of the TiN-based multilayer coating was about 10 μm. If made thicker, the coating fully collapsed within one hour of cooling. For interlayer thicknesses of 0.2 and 0.27 μm, the multilayer coating thicknesses were about 12 and 15 μm, respectively. Figure 3 illustrates the structure of a multilayer TiN-based coating.

The dressers improve their performance owing to multilayer TiN/Co and TiC/Co coatings: Both coatings contribute to a higher wear resistance of the multilayer dressing tool (see Table 1).

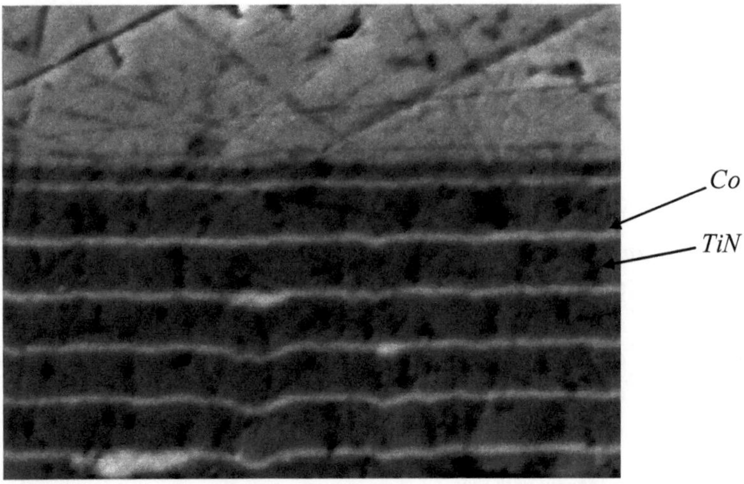

Fig. 3 Titanium nitride-based multilayer coating with cobalt interlayers

Considering that precision dressing rolls, which fall mainly into the category of single-layer dressing tools, are extensively used as well, we also studied the influence of multilayer ion plasma sprayed TiC/Co and TiN/Co coating on the performance of electroplated single-layer dressing tools.

The working surface of a dressing tool undergoes attrition due to abrasive wear of cutting elements (diamond grits) proper and pulling out of diamond grits from the bond. Analysis of the wear pattern of the working surface of diamond dressing rolls and dressers has revealed that in some cases, the second-mentioned mechanism prevails over the first-mentioned one. This is attributable to hydroabrasive erosion of the bond around a grit. The grit embedment depth decreases, and the grit retention in the bond fails. The specific volume of intergranular space ω, mm^3/mm^2 is the parameter that quantitatively represents the bond erosion (and, as a result, the grit protrusion from the bond) and the grit retention has place. The value of this parameter tends to its limit ω_{max}, which depends on the cutting conditions, while its growth rate is determined, among other things, by the bond hardness at the grit embedment location. The higher the hardness, the lower the ω value achievable over some period of time.

A hypothesis about the mechanism of reduction of diamond consumption in a dressing tool has been proposed recently, which states that an ion plasma sprayed coating increases the hardness and wear resistance of a nickel bond and thus hinders the bond hydroabrasive erosion and pulling out of the most loaded and protruding grits from the bond.

To verify this hypothesis for an electroplated single-layer dressing tool, we carried out similar studies of the influence of thick TiN and TiC coatings with thin Co interlayers on the tool wear resistance. The experiments were performed by the procedure as outlined above for multilayer dressing tools, by means of a dedicated

Fig. 4 Single-layer dressers before coating application

Table 2 Performance test results for single-layer diamond dressers

Coating	Abrasive consumption (kg)	Specimen weight change (mg)	Relative diamond consumption (mg/kg)
No coating	0.506	2.7	5.34
TiC/Co	1.185	2.7	2.29
TiN/Co	0.908	2.7	2.97

bench based on a Mod. 3B151 cylindrical grinding machine. The wheel to be dressed was a straight 1 (PP) 600 × 63 × 305 14A40PSM27K6 wheel. The dressing conditions were as follows: $v_w = 35$ m/s, traverse rate $S_{tr} = 1$ m/min, and infeed $S_{infeed} = 0.02$ mm.

The test specimens were prepared in the form of dressers, each containing five AS160T 630/500 synthetic diamond grits of total weight 2.7 mg which were taken from the same batch and fixed to the dresser surface by electroplating (Fig. 4).

Then, the specimens were nickel-plated under identical conditions, in pairs in the same modes. Three groups of specimens were identified: (i) uncoated, (ii) IPS TiC/Co-coated, and (iii) IPS TiN/Co-coated; the coatings were 10 ± 1 μm thick. The tests involved dressing of an abrasive wheel until diamonds were completely worn out (Table 2).

It has been found out (see Table 2) that the application of multilayer TiN coatings with compensating plastic metal (Co) interlayers improves the performance of a single-layer dressing tool by a factor of 1.8, while the TiC coatings provide a 2.3

Fig. 5 A test dresser made by electroforming: reference (uncoated) surface (1); coated surface (2)

times improvement. This finding is consistent with the data given in Table 1, which show that TiN and TiC coatings on a multilayer dressing tool raise its wear resistance 1.6- to 2.6-fold, respectively.

Thus, the deposition of thick wear-resistant TiN, and especially TiC (see Table 2), coatings with compensating interlayers of a plastic metal (Co) on a single-layer electroplated tool is promising method for improving efficiency of such tools.

The aim of the next stage of our investigation was to have an ion plasma sprayed TiN/Co coating on the working surface of diamond dressing tool (dresser) produced by electroforming. In this case, we choose titanium nitride for the coating because the optimal substrate temperature for this coating material is lower than that for titanium carbide (see above). A test dresser (Fig. 5) was prepared by a special process involving deposition of a thick nickel layer, thus eliminating the need to strengthen the shell structure by pouring wood metal whose melting point is below the technological ion plasma spraying temperature. The dresser was made multirib (see Fig. 5); therefore, the coating could be deposited on two ribs, while keeping the other two as reference ones; this enabled us to perform dressing of abrasive wheels in a single positioning, in the same mode, under identical conditions (the dresser was set crosswise to the machining direction) and monitor the status of coated and uncoated areas of the working surface. The coating was applied at the stage when the working surface was quite exposed. Thus, the surface of intergranular pockets was protected, which must hamper their further expansion.

The dresser working layer contained AS125T 500/400 diamond grits. The dressing operation was carried out on a Mod. ZE721VS surface grinder, with the dresser installed on the grinder table. The dressing conditions were the following: infeed $t_0 = 0.001$ mm/pass, $S_{tr} = 22.5$ m/min, and $v_w = 30$ m/s; the speed ratio of the wheel and dresser surfaces was 80. The dressing was performed wet, using a 10% soda solution. The dresser's working surface (both the coated and reference areas)

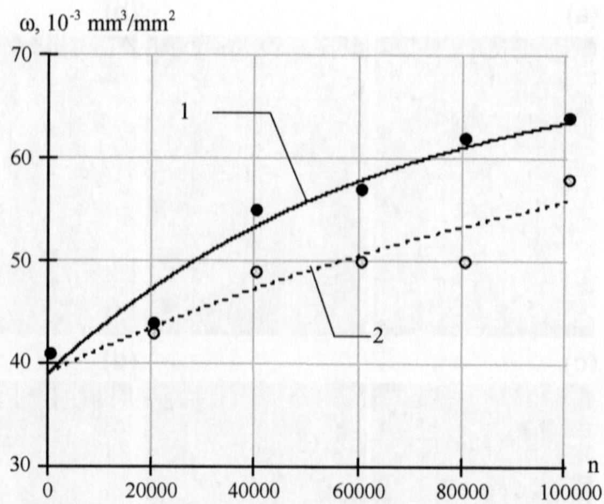

Fig. 6 Dynamics of intergranular pocket volume growth: reference working surface (1); surface protected with ion plasma sprayed coating (2)

was examined after every 20,000 double passes of the grinder table. The volume of intergranular pockets was measured by the procedure [14, 15]. The measurements are summarized in Fig. 6. It is evident that the intensity of growth of the pocket volume has been reduced due to the ion plasma sprayed surfaces, which is a positive factor and supports our earlier findings of the efficient use of ion plasma sprayed coatings on the dressers.

Visual studies of the surface also indicated this. The slowing down of the wear rate of the coated area in this case is not in doubt (Figs. 7 and 8).

5 Conclusion

Deposition of copper coatings (treated with ions Cu^+ and N^+, Cu^+ and Ti^+, and N^+) has failed to provide any improvement in the dresser's performance. It was only in the case of copper/titanium nitride + cobalt coatings that a positive effect was observed—the dresser's wear resistance increased 1.6-fold. This demonstrates that the benefits of tool coating should be sought for among "hard" coating rather than "soft" ones.

It has been found out that the application of multilayer TiN coatings with compensating interlayers of a plastic metal (cobalt) improves the performance of a single-layer dressing tool by a factor of 1.8, while the similar carbon titanium coatings provide a 2.3 times improvement. This is consistent with the data on coating deposition on multilayer dressers: The titanium nitride and carbide coating raise the tool wear resistance 1.6- and 2.6-fold, respectively. Thus, the deposition of thick wear-resistant TiN and especially TiC coatings with compensating interlayers of a plastic metal (Co) on a single-layer electroplated tool is a promising method for improving efficiency of such tools.

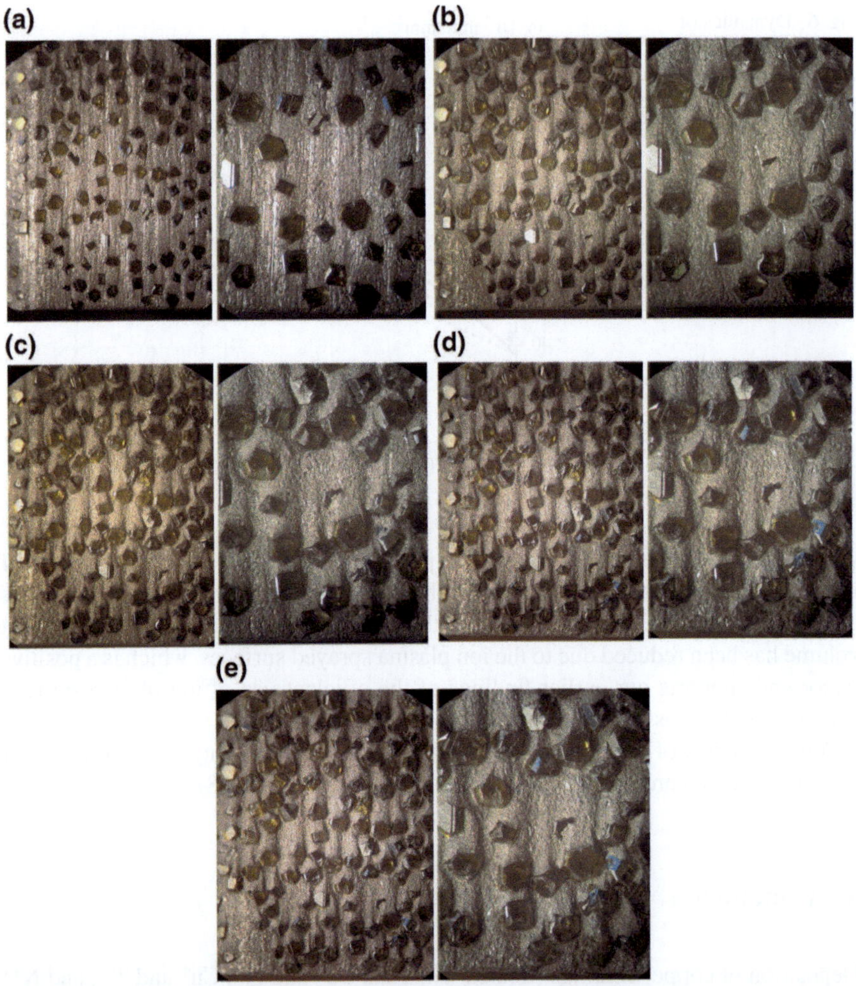

Fig. 7 Working surface (with magnifications that are twice as large) of the diamond dressing tool without coating at different stages of wear: **a** after opening; **b** after dispersing 4 conventional unit (c. u.) the volume of the grinding wheels in the straightening mode with feeding to the depth $t_0 = 0.001$ mm and dispersing 1.4 c. u. on the mode $t_0 = 0.005$ mm; **c** when the limit value is reached $\omega_{max} = 87 \times 10^{-3}$ mm; **d** after an additional dispersion of 0.6 s. u. on the mode $t_0 = 0.010$ mm; **e** after a total dispersion of 1.6. c. u. on the mode $t_0 = 0.010$ mm and reaching the limit value $\omega_{max} = 101 \times 10^{-3}$ mm

The use of wear-resistant TiN coatings with compensating interlayers of a plastic metal (Co) on a single-layer tool manufactured by electroforming (diamond dressers) has been demonstrated to greatly hinder hydroabrasive erosion of the nickel bond at the areas of the tool working profile.

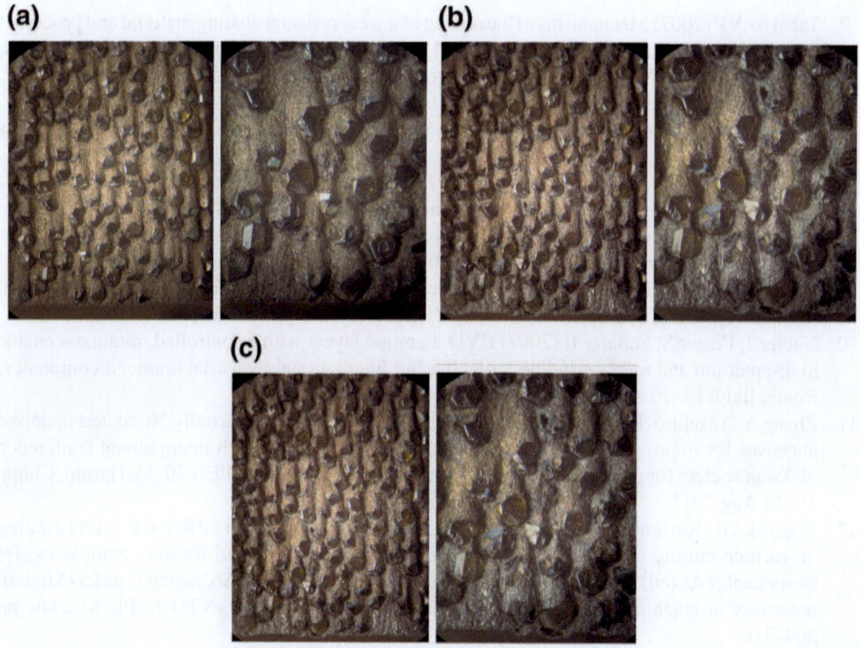

Fig. 8 Working surface (with magnifications differing by a factor of two) of the diamond ruling tool protected by the TiN/Co ion plasma coating, at different stages of wear: **a** after coating on the surface with "pockets" $\omega = 87 \times 10^{-3}$ mm and opening; **b** after dispersing 0.6 c. u. the volume of the grinding wheels in the straightening mode with feed to the depth $t_0 = 0.010$ mm; **c** after a total dispersion of 1.6. c. u. on the mode $t_0 = 0.010$ mm (the limit value $\omega_{max} = 101 \times 10^{-3}$ mm is still not reached)

References

1. Vereshchaka AA, Vereshchaka AS, Batako AD, Hoszhaev OH (2015) Development and research of nanostructured multilayer composite coatings for hard alloys with enhanced technological scope of application. In: Mat. XXIII Inter. Tech. Science Sem. High technologies: tendencies of development, NTU KhPI, Kharkov, pp 30–36
2. Kaidalov AA (2001) Plasma treatment of cutting and abrasive tools to improve their wear resistance. Instrument Svit 12:31–33
3. Lavrinenko VI, Novikov MV (ed) Nadtverdi abrasyvni matetially v mekhanoobrobtsi. Entsykloped. Dov. (Superabrasive materials in machining, Encyclopedic handbook). Bakul Institute for Superhard Materials, NAS of Ukraine, Kyiv
4. Boguslayev AV (2000) Characterization and application of thermal spray coating on aircraft engine parts and machinery components. Suchasne Mashinobud 3–4:28–33
5. Rozenberg OA, Tsarenko IV, Tsekhanov YA (1990) On performance of detonation-carbide-coated deformation broaches. In: Povyshenie effektivnosti protyagivaniya (kachestvo obrabotki), Sb. nauchn. tr. (Improving efficiency of broaching (Machining Quality). Collected Research Papers), Riga, pp 92–102
6. Garanin VM, Ivanenko AA, Klimenko SA (1999) Coating application by detonation. New potential in reconditioning of parts. Suchasne Mashinobud 2:96–99

7. Tabakov VP (2007) Mechanisms of hardening of a wear resistant coating material and processes for their implementation. In: Grabchenko AI (ed) Sovremennye tekhnologii v mashinostroenii, Sb. nauchn. satei (Modern processes in machine building. Collected research papers), vol 2, NTU KhPI, Kharkiv, pp 233–247
8. Belous VA (2005) Ion plasma surface treatment of structural materials. Oborud Instrum Profess 12:12–16
9. Akifyev VA, Polunina OV, Solopov AV (2007) Nanostructured droplet-free coatings for extending service life of cutting tools and dies. In: Effectivnost' realizatsii nauchnogo, resursnogo i promyshlennogo potentsiala v sovremennykh usloviyakh. Materialy VII Promyshl. konf. s mezdunar. uchastiyem (Efficiency of fulfillment of scientific, resource and industrial potential under present-day conditions. Proceeding of VII industrial conference with international attendance), Slavskoye, The Carpathians, UITs NTT, Kiev, pp 349–351, 12–16 Feb 2007
10. Gabler J, Pleger S, Schafer L (2007) CVD diamond layers with a controlled roughness enable high-precition and micro grinding tools. In: 2nd International industrial diamond conference, Rome, Italy, 19–20 Apr 2007. www.diamondat-work.com
11. Zhang Y, Yasuhiro T, Kawahata Y, Kirino O (2015) Proposal of partially Ni-coated diamond abrasives for improving grindabilities of diamond tools. In: The 6th international conference of Asian society for precision engineering and nanotechnology (ASPEN 2015), Harbin, China, 15–20 Aug 2015
12. Kostyuk GI, Kurinnyi AN (2007) A comparison of efficiency of 0.8ZRN + 0.2HFN coating on carbide cutting inserts made by Sandvik Koromant, USSR and Russian companies. In: Grabchenko AI (ed) Sovremennye technologii v mashinostroenii, Sb. nauchn. statei (Modern processes in machine building. Collected research papers), vol 2, NTU KhPI, Kharkiv, pp 304–316
13. Dabizha EV, Novikov NV, Bondar IV et al (2005) Development of modern vacuum coating processes. In: Sintez, specanie i svoistva sverkhtverdykh materialov. Sb. nauchn. tr. (Synthesis, sintering, and properties of superhard materials. Collected research papers), Kiev, ISM NANU, pp 95–107
14. Sheiko MN, Matskevich VP, Nemets VM, Skok VN (2006) Evolution of working surface of diamond sticks in dressing. In: Suchasni proctsesy mekhanichnoi obrobky instrumentamy z NTM ta yakist detalei mashin. Zb. nauk. prats' (Modern machining processes using tools of superhard materials, and surface quality of engineering components. Collected research papers), Kiev, INM NAN Ukrainy, pp 118–125
15. Sheiko MN, Borodavko DN, Skok VN (2007) Evolution of the working surface of diamond ruling bars at the stage preceding its destruction. Superhard Mater 2:65–72

Nitrogen-Vacancy Color Centers in Diamond Fabricated by Ultrafast Laser Nanomachining

Changkun Shi, Huihui Luo, Zongwei Xu and Fengzhou Fang

Abstract Nitrogen-vacancy (NV) color center is one kind of luminescent point defect in diamond. NV color center is a composite structure composed of substituted nitrogen atoms and adjacent carbon vacancies in diamond. It can be applied in many fields, such as super-resolution fluorescence imaging, high-sensitive detection, and quantum computing. In order to meet the requirements of NV color center's applications, many efforts have been devoted to study the manufacturing methods of NV color center. Nowadays, femtosecond (fs) laser technology has been widely used in the field of micro/nanomachining and gradually applied to the manufacturing of diamond NV color centers. In this chapter, the mechanism and characteristics of fs laser micro/nanomachining, the basic properties, and the applications of diamond NV color centers were concisely summarized. Moreover, the ultrafast laser processing of NV color center, the fluorescence detection of NV color center, and the anti-bunching analysis method of single NV color center are introduced and discussed in detail.

1 Introduction

Great progress has been made in the study of quantum mechanics and quantum subsystems, which largely depends on the experimental availability of the interaction between optical fields and atomic or atom-like systems. With the highly developed large-scale fabrication processes, a new research line of the 'artificial atom' has been prompted, such as the nitrogen-vacancy (NV) center in diamond, a solid-state 'artificial atom' with atom-like properties containing a paramagnetic triplet ground state, which can be sensitively interacted with both microwave and optical fields [1].

Nearly 40 years ago, the optical (Davies and Hamer in 1976) and spin properties (Loubser and Wyk in 1978) of the NV center in diamond were first discovered.

C. Shi · H. Luo · Z. Xu (✉) · F. Fang
State Key Laboratory of Precision Measuring Technology & Instruments,
Centre of MicroNano Manufacturing Technology, Tianjin University, Tianjin
300072, People's Republic of China
e-mail: zongweixu@tju.edu.cn

J. Zhang et al. (eds.), *Simulation and Experiments of Material-Oriented Ultra-Precision Machining*, Springer Tracts in Mechanical Engineering,
https://doi.org/10.1007/978-981-13-3335-4_11

Until 1997, after the advent of single-molecule spectroscopy [2, 3] in condensed matter, the optically detected magnetic resonance (ODMR) on single NV center was observed [4]. This single NV ODMR study was done at room temperature, while most of the pioneering experiments to detect magnetic resonance were done with aromatic molecules at low temperatures at that moment. This breakthrough work brought widespread attention to NV center because working with single spin is comparatively easier and more accessible to many groups without the need of the low-temperature setups. The ability to detect a single NV soon led to its investigation as a potential alternative to trapped ions for quantum information applications and precision magnetometry [1].

Nowadays, the unique properties and promising cutting-edge applications have maken NV center seen as a fascinating system; the center has a paramagnetic ground state with long electron spin coherence lifetime which could be extended to the millisecond range [5], and spin levels can be initialized and detected by optical excitation in a broad range of wavelengths 450–650 nm and manipulated on 1 GHz rate scales by microwave excitation, all at room temperature [6]. So by combining the millisecond coherent lifetime with the fast spin excitation rate with microwave microstructure, one million single-qubit operations can be generated in each coherent time. The strong optical coupling afforded by the NV center makes it of great interest as a potential interface between photons and other solid-state quantum computers. And NV center has also been explored for magnetic sensor applications for its optical readability.

Furthermore, the diamond lattice is an ideal host of NV center, as its excellent thermal and mechanical properties, biocompatibility, and nearly nuclear spin-free environment. These properties are highly favorable for NV operating over a very wide temperature range and in extremely harsh environments, so that it can be applied from high-sensitivity, high-resolution magnetometry to fluorescence probing of biological processes, solid-state quantum information processing, spin optomechanics, quantum optics, nano-photonics, and quantum plasmonics [7–13].

With the rapid progress of NV center application research, there is a pressing need to develop techniques which would allow us to engineer NV centers with tailored properties such as their position, their orientation in the crystal, their density. Therefore, to realize the technical application of the NV color center, it is a great challenge to obtain a high-quality NV color center at a specific position and to couple with the optoelectronic element. Higher requirements are imposed on the processing of NV color centers. Commonly used methods, like high-energy electron beam radiation, chemical vapor deposition, ion implantation, and other processing, generally have problems such as low yield and uncontrollable NV color center position. These methods also bring unnecessary residual damage and defects to the diamond lattice. Residual damage and defects can reduce the characteristics of NV color center through strain effect, which is not conducive to obtain NV color center of high-quality. Nowadays, the application of ultrafast laser micro–nano-processing technology has matured and has been widely used in the research of NV color center processing. Ultrafast laser has very little thermal effect in the process, especially in the processing of transparent materials; the Gaussian intensity distribution of the light field can achieve high-precision processing and can induce nonlinear multiphoton absorption to achieve processing inside transparent materials, ultrafast laser. These characteristics make it a powerful tool for NV color center processing.

This chapter first briefly introduced the research progress of NV color center, the research status, and the challenges faced by NV color center processing. The first section introduced the structure, electronic structure, and energy level structure of the NV color center and gave a brief overview of the application of the NV color center. The second section introduced the mechanism and characteristics of femtosecond laser micro–nano-processing, summarized the application of fs laser processing NV color center, and analyzed the parameters commonly used in fs laser processing in NV color center processing. In the third section, the typical case of fs laser processing NV color center was selected and analyzed. The positional accuracy of fs laser processing NV color center was then introduced. The fourth section introduced the confocal fluorescence scan of the NV color center and the anti-bunching test of the single NV color center and analyzed the fluorescence spectrum and anti-bunching test spectrum of the NV color center.

2 NV Color Center in Diamond

2.1 NV Color Center Structure

Diamond is known as a unique material of hardness, electrical non-conductivity, and large refractive index [14], these physical properties potentially meet the needs of next-generation optical and electronic devices. Apart from carbon, the diamond lattice can host various atomic impurities and multiple color centers can be formed such as nitrogen vacancies, silicon vacancies, nickel-nitrogen NE8 color center, and many others, which would change the properties of diamond [15]. Currently, strong attention is focused on the color centers in diamond, which can be widely used in quantum optics, quantum sensing, and other applications [16].

Nitrogen is the most abundant gas in the atmosphere, and it is the most common impurity found in diamond [17]. The single substitutional nitrogen presenting in natural diamond is called P1 center; its concentration in artificial crystals can reach several hundred ppm [18]. If a substitutional nitrogen atom is associated with an adjacent vacancy in the diamond lattice, the nitrogen-vacancy (NV) center is formed (shown in Fig. 1). As soon as a vacancy gets close to the P1 center, three electrons of the nitrogen atom connect with the neighboring carbon atoms through covalent bonds and the remaining electrons form a lone pair. It is an optically active point defect with C_{3v} symmetry orientated along the [111] crystalline direction [19]. The NV-axis lies along a C_3-axis of the crystal, but if one of the carbon atoms is a ^{13}C which does not lie on the symmetry axis of the center, the effective symmetry is reduced to a single mirror plane that contains the nitrogen, the vacancy, and the ^{13}C nucleus (shown in Fig. 1b). NV center widely exists in natural and artificial diamonds, but its existence is

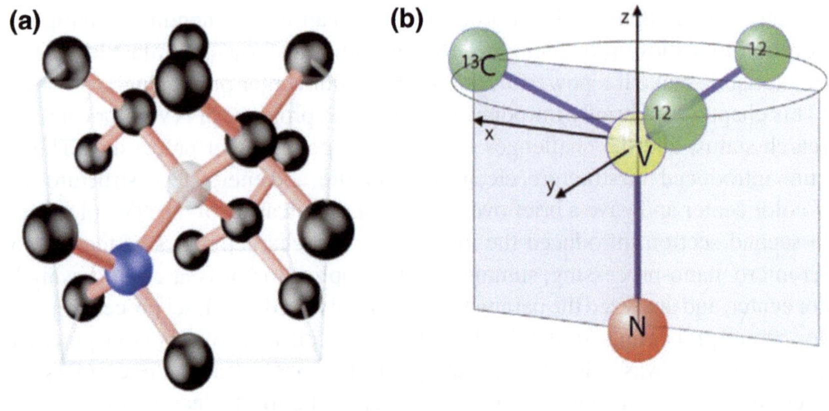

Fig. 1 **a** Atomic structure of a single nitrogen-vacancy (NV) defect in diamond. Substituted nitrogen atom (blue) bound to vacancy site (white) in a diamond lattice (black) [23]. **b** site symmetry of the NV center [18]

inhomogeneous and uncertain. Therefore, the controllable preparation of NV centers is necessary for further research. At present, NV center can be prepared in chemical vapor deposition (CVD) diamond as an 'ingrown' product of synthesis process or in bulk and nanocrystalline diamond through radiation damage or ion implantation and annealing [20].

Generally, diamonds are divided into natural and artificial diamonds. The diamonds commonly used in experiments are artificial diamonds, including nanocrystalline diamond and bulk diamond. Nanocrystalline diamond with particle size of 0–100 nm is made using high-temperature and high-pressure (HTHP) [21]. Bulk diamonds are usually classified into type I and type II according to nitrogen content, in which type I contains more nitrogen impurities and type II contains less nitrogen impurities. According to the optical transparency or optical absorption ability, they can be classified into type a and type b, in which the optical transparency of type a is better [22]. Samples of type Ib diamond and IIa diamond are mainly used in many studies.

Type Ib diamond is high-temperature and high-pressure (HTHP) diamond. It always appears yellow color with high nitrogen concentration. Such high N concentrations are excellent for creating large ensembles of NV but do not allow for isolated control at the single optical center level, as required by many quantum and sensing applications [24].while type IIa is prepared by chemical vapor deposition (CVD) and transparent with N concentration less than 10 ppm. Different diamond types will result in different photoelectric properties of NV color center, such as lifetime of excited state, zero-phonon line (ZPL), and spin coherence time. Some of the applications of NV centers are feasible only with nanocrystalline diamond; an obvious example is labeling for bio-imaging [25]. For quantum detection, the ultra-pure diamond samples with electrical grade are needed (Table 1).

Table 1 Typical type of diamond used in color center research

Type	Type Ib	Type IIa	Ultra-pure diamond

Nitrogen concentration	20–500 ppm	<10 ppm	<5 (typically 0.1–1) ppb
Synthesis method	HTHP	CVD	CVD
Typical size	2.6 mm × 2.6 mm × 0.5 mm	3 mm × 3 mm × 0.3 mm and 2.6 mm × 2.6 mm × 0.5 mm	2.0 mm × 2.0 mm × 0.5 mm

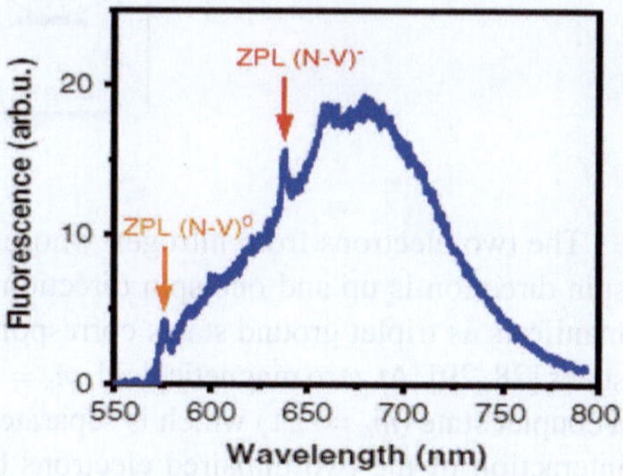

Fig. 2 Zero-phonon lines (ZPL) of NV^- and NV^0 in the fluorescence diagram [23]

2.2 NV Color Center Electronic Structure

In diamond, NV center occurs in two identified charge states with different spectral properties: the neutral center NV^0 with a zero-phonon line (ZPL) around 575 nm (2.16 eV), the negatively charged center NV^- with a ZPL at 637 nm (1.945 eV), and a phonon sideband extending to 800 nm (Fig. 2) [26]. Under the specific light field, NV^0 and NV^- can be converted to each other. As shown in Fig. 1, NV center contains a substitutional nitrogen atom and an adjacent vacancy in the diamond lattice. If the defect is neutral, it has five electrons (two from the N, three from the dangling carbon bonds) with a spin $S = 1/2$ of the ground state. In this state, NV^0 centers are not suitable for optically detected magnetic resonance, because the dynamic Jahn–Teller distortion of the ground state broadens the electron paramagnetic resonance (EPR) lines, dramatically reducing the detection sensitivity [27].

The NV^- center has been exclusively considered, and usually it is simply referred as NV center. As an important paramagnetic center in diamond crystal, it is a typical six-electron structure, in which two electrons are from nitrogen atom, the other three are from the dangling carbon bonds, and the sixth electron is captured by vacancy. Then the ground state has a spin $S = 1$ [18].

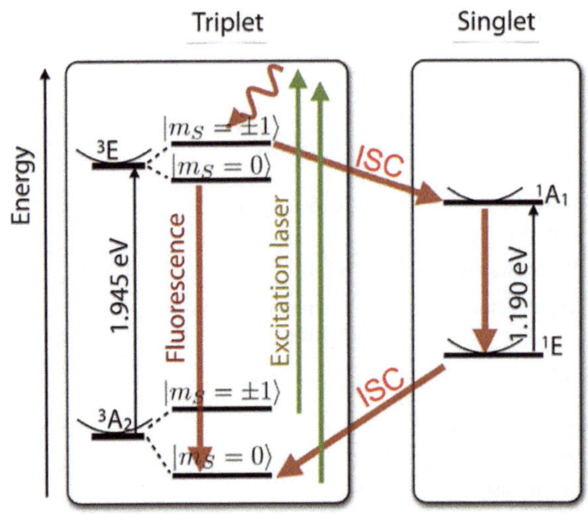

Fig. 3 Energy level structure of NV center. The numbers refer to the transition energies of the zero-phonon lines; the parabola indicates the continuum of vibrationally excited states. M_S refers to the electron spin. *ISC* = intersystem crossing [18]

The two electrons from nitrogen whose spin directions are up (or down) or one spin direction is up and one spin direction is down cause the electrons spin levels manifests as triplet ground states corresponding to $m_s = 0$, $m_s = +1$, and $m_s = -1$ states [28, 29]. At zero magnetic field, $m_s = +1$ and $m_s = -1$ states are degenerated as a couplet state ($m_s = \pm 1$) which is separated from $m_s = 0$ state due to the spin–spin interaction of the two unpaired electrons by 2.87 GHz. While under the effect of external magnetic field, the $m_s = +1$ and $m_s = -1$ states can be separated from the metastable $m_s = \pm 1$ state because of Zeeman splitting with an energy gap of ΔE. By adding a microwave signal with energy of ΔE to these electrons, the two spin states will interact with each other so that the electron spin coherent state is excited, which can be used for quantum manipulation [30].

2.3 NV Color Center Energy Level Structure

The optical properties of the NV center are determined by its unique three-level structure, which mainly consists of a triplet ground state ($^3A_2$), a triplet excited state (^3E), and two singlet metastable states ($^1A_1$ and ^1E). Figure 3 shows the energy level structure of NV center. There is a strong optical transition between $^3A_2$ and ^3E with a photon energy of 1.945 eV (637 nm). When the system is excited by a light less than 637 nm, such as the green laser, the photon energy greater than 1.945 eV drives the system well above the ^3E state. From there, it quickly relaxes nonradiatively down to the ^3E state, typically without changing its spin state [18].

In addition to the ground state, the excited state of NV center also has a zero field splitting between $m_s = 0$ and $m_s = \pm 1$ with 1.42 GHz. In this process, the electron in the ground $m_s = 0$ state transitions to the excited $m_s = 0$ state, and the electron in the ground $m_s = \pm 1$ state transitions to the excited $m_s = \pm 1$ state [31]. The electrons are not stable in the excited state, so they eventually relax back to the ground $m_s = 0$ state emitting red fluorescence (ZPL, 637 nm), while the $m_s = 0$ state most often (≈89%) relaxes directly to the ground state, the states often (≈45%) undergo intersystem crossing (ISC), and then decay to a long-lived singlet level [32], after which they transition predominately to the $m_s = 0$ sublevel of the $^3A_2$ ground state. The spin state is no longer conserved. Through this relaxation process, loop excitation method can be used to change spin of $m_s = \pm 1$ flip to $m_s = 0$ in order to realize the electronic spin polarization. As Wang et al. reported, the polarization rate has reached more than 90% at room temperature and has been close to 100% at low temperature [30]. Furthermore, since ISCs are nonradiative without red fluorescence, the photoluminescence (PL) intensity will decrease with the increase of this transition probability. Thus, fluorescent count is a viable method for determining the spin state of the NV center [33]. This spin-dependent transition is the basis for the detection of magnetic resonance by optical means.

2.4 Applications of NV Color Center

The NV defect in diamond has a spin triplet ground state, which can be optically initialized, coherently manipulated with microwave fields, and read out by optical means with long coherence time, even under ambient conditions. The results make the NV defect an outstanding candidate for the development of digital processing of quantum information [34]. Firstly, the NV defects can be used as single-photon emitters to produce single-photon pulses on demand, which is essential for the realization of quantum cryptographic protocols and optical quantum information processing schemes [35]. The quantum secure communication's error rate is less than 4.6% based on NV single-photon source in diamond, and the security is higher than the one based on coherent laser or entanglement photons [36]. Secondly, the key challenge in quantum information protocol is to create multipartite entangled states. So quantum mechanical systems with long coherence times that can be manipulated and coupled together in a scalable fashion are highly needed. In 2007, Dutt et al. described the coherent manipulation of an individual electron spin and nearby individual nuclear spins associated with the NV color center in diamond to create a controllable quantum register [37]. For example, the schematic representation of the NV center's interaction with two ^{13}C nuclear spins for nuclear spin-based quantum registers was shown in Fig. 4.

In 2013, Bernien et al. have realized the quantum entanglement of the NV color center at a distance of 3 m in experiment [38]. In the solid-state systems, the macroscopic number (ensemble) of NV centers in diamond is used as the memory element, while superconducting mesoscopic structures play the role of the operational element and quantum network [39]. Except these, as a solid single-spin quantum system, NV color center can also be used to study complex dynamic behaviors in quantum information, such as non-Markovianity [40].

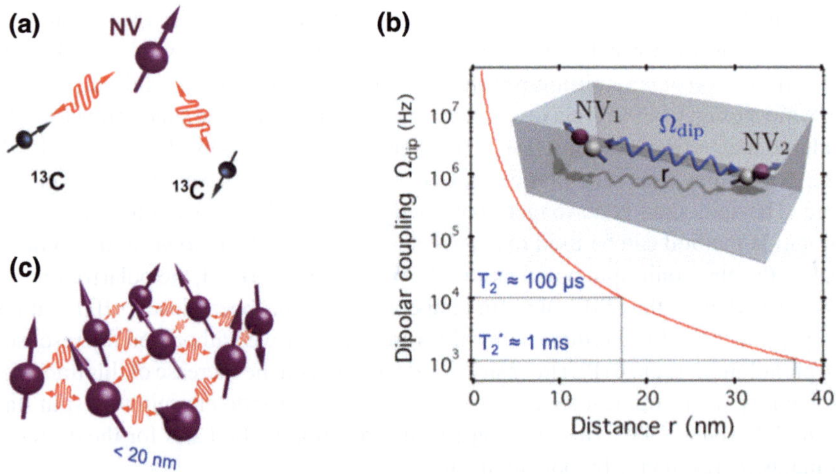

Fig. 4 **a** Schematic representation showing the interaction of the NV center and two ^{13}C nuclear spins. **b** Magnetic dipolar interaction between two electronic spin states associated with two NV centers as a function of their separating distance. **c** To achieve effective coupling between adjacent nodes, the distance between NV networks less than 20 nm which is associated to the coherence time of the spin [34]

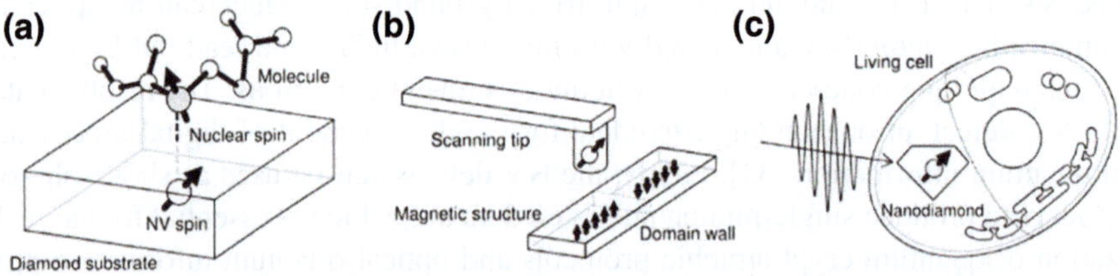

Fig. 5 Various modalities of NV-based magnetometers. **a** A near-surface NV center is used to detect the single nuclear spin of a single molecule. **b** High spatial resolution sensing of a magnetic structure using a scanning diamond probe with NV center. **c** Nanodiamond is implanted in a living cell [41]

On the other hand, the spin properties illustrated previously can also be used for magnetic fields measurements at nano-scale, which is one of the most important applications of NV color centers. The luminescence intensity from the NV center exhibits a sharp resonance when a microwave frequency that matches the splitting of the ground state levels of NV center is applied to the diamond. The ODMR signals are sharp even at room temperature and can be used in miniature sensors. The NV-based magnetometer can take a variety of modalities (shown in Fig. 5).

In 2008, Maze et al. reported that the NV magnetometer sensors can detect magnetic fields of a few nanotesla at kilohertz frequencies after averaging 100 s [42], as shown in Fig. 6. In 2011, Dolde et al. demonstrated that the electric fields about 10 V/cm were detected by NV center [43]. Recently, DR Glenn et al. demonstrated a measurement technique that used a solid-state spin sensor (a magnetometer) consist-

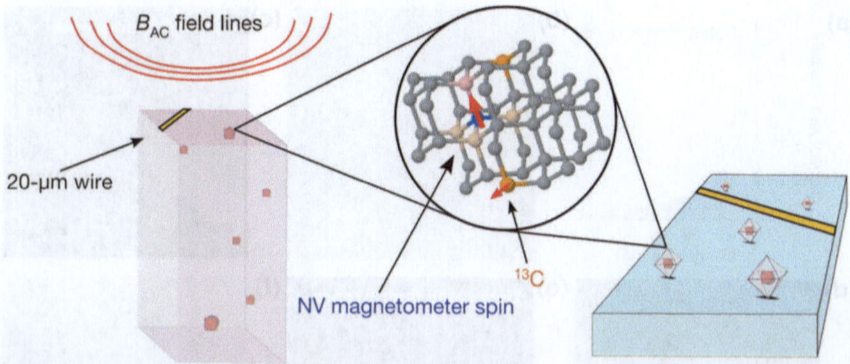

Fig. 6 Schematic of the magnetic sensor, which is based on individual nitrogen-vacancy electronic spins in diamond. A single NV defect proximal to the surface of an ultra-pure bulk single-crystal diamond sample (left) or localized within a diamond nanocrystal (right) is used to detect an AC magnetic field for external applications (BAC, top left). The electronic spin states is manipulated by a 20-mm-diameter wire (yellow) which generates microwave pulses [42]

ing of an ensemble of nitrogen-vacancy centers in combination with a narrowband synchronized readout protocol to obtain nuclear magnetic resonance (NMR) spectral resolution about one hertz [44]. This technique fills the gap between the nano-scale NV-NMR experiments and microcoils and enables analytical NMR spectroscopy at the scale of single cells.

As an optically readable sensor, the NV center could measure variety of physical quantities with very high precision. During the last few years, the precision detections of magnetic field, electric field, pressure, temperature, and even nuclear spin have been realized [43, 45–47].

In addition, the robust photostability and the low cytotoxicity of the NV color center in nanodiamond make it attractive in both bio-imaging and bio-sensing. Earlier, Fu et al. and Neugart et al. demonstrated the first bio-application of fluorescent nanodiamond (FND) particles for single-particle tracking in HeLa cells [48, 49]. For in vitro biological applications, FND has been demonstrated to be a long-term cell marker. In 2011, Fang et al. used the FND-based platform to track cell division and proliferation in vitro [50]. In vivo studies, FND-cell labeling technique has been applied to model organisms like Caenorhabditis elegans and Drosophila melanogaster [51, 52]. To eliminate the autofluorescence background signal which interferes the FND signal during fluorescence detection, time-gating fluorescence imaging, and fluorescence lifetime imaging (FLIM) was developed [53, 54]. The different fluorescence signals can be seen in Fig. 7.

And then optical super-resolution nanoscopy, magneto-optical (spin-assisted) subwavelength localization and imaging were developed based on NV center. Han et al. found that there is no observable photobleaching when NV centers in diamond are used as luminescence sources in STED microscopy [55]. Eva Rittweger et al. demonstrated that the STED technology can break the optical diffraction limit of optical

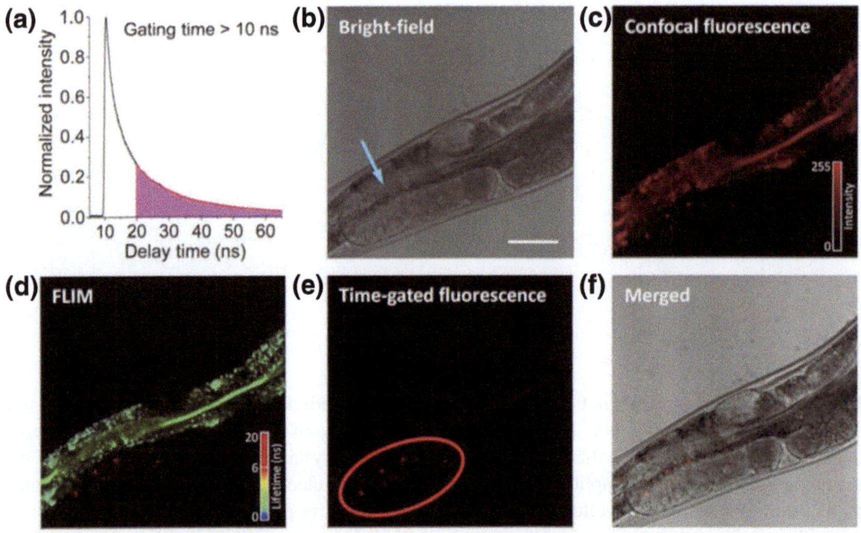

Fig. 7 Observation of GFP: YLC-FNDs in C. elegans by FLIM. **a** 100-nm FNDs' fluorescence decay time trace in water. The area shaded in magenta represents the fluorescence signal collected at the gating time (s) longer than 10 ns. **b** Bright field. **c** Confocal fluorescence. **d** FLIM. **e** Time-gated fluorescence image at gating time > 10 ns, and **f** merged bright field and time-gated fluorescence images of a worm microinjected with GFP: YLC-FNDs at the distal gonad. A blue arrow indicates the site of injection. Anterior is left and dorsal is up in all figures. Scale bar is 50 mm [53]

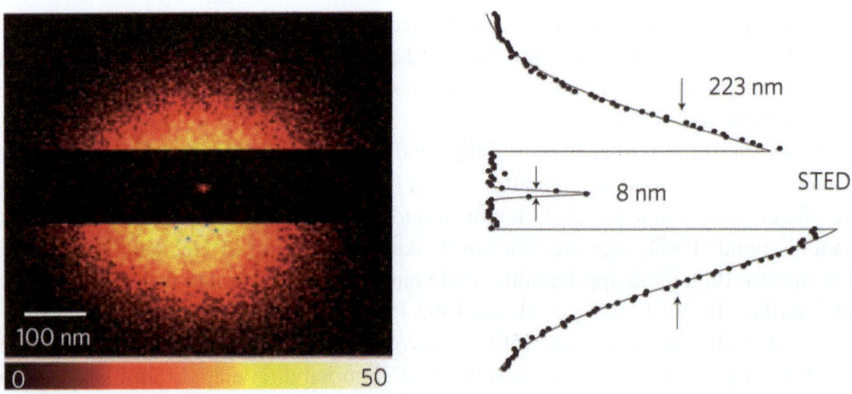

Fig. 8 STED improve the resolution to 8 nm by taking NV color center as the sample [56]

molecules and shrinks a confocal spot of 223 nm diameter (FWHM) down to 8 nm by taking NV center as the sample [56]. Combined with solid immersion lenses, the resolution can even reach 2.4 nm [57]. Now the applications such as single-molecule spin imaging, nano-scale imaging of biomagnetic fields, sensing molecular fluctuations, and temperatures in live cellular environments are on the rise (Fig. 8).

3 Ultrafast Laser Processing NV Color Center in Diamond

3.1 Ultrafast Laser Processing Mechanism

In recent years, ultrafast laser processing technology has been widely used in different field, especially for the femtosecond (fs) laser micro–nano-processing method, which has attracted more and more attention.

Therefore, in order to fully exploit the fs laser micro–nano-processing, it is necessary to understand machining mechanism of the fs laser [58–62]. During the process of fs laser-material interactions, the energy of laser is deposited in a very short time and has a very small thermal effect. This process is extremely short and accompanied by a variety of nonlinear optical effects.

In general laser processing, the absorption of photons is mainly linear absorption, when the energy of the photon is large enough, the electron of the material can absorb the energy of single photon to realize the transition from valence band to conduction band. When the energy of the photon is insufficient to meet the needs of the electron transition, the bandgap transition of electron will not occur. Due to the high peak power of fs laser, the photon density of fs laser is high enough to make the electron to absorb the energy of multiple photons to realize the bandgap transition. This process is called the nonlinear absorption effect of photons [58, 59, 63, 64]. The rate of nonlinear multiphoton absorption is mainly determined by the laser intensity and the nonlinear multiphoton absorption coefficient of the material, as described in the following formula.

$$P(I) = \sigma_k I^k \tag{1}$$

where I is the intensity of the laser, σ is the nonlinear absorption coefficient of the material, and k is the number of photons absorbed. In the process of fs laser micro–nano-processing, the transition of electrons from valence band to conduction band is called ionization, which includes photoionization and avalanche ionization. The process in which electrons are directly excited by the fs laser is called photoionization, which includes the process of nonlinear photoionization and the process of tunneling ionization, respectively [58, 60, 62, 65]. More concretely, in the process of nonlinear photoionization, electrons absorb the energy of multiple photons to realize bandgap transition and become free-moving active electrons. It occurs mainly in the process of multiphoton nonlinear absorption, which mainly existed in the process of fs laser action of high-frequency ultrashort pulses. The rate of photoionization mainly depends on the intensity of the fs laser. The greater the intensity, the faster the photoionization rate is, as shown in Fig. 9c.

While in the process of strong low-frequency fs laser processing, under the condition of the strong electric field generated by the fs laser, the electron of materials get rid of the atomic electric field directly and pass through the short barrier to be a free-moving electron. This process is described as tunneling ionization, as shown in

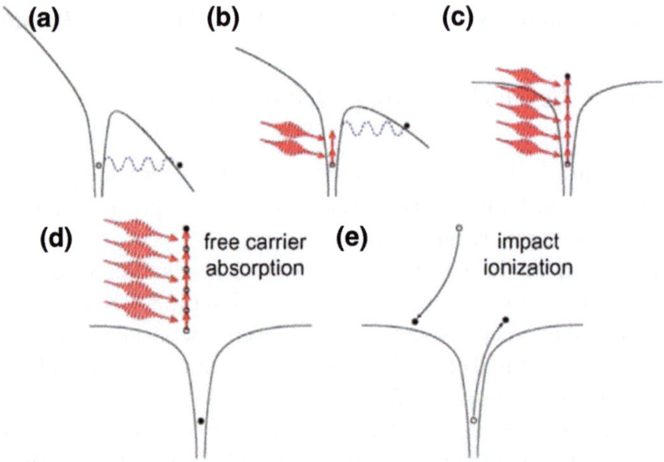

Fig. 9 Schematic diagrams of the photoionization excited by fs laser. **a** Tunneling ionization, **b** mixture of tunneling and multiphoton ionization, **c** multiphoton ionization, and **d**, **e** avalanche ionization (Reproduced with permission from [62])

Fig. 9e. It is noteworthy that the transition between nonlinear photoionization and tunnel ionization can be described by Keldysh parameters.

$$\gamma = \frac{\omega}{e}\left[\frac{mcn\varepsilon_0 E_g}{I}\right]^{\frac{1}{2}} \tag{2}$$

where ω is the laser frequency, m and e are the reduced mass and electron charge, respectively. I is the laser intensity at the focus, m and c is the speed of light, and n is the refractive index of the material, E_g is the bandgap of the material, and ε_0 is the permittivity of free space [58, 62]. When the value of the Keldysh is less than 1.5, the tunneling ionization process will take place. However, there still exists an intermediate region that both the tunneling ionization and the nonlinear photoionization occur simultaneously (shown in Fig. 9b).

In addition, as shown in Fig. 9e, the active electrons reaching the conduction band can continuously absorb the energy of photon and then became the high-energy ionized electrons. At this time, the high-energy ionized electrons will release excess energy, collide with the surrounding electrons, and realize energy conversion, so that other electrons can be converted. Electrons are shifted from valence band to conduction band. Thereafter, in order to ionize more electrons, electrons in the conductive band will continue to repeat this process, which is called avalanche ionization. The avalanche ionization can generate the plasma with high density, which is beneficial to improve the efficiency of laser energy conversion. Therefore, in the process of fs laser machining NV color center in diamond, the high-density plasma accelerated by the strong electric field generated by the fs laser will bombard the lattice of the diamond, which will result in the generation of vacancy and realize the fabrication of NV color centers in diamond.

3.2 *Ultrafast Laser Processing Features*

Fs laser micro–nano-fabrication technology has been widely used in the fields of micro–nano-photonics and information optics. The interaction between fs laser and materials at micro–nano-scale can induce optical nonlinear absorption effect, which can produce strong optical absorption during the processing of transparent materials. Due to the optical nonlinear absorption effect, fs laser can be processed inside the transparent material, and the processing precision of fs laser can be improved. This provides a possibility for ultrafast laser processing of NV color centers in diamond. In detail, the advantages of fs laser in material micro–nano-processing mainly include the following aspects:

(1) **Minimal heat-affected zone function**

The ablation effect was usually used to explain in the process of continuous laser interaction with the material. The fs laser micro–nano-processing is often called 'cold processing' because the fs laser can effectively suppress the thermal effect of the active area in material processing. Taking metal with high thermal conductivity processing as an example, the laser pulse duration is significantly smaller than the electron–phonon coupling time (1–100 ps) in the interaction of laser and metal. The electrons absorb the energy of the photon and transmits it to the crystal lattice in a short time, this process is accompanied with a small thermal diffusion effect [66]. Since the duration of the fs laser pulse is shorter than the electron–phonon coupling time (picosecond) of most metals, the relationship between the thermal diffusion length l_d and the time required for the material to heat to the melting point T_{im} is shown as the formula (3).

$$l_\mathrm{d} = \left(\frac{128}{\pi}\right)^{1/8} \left(\frac{DC_i}{T_{im}\gamma^2 C_\mathrm{e}'}\right)^{1/4} \quad (3)$$

where D is the thermal conductivity of the metal, C_i is the heat capacity of the metal lattice, γ is the electron–phonon coupling coefficient of the metal, and C_e' is the ratio of the electron heat capacity to the electron temperature [67, 68]. Therefore, fs laser can significantly reduce the effect of heat-affected zone (HAZ) and the area of HAZ in the fabrication of the materials, which can ensure the quality of the processing structure and improve the processing accuracy.

(2) **Nonlinear multiphoton absorption effect**

An important characteristic of fs laser processing is the nonlinear multiphoton absorption effect produced by the reaction process. Generally, the absorption of light is dominated by linear single-photon absorption. When the energy h_v of a single photon is greater than the transition energy E_g of the material from valence band to conduction band, the electron in the material absorbs the energy carried by a single photon in the laser and then transits from the valence band to the conduction band. Because of its high peak power, fs laser irradiates the material with extremely high photon density, so it is easy to realize nonlinear multiphoton absorption effect. When the photon energy is less than the bandgap, the electron can not excite the electron to realize the transition, so the absorption does not occur in the static state [69, 70]. However, when high-density photons are incident on materials, electrons in transparent materials can absorb the energy of multiple photons and excited by multiple photons to produce a strong optical absorption effect, which has been widely used in the fabrication of microlenses. The transition processes of electrons in a material have been shown in Fig. 10.

(3) **Effective reduction of the plasma effect**

In the laser ablation process, the material will produce a large number of ablation plasma; this process is only a few hundred picoseconds. The plasma is mainly controlled by the electromagnetic field generated by laser ionization, showing obvious agglomeration. The existence of plasma can shield the laser pulse behind, thus causing the loss of laser energy, affecting the efficiency and accuracy of processing. However, the ultrashort laser pulse of fs laser has very high power, so that the material is processed before the plasma is produced, which can effectively reduce the plasma shielding effect.

(4) **High processing spatial resolution**

Femtosecond laser pulse can inhibit the diffusion of thermal energy to the surrounding of the processing area. The thermal diffusion can be neglected in the fs laser processing. The processing area is limited to the size of the spot in the focus; thus, it can effectively improve the spatial resolution of material processing. In ideal condition, the intensity of fs laser is Gaussian distribution. In the linear absorption process, the area of interaction between the laser and the material is the same as the area irradiated by the laser. But for the process of multiphoton nonlinear absorption, the spatial distribution of absorbed energy decreases with the increase of multiphoton absorption order *n*.

$$\omega = \frac{\omega_0}{\sqrt{n}} \tag{4}$$

where ω_0 is the actual size of the focused beam. In addition, there is a threshold for the dependence of laser intensity on the changes of material physical and chemical properties induced by laser. Laser processing can be achieved when the laser energy is above the ablation threshold of the material, so the processing accuracy of the fs laser can be further improved by adjusting the laser power.

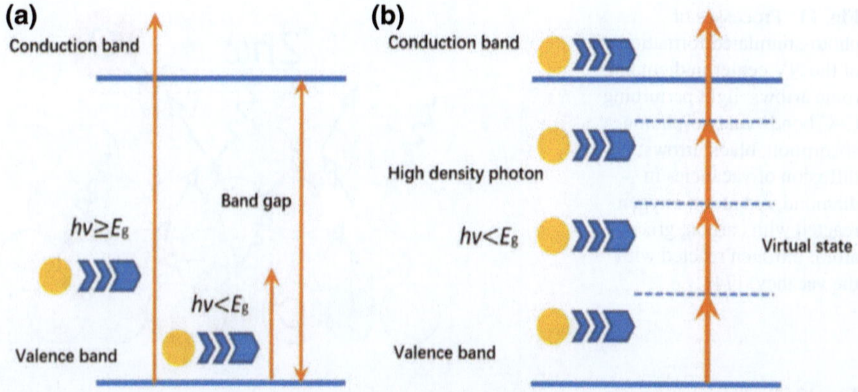

Fig. 10 Electron transitions in materials. **a** Single-photon absorption process; **b** multiphoton absorption process

3.3 *Ultrafast Laser Processing NV Color Center*

Nowadays, many efforts have been devoted to ultrafast laser processing of NV color centers in diamond. In 2004, Dumeige et al. found the photo-creation of color centers in diamond nanocrystals under fs illumination. In the presented work, a frequency-doubled Ti:Sa laser with a central wavelength at 500 nm was used. The pulse duration and repetition rate were 150 fs and 8.2 MHz, respectively. The maximum energy per pulse was about 55 μJ, which is close to the saturation level for the emission of the NV center [3]. Such pulse energy insures single-photon emission for each excitation pulse with a peak intensity of 4.5×10^{10} W/cm^2 [71]. In 2013, Wu et al. successfully fabricated the NV color centers in a diamond sample by the fs laser pulse illumination. The diamond sample used in the experiment was a $3 \times 3 \times 1$ mm^3 synthetic Ib bulk diamond which has a high nitrogen concentration. Then a 800 nm Ti:sapphire pulsed laser of 250 kHz repetition rate and 80 fs pulse duration was used to illuminate the diamond sample. During the processing, the average power of the laser was 1 W and the energy of the single fs laser pulse was 4 μJ. In the focus point, the peak intensity of the laser pulse was 2.8×10^{13} W/cm^2 which could incite both multiphoton ionization ($I < 10^{12}$ W/cm^2) and tunneling ionization ($I > 10^{14}$ W/cm^2) of oxygen and nitrogen molecules in air to a relatively high ionization rate [72]. After this, Chen et al. realized the controllable generation of NV centers in diamond by using fs laser direct writing technology. By means of the aberration correction, it can precisely position the vacancies within the diamond crystal when writing in the optics, and after annealing (1000 °C, 3 h) the diamond sample, the production rate of the single NV centers was up to 45% and the position accuracy was within 200 nm of the desired position. In the processing of the NV centers, the wavelength of the laser they used was 790 nm and the duration of the single laser writing pulse was 300 fs; the laser writing pulses were delivered to each site in 25×20 square grids

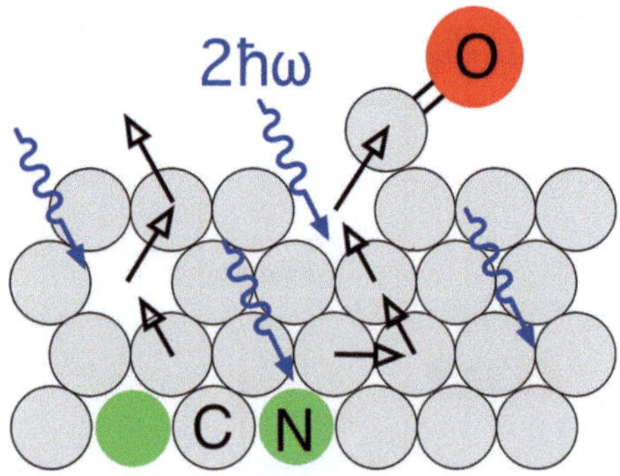

Fig. 11 Processes of photo-stimulated formation of the NV center in diamond: (blue arrows: light perturbing C–C bonds via two-photon absorption, black arrows: diffusion of vacancies in diamond, red atom: oxygen reacted with carbon, green atom: nitrogen reacted with the vacancy) [74]

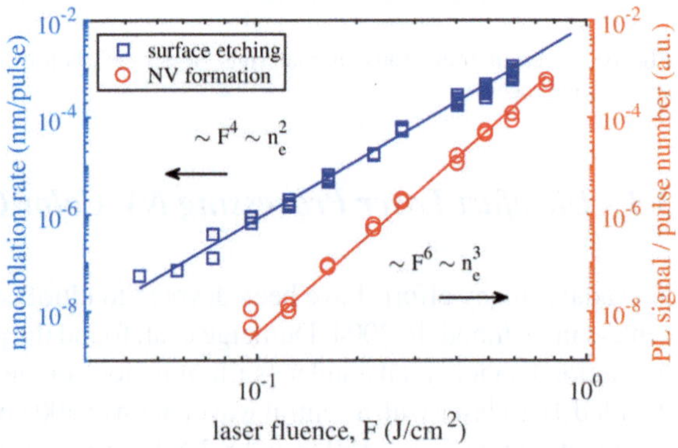

Fig. 12 Surface etching rate and the yield of NV centers in the nanoablation regime versus laser fluence [74]

with a separation of 5 μm at a depth of 50 μm. When the laser pulse energy was 25.7 nJ, they found that the generation of the single NV at a lattice site was as high as 45% [73].

In 2017, Kononenko et al. put forward a novel strategy for NV centers production that can generate the NV centers in the regime of nanoablation of a diamond surface effectively and controllably. The process of photo-stimulated formation of the NV center in diamond was shown in Fig. 11. The laser used in the experiment platform consisted of a Ti:sapphire oscillator and a spitfire regenerative amplifier; the duration of the laser pulse was 100 fs, and the wavelength and the repetition rate of the laser were 800 nm and 1 kHz, respectively. By means of the two crystal β-BaB_2O_4, frequency-tripled radiation was obtained. Then a fs laser pulse with 266 nm wavelength was used to irradiate the diamond sample; it was focused on the surface with the help of an aspherical lens. As shown in Fig. 12, the nanoablation rate is proportional to the fourth power of the laser fluence, the formation rate of NV centers is proportional to the sixth power, and the TPA-induced ionization is proportional to the second and third powers of the plasma concentration. After processed by the laser nanoablation, the diamond sample was annealed in vacuum at 700 °C for 2 h [74].

Table 2 Laser processing parameters

	Yan Liu	Yuchen Chen	Vitali V. Kononenko	Belén Sotillo
Wavelength (nm)	800	790	266	515
Duration of pulse (fs)	80	250	100	230
Annealing temperature (°C)	680	1000	700	1000
Annealing time	15 min	3 h	2 h	3 h

In the same year, Sotillo et al. showed the fabrication of the optical waveguides in diamond by the high repetition rate fs laser pulses and the laser writing of the single NV centers in the diamond sample. Their research showed for the first time that the fabrication of optical waveguides in diamond sample can address single NV center in quantum information systems. The wavelength of the fs used in the experiment was 515 nm (frequency double), the duration of the fs pulse was 230 fs, and the repetition rate of the laser was 500 kHz. The diamond sample was set on the three-axis motion stages and controlled by the computer, then the fs laser was focused by an oil immersion lens (NA = 1.25) and illuminated on the diamond sample to realize the writing of the optical waveguides in the diamond. After the treatment of fs laser pulses, the diamond sample finally annealed in the nitrogen atmosphere at 1000 °C for 3 h [75].

By comparing the four groups of experimental data from Table 2, for the study of NV color center in fs laser processing diamond, the wavelength of fs laser is more than 800 nm, which shows that diamond has very good absorption of near-infrared laser. However, other processing parameters, such as pulse duration, repetition rate, pulse energy, and peak pulse energy, are influenced by the experimental instruments and processing techniques used. In addition, it is noteworthy that in the annealing process of diamond, different from the other three groups, the annealing process used in [72] firstly puts illuminated diamond sample into a ceramic stove at 300 °C and then started heating to 680 °C and maintained for 5 min. The purpose of this treatment is to oxidize the impurities on the surface of the diamond sample with oxygen in the air and to minimize the oxidation of the diamond itself. The high-temperature treatment could also enhance vacancies moving toward nitrogen atoms. Time of the whole oxidation procedure including annealing to room temperature was about 15 min. Generally speaking, the annealing temperature used is more than 1000 °C, and the annealing time is generally 2–3 h. This shows that the vacancy movement in diamond samples can be effectively promoted under the action of temperature and time, and the carbonization of diamond can be limited to the greatest extent, the damage to diamond lattice can be reduced, and the quality of NV color center can be improved effectively.

3.4 *Experimental Configuration and Annealing*

3.4.1 Laser Processing

NV color center is a lattice structure of point defect in diamond. In the common preparation methods, whether it is electron beam etching, ion beam implantation, or laser irradiation effect, the purpose is to produce the vacancies structure in diamond and then through annealing treatment to obtain NV color center structure. Therefore, the study of the mechanism between fs laser and diamond is helpful to solve the specific problems in the research of fs laser processing NV color centers in diamond [73].

As we all know, fs laser pulse width is very short which is generally dozens of femtoseconds, so that it has a very high peak power pulse. Photons in laser pulse have high coherence, which can effectively induce photoionization and avalanche ionization. Therefore, in the process of fs laser interacting with diamond samples, the strong electromagnetic field generated by fs laser can ionize nitrogen and oxygen molecules in the air to produce plasma, thus obtaining a large number of high-density electrons. The strong electric field generated by the electric vector superposition can make the free electrons and charged ions get a high speed [76–82]. High-speed charged particles generated by fs laser ionization will bombard the lattice of diamond along the plasma channel. The vacancies are eventually produced in diamond samples. After annealed at high temperature, the vacancies move freely in the sample and were captured by the nitrogen atoms; thus, the NV color centers were generated in the diamond. In summary, based on the principle that fs laser fabrication generating the vacancies in the diamond, it is possible to fabricate NV color centers in diamond samples by fs laser direct writing technology.

Therefore, for example, the Oxford University team (Chen et al.) [73] realized the fabrication of NV color centers in diamond samples by fs laser direct writing technique. This part will introduce the experimental setup of fabricating NV color centers in diamond with fs laser in detail. In the experimental platform, the laser used by Chen et al. is a Ti:sapphire laser. The laser beam was firstly modulated by a half-wave plate and a Gran laser prism; then the laser beam was expanded and modulated by a space light modulator (SLM) to generate the optical field array. By means of the dichroic lens, the laser beam was divided into two beams, one of the beams is incident into the CCD through the mirror to observe the machining process, and the other one modulated by a 4f system was focused by the 60×1.4 oil-immersed lens to fabricate the diamond sample. During the process of the fs laser, the diamond sample was placed on a three-dimensional displacement platform which was controlled by the computer and illuminated visually by LED illumination transmission microscope; then the processing of the NV color centers realized [73] (Fig. 13).

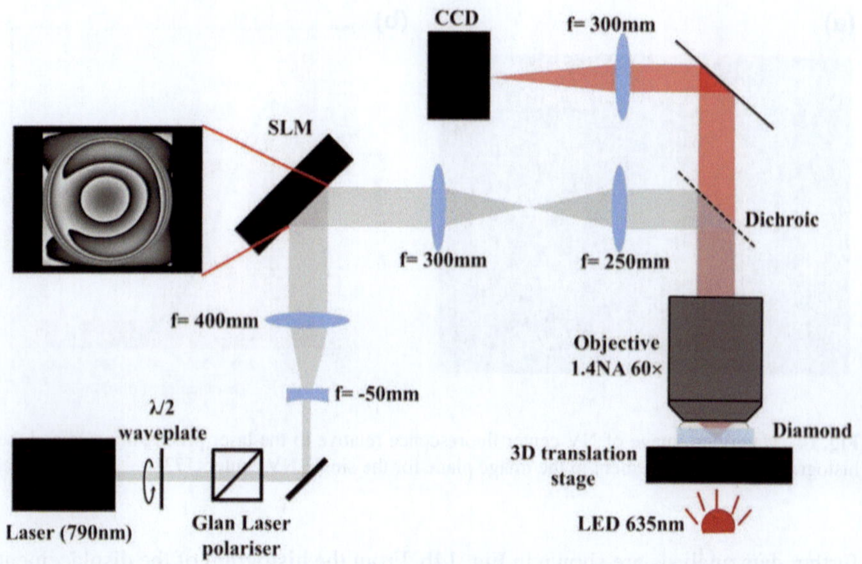

Fig. 13 Schematic diagram of NV color center in diamond processed by fs laser direct writing [73]

In order to improve the accuracy of fs laser processing, a high numerical aperture oil immersion objective is used to focus the laser to limit the size of the laser spot, so that the super-resolution spot is obtained and the NV color center could be fabricated in a small region. The focus size of the laser in the diamond is close to the optical diffraction limit, the lateral and axial dimensions (full width at half maximum) are 350 nm and 2 μm, respectively, and the focal volume is about 0.12 μm^3. However, due to the strong nonlinear optical effect during fs laser and diamond interaction, the volume of vacancy may be much smaller than the theoretical value [83]. This is because high NA optical devices focusing below the diamond surface introduce a large amount of spherical aberration for the refraction at the oil/diamond interface where the refractive index has a large step. It is necessary to correct the aberrations to restore the best optical performance, which has also been shown to enhance the photo-material interaction [84] when the laser is written into the subsurface graphite orbitals. Aberration correction can significantly reduce the size of the laser focus. The use of a high numerical aperture oil immersion objective can effectively improve the controllability of the processing position of the NV color center, limiting it to a very small local range, avoiding the NV cluster formed around the graphitized region induced by the laser in the diamond material. It is beneficial to increase the yield of single NV color center.

As shown in Fig. 14a, it is the superimposed grid map of the NV color center array and the target position. The red circle in the figure shows the target position of the NV color center. It can be seen that the position of the NV color center produced by ultrafast laser processing almost overlaps with the target position. The results of

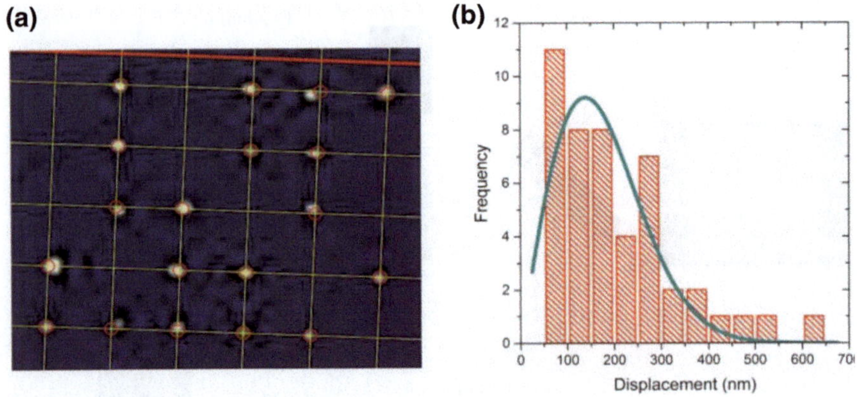

Fig. 14 Magnified image of NV center fluorescence relative to the laser processing grid and the histogram of the displacement in the image plane for the single NV centers [73]

further data analysis are shown in Fig. 14b. From the histogram of the displacement of the NV color center measured from the target point, it can be clearly seen that the position error of the NV in ultrafast laser processing mainly concentrates in the range of 200 nm, which indicates that the ultrafast laser processing of NV color center can effectively improve the position accuracy of NV color center, which is conducive to the coupling of NV color center and optoelectronic components; it is of great significance in the practical technical applications of NV color center.

3.4.2 Annealing

Diamond annealing treatment was carried out in an annealing tube furnace, the diamond sample was placed in an alumina boat and buried in the diamond sand, then annealing furnace in vacuum or dry nitrogen environment gradually heated to the appropriate temperature. In the annealing process, the vacuum and the dry nitrogen environment are to reduce the oxidative damage and graphitization of the diamond surface, so as to improve the NV color center yield. It is well known that different annealing temperatures and times directly affect the number of individual NV color centers in a diamond sample. Generally, the time of annealing treatment is 2–3 h. Chen et al. and his members have studied the presence of NV color centers in laser-processed diamond sample at the annealing temperature of 900, 1000, and 1200 °C through experiment. They firstly detected and recorded the distributions in the diamond sample; the results shown that a part of the NV color centers are generated at the laser ablation position in the diamond sample annealed at 900 °C, indicating that under this action, some vacancies moved and were captured by the nitrogen atoms to form color centers, but the defects and damages at most ablation sites still existed. The yield of NV color centers is very low at 900 °C. Next, they processed sample at 1000 °C [73, 85]. Compared with the sample annealed at 900 °C, a plurality of single NV color center defect structures was found in the diamond sample after the annealing treatment. Since then, they continued to raise the temperature to 1200 °C and annealed for 24 h. It was found by comparing the fluorescence detection results before and after annealing that the NV color center structure presented in the previous process disappeared and no new one is formed [73, 86]. Based on the experimental results, the best annealing temperature and time for NV color center preparation in diamond by fs laser direct writing are 1000 °C and 3 h, respectively [73].

4 Spectral Characterization of NV Color Center in Diamond

4.1 Confocal Fluorescence Scanning

In order to observe the distribution of the NV color centers in diamond samples and make a spectrum analysis for the NV color centers, a confocal fluorescence scanning microscopy imaging system (CFSMI) was used to detect the NV color centers in diamond. The functions of CFSMI system mainly include fluorescence detection, fluorescence spectrum acquisition, and photon second-order correlation coefficient measurement. Especially, the measurement of photon second-order correlation coefficient plays an important role in the detection of single NV color center. The CFSMI system is essential for us to understand the properties of the NV color center; the components of the system are shown in Fig. 15.

The CFSMI system consists of a 532 nm laser, a dichroic beam splitter, a spectrometer, an avalanche photodiode detector (APD) single-photon counting module, a time-correlated single-photon counter (TCSPC) time-dependent single-photon counter, a piezoelectric displacement controller, a microscope objective, and a filter [72, 87]. The CFSMI system can be used for fluorescence excitation, fluorescence collection, and optical performance analysis. In order to realize the function of filtering, the conjugation of the objective lens focus in CFSMI system can be modulated by the keyhole and the lens in front of the keyhole. In detail, only the light in the focus of the objective lens can pass through the small aperture, while some optical signals can be shield due to the small aperture. Therefore, the selection of the keyhole mainly depends on the numerical aperture of the objective lens and the measurement requirements. After filtering, the fluorescence of the sample can be reflected into the spectrometer, to realize the function of spectral detection. In order to collect the fluorescence information of NV color center in real time, the output terminal of the photon counter was connected with the TCSPC and the computer, respectively; in addition, the luminescence intensity information of the sample was synchronously monitored by the computer. The movement of the samples is also controlled by computers [72, 87]. One set of piezoelectric ceramic platforms was used to place

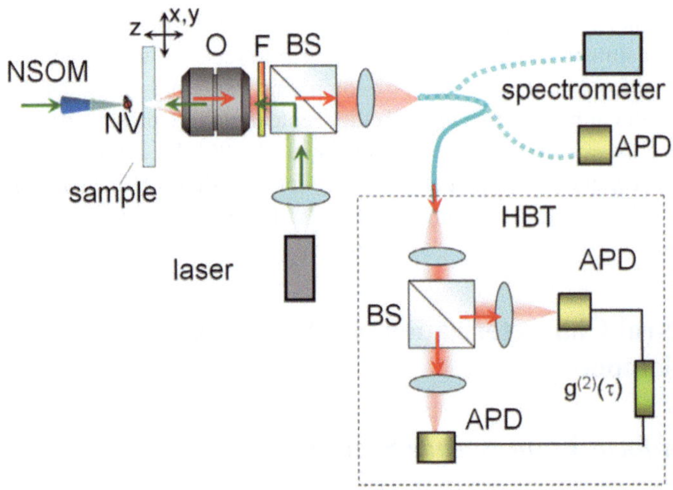

Fig. 15 Scheme of NV color center confocal fluorescence scanning and anti-bunching optical device; (O: microscope objective, F: filter and dichroic mirror, BS: beam splitter, APD: avalanche photodiode in single-photon counting mode). The 532 nm laser is excited from the polymer-coated optical tip. The NV center fluorescence is collected with a high NA objective lens and filtered and injected into the multimode fiber. The latter can be connected to an APD, spectrometer, or HBT correlator involving 50/50 BS, two APDs, and a time-dependent single-photon counting module [88]

diamond samples and achieve accurate displacement in the two-dimensional direction of X-Y, and the other set of piezoelectric ceramic platforms was used to connect the objective lens so that the objective lens can move along the Z-direction to adjust the focal length of the objective lens.

As shown in Fig. 16, the image of confocal scanning and NV color center fluorescence spectrogram [74, 89]. It can be seen from Fig. 16a, c that the fluorescence distribution of the single NV color center in the diamond sample is a regular circular shape and is independent of other fluorescent distribution regions. In addition, as is shown in Fig. 16b, d, the fluorescence spectra of the reflecting points are also different for different valence states of the single NV color center. The fluorescence spectra of the negative valence state of the single NV color center are 637 nm, and the fluorescence spectrum of a single NV color center at zero valence is at 575 nm, both show obvious narrow zero-phonon lines.

4.2 *Anti-bunching Detection*

In order to identify whether the single emitting point is a single NV color center, the anti-bunching detection was carried out to make sure the generation of single NV color center. In the Hanbury Brown and Twiss (HBT) system shown in Fig. 15,

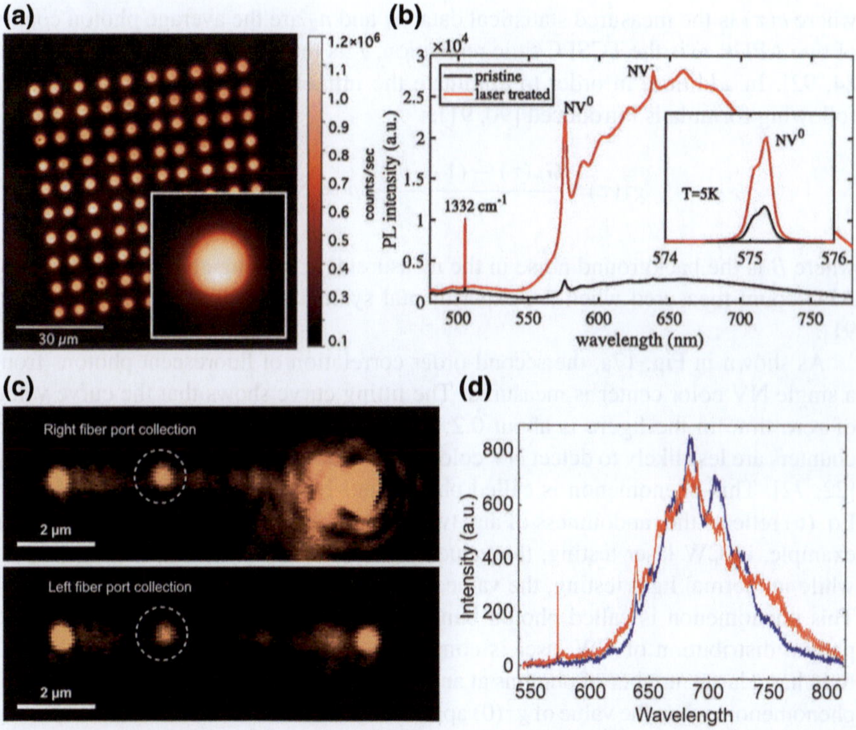

Fig. 16 Image of confocal scanning and NV color center fluorescence spectrogram. **a** Image of the single NV centers array with a period of 8 μm; **c** confocal scans of fluorescence detected from two fiber ends (light originating from a single NV center is shown in the dotted circles); **b, d** NV color center fluorescence spectrogram [74, 89]

the photons excited by the sample arrive at the PBS which is the first element of the measurement system. Due to the indivisibility of the single photon, the photons excited by the sample are reflected or transmitted to another APD. When the first APD receives the signal, it will send an electrical pulse to the TCSPC; then the TCSPC starts counting. While the other APD receives the photon and sends an electrical pulse to the TCSPC, the TCSPC will store the information of the time difference and transmits it to the computer for analyzing. By counting the information of the time difference that collected with the TCSPC, the information of fluorescent photons emitted by the sample was recorded and calculated, which was used to predict the correlation of fluorescent photons. The statistical data can be completed by the following formula [90, 91].

$$G_n(\tau)=\frac{c(\tau)}{n_1 n_2 \omega T} \tag{5}$$

where $c(\tau)$ is the measured statistical data, n_1 and n_2 are the average photon counts of two APDs, ω is the TCSPC time resolution, T is the total measurement time [62, 84, 92]. In addition, in order to eliminate the influence of background noise, the following formula is introduced [90, 91].

$$g^2(\tau)=\frac{G_n(\tau)-(1-\rho^2)}{\rho^2}, \rho = S/(S+B) \tag{6}$$

where B is the background noise in the measurement system, and $S + B$ is the total APD count measured when the experimental system aims at the color center [90, 91].

As shown in Fig. 17a, the second-order correlation of fluorescent photons from a single NV color center is measured. The fitting curve shows that the curve value of zero time in the figure is about 0.2; it indicates that both of the single-photon counters are less likely to detect NV color center fluorescent photons simultaneously [22, 72]. This phenomenon is called photon anti-bunching phenomenon. $g^2(\tau)$ in Eq. (6) reflects the randomness of any two photons in the beam in time domain. For example, in CW laser testing, the value of $g^2(\tau)$ is almost always the constant 1, while in thermal light testing, the value of $g^2(0)$ is greater than the value of $g^2(\tau)$. This phenomenon is called photon bunching phenomenon for the reason that the photon distribution of CW laser is completely random and the source of thermal light has a large number of photons at any moment. Therefore, for the anti-bunching phenomenon, when the value of $g^2(0)$ approaches zero, it is indicated that the spacing between any two photons is impossible to be zero in the time domain, that is, the light source only emits one photon at a time, so we can make sure the existence of a single NV color center. Generally, when the value of $g^2(\tau)$ is less than 0.5, we can think that there is a single NV color center in this position [22, 72, 87]. After the position of the single NV color center was determined, the fluorescence spectra of the NV color center can be further analyzed to make sure whether it is a single NV color center or not. Figure 17b shows the histogram of $g^2(0)$ values from all measurement points. It is obvious that the $g^2(0)$ values of all measuring points are divided into three parts. The measurement points of $g^2(0) < 0.32$ and $0.32 < g^2(0) < 0.65$ occupy the majority of all the measurement points, which are attributed to the existence of one and two NV centers, respectively. In addition, some sites show that $g^2(0)$ is between 0.65 and 0.9, which is attributed to the existence of three NV centers. Figure 17c shows the spatial distribution of the locations of single, double, and three NV centers.

5 Summary

In order to obtain the high position accuracy processing method of NV color center and realize the technical application of the NV color center in diamond, it is necessary to understand the structure of the NV center and the machining mechanism of the

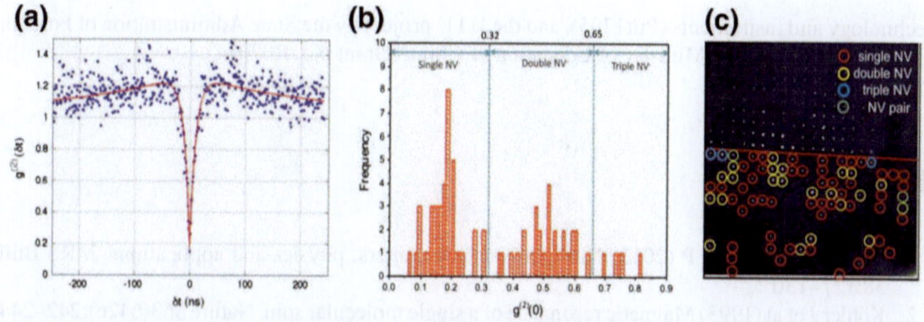

Fig. 17 Anti-bunching test and fluorescence spectrum analysis of single NV color center. **a** The second-order correlation of fluorescent photons from a single NV color center; **b** histogram of $g^2(0)$ values from all measurement NV centers; **c** the spatial distribution of the locations of single, double, and three NV centers [73]

ultrafast laser. More importantly, the quality of NV color center obtained by different processing method can reflect the advantages and disadvantages of the processing method. The ultrafast laser processing method has been proved to be a powerful tool for the fabrication of NV center. It is of great importance to study the ultrafast laser processing scheme which includes the parameters of the ultrafast laser used in the fabrication, annealing treatment, and confocal fluorescence detection.

This review has firstly illustrated the structures of NV color center, which includes the electronic structure and the energy level structure; and then, based on the study of NV color center, the applications of NV color center in different fields were introduced. Subsequently, the mechanism and characteristics of ultrafast laser processing were described, and the parameters of the laser and annealing treatment used in different ultrafast laser processing schemes were analyzed. At last, a typical scheme of ultrafast laser processing NV color centers was chosen, then the optical path of ultrafast laser processing, the annealing treatment, and the confocal fluorescence detection were analyzed in detail. The position error of single NV color center prepared by fs laser can be about 200 nm around the target point, and the yield of single NV color center after annealed at 1000 °C for 3 h was up to 45%. The NV color center in diamond prepared by ultrafast laser processing is promising method with high position accuracy and high yield.

To realize the technical application, single NV color center is desired to place in a fixed position in the diamond sample and couple it with optical or electronic devices. However, it is still a great challenge for the current NV color center processing methods [86, 92–95]. More efforts should be done to improve the machinability of the atomic scale color center structure and push forward its application in different areas.

Acknowledgements The study was supported by the National Natural Science Foundation of China (No. 51575389, 51511130074), National Natural Science Foundation of China (NSFC)-German Research Foundation (DFG) International Joint Research Programme (51761135106), the Natural Science Foundation of Tianjin (15JCYBJC19400), State key laboratory of precision measuring technology and instruments (Pilt1705), and the '111' project by the State Administration of Foreign Experts Affairs and the Ministry of Education of China (Grant No. B07014).

References

1. Acosta V, Hemmer P (2013) Nitrogen-vacancy centers: physics and applications. MRS Bull 38:127–130
2. Köhler J et al (1993) Magnetic resonance of a single molecular spin. Nature 363(6426):242–244
3. Wrachtrup J et al (1993) Optical detection of magnetic resonance in a single molecule. Nature 363(6426):244–245
4. Gruber A (1997) Scanning confocal optical microscopy and magnetic resonance on single defect centers. Science 276(5321):2012–2014
5. Balasubramanian G et al (2009) Ultralong spin coherence time in isotopically engineered diamond. Nat Mater 8(5):383–387
6. Acosta VM et al (2010) Optical properties of the nitrogen-vacancy singlet levels in diamond. Phys Rev B 82(20):2889–2898
7. Rondin L et al (2015) Magnetometry with nitrogen-vacancy defects in diamond. Cheminform 45(42):056503
8. Mcguinness LP et al (2011) Quantum measurement and orientation tracking of fluorescent nanodiamonds inside living cells. Nat Nanotechnol 6(6):358–363
9. Schuster DI et al (2010) High-cooperativity coupling of electron-spin ensembles to superconducting cavities. Phys Rev Lett 105(14):140501
10. Maletinsky P et al (2012) A robust scanning diamond sensor for nanoscale imaging with single nitrogen-vacancy centres. Nat Nanotechnol 7(5):320
11. Kubanek A et al (2012) Quantum interference of single photons from two remote nitrogen-vacancy centers in diamond. In: Meeting of the APS division of atomic, molecular and optical physics
12. Schell AW et al (2014) Scanning single quantum emitter fluorescence lifetime imaging: quantitative analysis of the local density of photonic states. Nano Lett 14(5):2623–2627
13. Cuche A et al (2010) "Deterministic" quantum plasmonics. Nano Lett 10(11):4566
14. Zaitsev SAM (2001) Optical Properties of Diamond. Springer, Berlin, Heidelberg, pp 5090–5097
15. Koizumi S, Nebel CE, Nesladek M (2008) Physics and applications of CVD diamond. Wiley, Weinheim
16. Jelezko F, Wrachtrup J (2010) Single defect centres in diamond: a review. Phys Status Solidi 203(13):3207–3225
17. Mildren, Assoc. Rich P, Rabeau AJR (2013) Optical quality diamond grown by chemical vapor deposition. Optical engineering of diamond. Wiley-VCH Verlag GmbH & Co. KGaA
18. Suter D, Jelezko F (2017) Single-spin magnetic resonance in the nitrogen-vacancy center of diamond. Prog Nucl Magn Reson Spectrosc 98–99:50
19. Davies G, Hamer MF (1976) Optical studies of the 1.945 eV vibronic band in diamond. Proc R Soc Math Phys Eng Sci 348(1653):285–298
20. Doherty MW et al (2013) The nitrogen-vacancy colour centre in diamond. Phys Rep 528(1):1–45
21. Wang J (2016) Preparation and spin coherence of NV centers in diamond and temperature detection application. University of Science Technology of China
22. Liu Y (2016) Generation of single photons based on color centers in diamond and investigation on its fluorescence dynamics. East China Normal University
23. Balasubramanian G et al (2014) Nitrogen-vacancy color center in diamond-emerging nanoscale applications in bioimaging and biosensing. Curr Opin Chem Biol 20(20):69–77

24. Orwa JO et al (2011) Engineering of nitrogen-vacancy color centers in high purity diamond by ion implantation and annealing. J Appl Phys 109(8):3207
25. Plakhotnik T, Aman H (2017) NV-centers in nanodiamonds: how good they are. Diam Relat Mater 82:87–95
26. Berthel M et al (2016) Photophysics of single nitrogen-vacancy centers in diamond nanocrystals. Phys Rev B 91(3):035308
27. Felton S et al (2008) Electron paramagnetic resonance studies of the neutral nitrogen vacancy in diamond. Phys Rev B: Condens Matter 77(77):439–446
28. Lenef A, Rand SC (1996) Electronic structure of the N-V center in diamond: theory. Phys Rev B: Condens Matter 53(56):13427–13440
29. Manson NB, Harrison JP, Sellars MJ (2006) Nitrogen-vacancy center in diamond: model of the electronic structure and associated dynamics. Phys Rev B Cond matter 74(10):104303
30. Fuchs GD et al (2011) A quantum memory intrinsic to single nitrogen–vacancy centres in diamond. Nat Phys 7(10):789–793
31. Kong X (2015) Magnetic resonance towards single nuclear spin sensitivity based on single solidstate spin in diamond. University of Science and Technology of China
32. Dumeige Y et al (2014) Magnetometry with nitrogen-vacancy ensembles in diamond based on infrared absorption in a doubly resonant optical cavity. Phys Rev B 87(15):155202
33. Nizovtsev AP et al (2001) Modeling fluorescence of single nitrogen–vacancy defect centers in diamond. Phys B Phys Condens Matter 308(308):608–611
34. Lesik M (2015) Engineering of NV color centers in diamond for their applications in quantum information and magnetometry
35. Gisin N et al (2001) Quantum cryptography. Rev Mod Phys 74(1):145–195
36. Groeblacher S et al (2005) Experimental quantum cryptography with qutrits. New J Phys 8(5):75
37. Dutt MVG et al (2007) Quantum register based on individual electronic and nuclear spin qubits in diamond. Science 316(5829):1312–1316
38. Bernien H et al (2012) Heralded entanglement between solid-state qubits separated by 3 meters. In: APS division of atomic, molecular and optical physics meeting
39. Tsukanov AV (2013) Quantum memory based on ensemble states of NV centers in diamond. Russ Microlectron 42(3):127–147
40. Rivas A, Huelga SF, Plenio MB (2010) Entanglement and non-markovianity of quantum evolutions. Phys Rev Lett 105(5):050403
41. Abe E, Sasaki K (2018) Tutorial: magnetic resonance with nitrogen-vacancy centers in diamond—microwave engineering, materials science, and magnetometry. J Appl Phys 123(16):161101
42. Maze JR et al (2008) Nanoscale magnetic sensing with an individual electronic spin in diamond. Nature 455(7213):644–647
43. Dolde F et al (2011) Electric-field sensing using single diamond spins. Nat Phys 7(6):459–463
44. Glenn DR et al (2018) High-resolution magnetic resonance spectroscopy using a solid-state spin sensor. Nature 555(7696):351
45. Grazioso F et al (2013) Measurement of the full stress tensor in a crystal using photoluminescence from point defects: the example of nitrogen vacancy centers in diamond. Appl Phys Lett 103(10):133
46. Kucsko G et al (2013) Nanometer scale quantum thermometry in a living cell. Physics 500(7460):54–58
47. Ajoy A et al (2015) Atomic-scale nuclear spin imaging using quantum-assisted sensors in diamond. Phys. Rev X 5(1):011001
48. Fu CC et al (2007) Characterization and application of single fluorescent nanodiamonds as cellular biomarkers. Proc Natl Acad Sci U S A 104(3):727–732
49. Neugart Felix et al (2007) Dynamics of diamond nanoparticles in solution and cells. Nano Lett 7(12):3588–3591
50. Fang CY et al (2011) The exocytosis of fluorescent nanodiamond and its use as a long-term cell tracker. Small 7(23):3363–3370

51. Mohan N et al (2010) In vivo imaging and toxicity assessments of fluorescent nanodiamonds in Caenorhabditis elegans. Nano Lett 10(9):3692–3699
52. Simpson DA et al (2014) In vivo imaging and tracking of individual nanodiamonds in drosophila melanogaster embryos. Biomed Opt Express 5(4):1250–1261
53. Kuo Y et al (2013) Fluorescent nanodiamond as a probe for the intercellular transport of proteins in vivo. Biomaterials 34(33):8352–8360
54. Wu TJ et al (2013) Tracking the engraftment and regenerative capabilities of transplanted lung stem cells using fluorescent nanodiamonds. Nat Nanotechnol 8(9):682–689
55. Han KY et al (2009) Optimizing fluorophores for super-resolution fluorescence STED microscopy. Biophys J 96(3):637a–637a
56. Rittweger E et al (2015) STED microscopy reveals crystal colour centres with nanometric resolution. Nat Photonics 3(3):144–147
57. Wildanger D et al (2012) Solid immersion facilitates fluorescence microscopy with nanometer resolution and sub-ångström emitter localization. Adv Mater 24(44): OP309–OP313
58. Mazur E (2008) Femtosecond laser micromachining in transparent materials. Nat Photonics 2(4):219–225
59. Schaffer CB, Brodeur A, Mazur E, Laser-induced breakdown and damage in bulk transparent materials induced by tightly focused femtosecond laser pulses. Measur Sci Technol 12(11):1784–1794
60. Ams M et al (2008) Investigation of ultrafast laser-photonic material interactions: challenges for directly written glass photonics. IEEE J Sel Top Quant Electron 14(5):1370–1381
61. Mao SS et al (2004) Dynamics of femtosecond laser interactions with dielectrics. Appl Phys A 79(7):1695–1709
62. Tan D et al (2016) Femtosecond laser induced phenomena in transparent solid materials: fundamentals and applications. Prog Mater Sci 76(1):154–228
63. Qiu J, Miura K, Hirao K (2008) Femtosecond laser-induced microfeatures in glasses and their applications. J Non-Cryst Solids 354(12–13):1100–1111
64. Mysyrowicz A et al (2003) Femtosecond laser irradiation stress induced in pure silica. Opt Express 11(9):1070–1079
65. Joglekar AP et al (2004) Optics at critical intensity: applications to nanomorphing. Proc Natl Acad Sci U S A 101(16):5856–5861
66. Macandrew JA (1988) The programme. Optica Acta Int J Opt 303–316
67. Kane DJ, Trebino R (1993) Using phase retrieval to measure the intensity and phase of ultrashort pulses: frequency-resolved optical gating. J Opt Soc Am A 10(5):1101–1111
68. Trebino R et al (1997) Measuring ultrashort laser pulses in the time-frequency domain using frequency-resolved optical gating. Rev Sci Instrum 68(9):3277–3295
69. Meijer J (2004) Laser beam machining (LBM), state of the art and new opportunities. J Mater Process Technol 149(1):2–17
70. Küper S, Stuke M (1988) Ablation of uv-transparent materials with femtosecond UV excimer laser pulses. MRS Proc 129(1–4):475–480
71. Dumeige Y et al (2004) Photo-induced creation of nitrogen-related color centers in diamond nanocrystals under femtosecond illumination. J Lumin 109(2):61–67
72. Wu B et al (2013) Fabrication of nitrogen vacancy color centers by femtosecond pulse laser illumination. Opt Express 21(10):12843–12848
73. Chen YC et al (2016) Laser writing of coherent colour centres in diamond. Nat Photonics 11:77
74. Kononenko VV et al (2017) Nitrogen-vacancy defects in diamond produced by femtosecond laser nanoablation technique. Appl Phys Lett 111(8):081101
75. Sotillo B et al (2017) Visible to infrared diamond photonics enabled by focused femtosecond laser pulses. Micromachines 8(2):60
76. Kasparian J, Sauerbrey R, Chin SL (2000) The critical laser intensity of self-guided light filaments in air. Appl Phys B 71(6):877–879
77. Tzortzakis S et al (2002) Femtosecond laser-guided electric discharge in air. In: Conference digest: 2000 international quantum electronics conference

78. Kasparian J et al (2003) White-light filaments for atmospheric analysis. Science 301(5629):61–64
79. Sprangle P, Peñano JR, Hafizi B (2002) Propagation of intense short laser pulses in the atmosphere. Phys Rev E Stat Nonlinear Soft Matter Phys 66(2): 046418
80. Couairon A, Mysyrowicz A (2007) Femtosecond filamentation in transparent media. Phys Rep 441(2):47–189
81. York AG et al (2008) Direct acceleration of electrons in a corrugated plasma waveguide. Phys Rev Lett 100(19):195001
82. Shi L et al (2011) Generation of high-density electrons based on plasma grating induced Bragg diffraction in air. Phys Rev Lett 107(9):095004
83. Lagomarsino S et al (2016) Photoionization of monocrystalline CVD diamond irradiated with ultrashort intense laser pulse. Phys Rev B. 93(8):085128
84. Sun B, Salter PS, Booth MJ (2014) High conductivity micro-wires in diamond following arbitrary paths. Appl Phys Lett 105(23):397
85. Yamamoto T et al (2013) Extending spin coherence times of diamond qubits by high temperature annealing. Phys Rev B 88(7):4049–4056
86. Chu Y et al (2014) Coherent optical transitions in implanted nitrogen vacancy centers. Nano Lett 14(4):1982–1986
87. Cuche A et al (2009) Near-field optical microscopy with a nanodiamond-based single photon tip. Opt Express 17(22):19969–19980
88. Drezet A et al (2015) Near-field microscopy with a scanning nitrogen-vacancy color center in a diamond nanocrystal: a brief review. Micron 70:55–63
89. Patel RN et al (2015) Efficient photon coupling from a diamond nitrogen vacancy center by integration with silica fiber. Light Sci Appl 5(2):e16032
90. Gaebel T et al (2004) Stable single-photon source in the near infrared. New J of Phys 6(1):98
91. Kurtsiefer C et al (2002) Stable solid-state source of single photons. Phys Rev Lett 85(2):290–293
92. Englund D et al (2010) Deterministic coupling of a single nitrogen vacancy center to a photonic crystal cavity. Nano Lett 10(10):3922
93. Faraon A et al (2011) Resonant enhancement of the zero-phonon emission from a colour centre in a diamond cavity. Nat Photonics 5(5):301–305
94. Riedrichmöller J et al (2011) One- and two-dimensional photonic crystal microcavities in single crystal diamond. Nat Nanotechnol 7(1):69
95. Mclellan CA et al (2015) Patterned formation of highly coherent nitrogen-vacancy centers using a focused electron irradiation technique. Nano Lett 16(4):2450–2454

MIX
Papier aus verantwortungsvollen Quellen
Paper from responsible sources
FSC® C105338

If you have any concerns about our products,
you can contact us on
ProductSafety@springernature.com

In case Publisher is established outside the EU,
the EU authorized representative is:
Springer Nature Customer Service Center GmbH
Europaplatz 3, 69115 Heidelberg, Germany

Printed by Libri Plureos GmbH
in Hamburg, Germany